Choquet–Deny Type Functional Equations with Applications to Stochastic Models

C. Radhakrishna Rao
Pennsylvania State University
University Park, PA 16802, USA

D. N. Shanbhag
University of Sheffield,
Sheffield S10 2TN, UK

JOHN WILEY & SONS
Chichester • New York • Brisbane • Toronto • Singapore

Other Wiley Editorial Offices

John Wiley & Sons, Inc., 605 Third Avenue,
New York, NY 10158-0012, USA

Jacaranda Wiley Ltd, 33 Park Road, Milton,
Queensland 4064, Australia

John Wiley & Sons (Canada) Ltd, 22 Worcester Road,
Rexdale, Ontario M9W 1L1, Canada

John Wiley & Sons (SEA) Pte Ltd, 37 Jalan Pemimpin #05-04,
Block B, Union Industrial Building, Singapore 2057

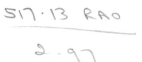
Library of Congress Cataloging-in-Publication Data

Rao, C. Radhakrishna (Calyampudi Radhakrishna), 1920–
    Choquet-Deny type functional equations with applications to
stochastic models / C. Radhakrishna Rao, D.N. Shanbhag.
        p.   cm. — (Wiley series in probability and mathematical statistics
    Includes bibliographical references and index.
    ISBN 0 471 95104 8
    1. Functional equations.   2. Stochastic processes.   I. Shanbhag
D. N.   II. Title.   III. Series.
QA431.R36   1994
519.2 — dc20   94-10080
                                                                   CIP

British Library Cataloguing in Publication Data

A catalogue record for this book is available from the British Library

ISBN 0 471 95104 8

Typeset in 10/12pt Times by Laser Words, Madras
Printed and bound in Great Britain by Biddles Ltd, Guildford and Kings Lynn

Dedicated

to

our parents

C.D. Naidu
A. Laxmikanthamma

and

Nagesh G. Shanbhag
Rama Shanbhag

Contents

Preface

The present volume is the product of a collaborative effort by the authors over a period of several years. It all started with the solution to what appeared to be an interesting problem in probability. Let $X \geqslant 0$ be an integer valued random variable (r.v.) and suppose that a damage process reduces X to Y according to a binomial damage model, i.e., for some $0 < \pi < 1$

$$P(Y = r | X = n) = \binom{n}{r} \pi^r (1 - \pi)^{n-r}, \quad r = 0, \ldots, n; n = 0, 1, 2, \ldots$$

Rao (1963) observed that if X is a Poisson r.v., then

$$P(Y = r | X \text{ is damaged}) = P(Y = r) = P(Y = r | X \text{ is undamaged}) \quad (1)$$

and raised the question whether the property (1) characterizes a Poisson distribution. He showed that it is so if (1) holds for all π, the parameter of the binomial distribution. Later in collaboration with Rubin (Rao and Rubin, 1964), it was shown that the result holds if (1) is satisfied for just one value of π. It was pointed out by Shanbhag (1977) that the Rao–Rubin result can be deduced from the solution to a general recurrence relation of the form

$$v_m = \sum_{n=0}^{\infty} p_n v_{m+n}, \quad m = 0, 1, \ldots \quad (2)$$

where $p_0, p_1, \ldots$ is a given sequence of nonnegative real numbers, not all zero, and $v_0, v_1, \ldots$ is a sequence of nonnegative real numbers to be determined. Shanbhag obtained a complete solution to problem (2), which provided a unified approach to a variety of characterizations of discrete distributions.

In a series of research seminars given at the University of Pittsburgh in 1979–80, Rao formulated a more general version of Shanbhag's problem (2) in the following form.

Let μ be a σ-finite Borel measure on $\mathbb{R}_+$, nondegenerate at zero and $H : \mathbb{R}_+ \to \mathbb{R}_+$ be a locally integrable (with respect to Lebesgue measure) Borel measurable function on $\mathbb{R}_+$ satisfying the integral equation

$$H(x) = \int_{\mathbb{R}_+} H(x + y) \mu(dy), \quad \text{for a.a. } (L) x \geqslant 0. \quad (3)$$

Under what conditions does a nontrivial H exist and what is its functional form? If $\int_{\mathbb{R}_+} H(y)\mu(dy) = c \neq 0$, then (3) can be written in the form

$$\int_{\mathbb{R}_+} (h(x+y) - h(x)h(y))v(dy) = 0, \quad \text{for a.a. } (L)x \geqslant 0 \tag{4}$$

where $H = ch$ and $v = c\mu$. Since (4) involves explicitly the well known Cauchy equation $h(x + y) - h(x)h(y) = 0$, Rao named (3) as the ICFE (Integrated Cauchy Functional Equation). In collaboration with Lau (Lau and Rao, 1982), a complete solution to (3) was obtained, which provided solutions to a variety of characterization problems involving continuous distributions. Since 1982, a number of research papers have appeared on the ICFE and its applications.

Encouraged by the unification of several characterization problems provided by Lau–Rao–Shanbhag theorems, attempts were made by several authors to formulate the ICFE in a general form

$$H(x)(1 - g(x)) = \int_S H(x+y)(\mu - v)(dy) \; \forall x \in T. \tag{5}$$

The original formulations of Shanbhag and Rao may be described as ICFE on $\mathbb{N}_0$ and $\mathbb{R}_+$. The general formulation (5) involves the domain of integration S, the range of validity of the equation T, the signed measure $\mu - v$ and the error term g. Some special cases of (5) have been solved, while others are still under investigation.

Historically, a special case of (5) was considered by Choquet and Deny (1960). They showed that if S is a locally compact Abelian topological group, then a bounded continuous function $H : S \to \mathbb{R}$ satisfies

$$H(x) = \int_S H(x+y)\mu(dy), \quad x \in S \tag{6}$$

with μ as a probability measure, only if $H(x+y) = H(x)$ for each $y \in \text{supp}[\mu]$ and each $x \in S$. With an additional condition that S is second countable, Deny (1961) gave an extended version of this result. He showed that if $H : S \to \mathbb{R}_+$, with H not necessarily bounded, satisfies (6) with μ as a Radon measure (that is not necessarily a probability measure) such that the smallest closed subgroup of S generated by $\text{supp}[\mu]$ equals S itself, then H has an integral representation as a weighted average of μ-harmonic exponential functions. Choquet and Deny (1960) and Deny (1961) have also given related results on the convolution equation $v * \mu = v$ for measures. Applications of the results of Choquet and Deny, and Deny to probability theory have appeared in Doob (1953), Meyer (1966) and Feller (1966). The convolution equation $v * \mu = v$ was also studied on semisimple Lie groups by Furstenberg (1963) and Azencott (1970) in the restricted case when μ is a probability measure and v is bounded.

The early papers which extended Deny's theorem to semigroups are those of Shanbhag (1977), Lau and Rao (1982), Davis and Shanbhag (1987), Alzaid *et al* (1988), Rao and Shanbhag (1989b, 1991, 1992a) and Lau and Zeng (1990). A recent

book by Ramachandran and Lau (1991) summarizes some of the contributions
to this area. In the present monograph, we concentrate on the discussion of the
solutions of (5) under different conditions and discuss a number of applications.
We show that the results contained in several recent papers on characterizations
of discrete and continuous distributions could have been obtained more simply by
using the solution of an appropriate ICFE.

A brief outline of the material covered in this book is as follows.

Chapter 1 gives the probability tools used in the rest of the monograph. A
rigorous treatment of the inversion theorem of the characteristic function is given
along with some results on probability measures and stochastic processes.

Chapter 2 covers the early work of Lau, Rao and Shanbhag on the ICFE

$$H(x) = \int_S H(x + y)\mu(\mathrm{d}y) \quad \text{for a.a. } (L)x \in T \tag{7}$$

where $S = T = \mathbb{N}_0$ or $S = T = \mathbb{R}_+$ or $S = \mathbb{R}$ and $T = \mathbb{R}_+$. Chapter 3 deals
with the case where $S = T$ is an Abelian topological semigroup, H is a bounded
nonnegative Borel measurable function on S and μ is a subprobability measure
on S, and a few other cases. It is shown that some of the famous theorems of
Bernstein and Bochner on the representations of completely mononotic functions
follow from the solution of an ICFE.

Chapter 4 considers multiple integral equations of the type

$$H_i(x) = \sum_{j=1}^{k} \int_S H_{i+j-1}(x + y)\mu_j(\mathrm{d}y), \quad x \in S, i = 1, \ldots, k. \tag{8}$$

Shimizu (1978) and Ramachandran *et al* (1988) studied the above problem when
$S = \mathbb{R}$ or $S = \mathbb{R}_+$. We consider the more general case when S is a Polish Abelian
semigroup. We also show that the solution to the ICFE with an error term follows
from the results in Chapter 3 in conjunction with some limit arguments. Some
applications to characterization problems are considered.

Chapter 5 deals with numerous characterizations of the distribution of a r.v.
or a vector r.v. based on the mean residual life function or the hazard measure.
Chapter 6 discusses characterizations of stable distributions and other problems
based on properties of Fourier and Mellin transforms. Chapter 7 is devoted to
characterizations based on damage models with generalizations of the earlier results
beginning with the papers by Rao and Rubin (1964) and Shanbhag (1977).

Chapter 8 discusses numerous results based on properties of order statistics,
record values and stochastic processes. Finally, Chapter 9 deals with characterizations based on regression and related statistical properties providing generalizations
of earlier results of Laha and Lukacs (1960), Kagan *et al* (1965) and others.

In honor of the original contributions made by Choquet and Deny, we have
referred to 'Choquet–Deny type equations' in the title of our monograph instead
of the ICFE, which is our main subject matter. Our arguments are mostly based
on properties of exchangeable random elements martingales and certain Markov

processes, in addition to those of transforms. We have mainly considered in our monograph applications in characterizing stochastic models. The ICFE also plays an important role in renewal processes, potential theory and other applications of stochastic processes. We hope our monograph devoted to a special study of the ICFE will be useful to research workers in pure and applied probability.

We wish to thank our wives Bhargavi Rao and Vibha Shanbhag for the help and encouragement they have given during the years we were engaged on drafting and redrafting the various chapters of the monograph, and continuing our research on the ICFE. We also wish to record that Rao's grandsons Amar and Rohit and Shanbhag's children Sharayu, Sandeep and Anita played their role in trying to entertain us to relieve us from the stresses involved in writing a book.

The work of the authors was supported by the US Army Research Office under the grant DAAH04-93-G-0030.

CHAPTER 1

Probability Tools and Preliminary Results

We devote this chapter to the presentation of the basic tools and preliminary results that are to be used in the derivations of the main results appearing in the monograph. If a result is well known we omit its proof unless we have an alternative or an improved proof. On the other hand, for a result that is new or not very well known we give a proof.

1.1 THE INVERSION, UNIQUENESS AND CONTINUITY THEOREMS FOR FOURIER TRANSFORMS

If v is a real signed measure on (the Borel σ-field of) $\mathbb{R}^n$, where n is a positive integer, then the function v^* on $\mathbb{R}^n$ defined by

$$v^*(t) = \int_{\mathbb{R}^n} \exp\{i\langle t, x\rangle\} v(\mathrm{d}x), \quad t \in \mathbb{R}^n$$

is called the Fourier transform of v. (This function is easily seen to be uniformly continuous on $\mathbb{R}^n$.) The following are amongst the most frequently used results concerning Fourier transforms.

1.1.1 The inversion theorem

Let v^* be the Fourier transform of a real signed measure v on $\mathbb{R}^n$ and let $a = (a_1, a_2, \ldots, a_n)$ and $b = (b_1, b_2, \ldots, b_n)$ be points of $\mathbb{R}^n$ such that $a < b$ and

$$|v|((a, b)) = |v|([a, b])$$

(i.e. for which (a, b) is a $|v|$-continuity set), where $|v|$ is the total variation of v. Then

$v((a, b])$

$$= \lim_{\tau_r \to \infty, r=1,2,\ldots,n} \frac{1}{(2\pi)^n} \int_{-\tau_n}^{\tau_n} \cdots \int_{-\tau_1}^{\tau_1} \left\{ \prod_{r=1}^{n} \left(\frac{e^{-it_r a_r} - e^{-it_r b_r}}{i t_r} \right) \right\} v^*(t) \, \mathrm{d}t_1 \ldots \mathrm{d}t_n,$$

where, obviously, the notation t stands for the vector $(t_1, t_2, \ldots, t_n)$.

Proof   As pointed out on pages 42 and 43 of Section 1.76 in Titchmarsh (1978), $\int_0^v (\sin u)/u\, du$, $v > 0$ is a positive bounded function tending as $v \to \infty$ to $\pi/2$. Define

$$J(\tau_1, \ldots, \tau_n) = \frac{1}{(2\pi)^n} \int_{-\tau_n}^{\tau_n} \cdots \int_{-\tau_1}^{\tau_1} \left\{ \prod_{r=1}^{n} \left(\frac{e^{-it_r a_r} - e^{-it_r b_r}}{it_r} \right) \right\} v^*(t)\, dt_1 \ldots dt_n.$$

Substituting $\int_{\mathbb{R}^n} \exp\{i\langle t, x\rangle\} v(dx)$ for $v^*(t)$ and noting that

$$\left| \left\{ \prod_{r=1}^{n} \left(\frac{e^{-it_r a_r} - e^{-it_r b_r}}{it_r} \right) \right\} \exp\{i\langle t, x\rangle\} \right| \leqslant \prod_{r=1}^{n} (b_r - a_r),$$

we get in view of Fubini's theorem and a minor manipulation that

$$J(\tau_1, \ldots, \tau_n) = \frac{1}{\pi^n} \int_{\mathbb{R}^n} \left(\prod_{r=1}^{n} \int_0^{\tau_r} \frac{\sin(t_r(x_r - a_r)) - \sin(t_r(x_r - b_r))}{t_r}\, dt_r \right) v(dx).$$

From the cited result in Titchmarsh (1978), it is obvious that the function

$$g(x, \tau_1, \ldots, \tau_n)$$

$$= \frac{1}{\pi^n} \prod_{r=1}^{n} \int_0^{\tau_r} \frac{\sin(t_r(x_r - a_r)) - \sin(t_r(x_r - b_r))}{t_r}\, dt_r, \quad x \in \mathbb{R}^n, \tau_1, \ldots \tau_n > 0$$

is bounded and is such that

$$\lim_{\tau_r \to \infty, r=1,2,\ldots,n} g(\bullet, \tau_1, \ldots, \tau_n) = I_{(a,b]}(\bullet) \quad \text{a.e. } [|v|].$$

Consequently, we have the theorem on appealing to the Lebesgue dominated convergence theorem. ∎

1.1.2   Corollary (the uniqueness theorem)

Two real signed measures v_1 and v_2 on $\mathbb{R}^n$ are identical if and only if their Fourier transforms v_1^* and v_2^* are identical.

Proof   The 'only if' part is straightforward and the 'if' part follows because if $v_1^* = v_2^*$ then the inversion theorem gives (on noting that, for each interval $(a, b] \subset \mathbb{R}^n$, there exists a sequence $\{(a_m, b_m) : m = 1, 2, \ldots\}$ of intervals in $\mathbb{R}^n$ converging to it such that for each of the intervals $(a_m, b_m]$ the assertion of the theorem is valid for both v_1 and v_2) that $v_1((a, b]) = v_2((a, b])$ for each $a, b \in \mathbb{R}^n$ with $a < b$ and hence that $v_1(B) = v_2(B)$ for each Borel set $B \subset \mathbb{R}^n$. ∎

1.1.3   Corollary (Rao, 1947; Laha and Lukacs, 1960)

Let $U, V_1, \ldots, V_n$, where $n \geqslant 1$, be (real) random variables defined on a probability space $(\Omega, \mathcal{E}, P)$ such that U is integrable. Then

$$E(U|V_1, \ldots, V_n) = 0, \quad \text{a.s. } [P] \tag{1.1.1}$$

if and only if

$$E(U \exp\{it_1 V_1 + \cdots + it_n V_n\}) = 0, \quad (t_1, \ldots, t_n) \in \mathbb{R}^n. \tag{1.1.2}$$

Proof   As

$$E(U \exp\{it_1 V_1 + \cdots + it_n V_n\}) = E\{\exp\{it_1 V_1 + \cdots + it_n V_n\} E(U | V_1, \ldots, V_n)\},$$

it is immediate that (1.1.1) implies (1.1.2). Also, it now follows that if (1.1.2) is valid then, in view of Corollary 1.1.2, we have

$$\int_{\{(V_1, \ldots, V_n) \in B\}} E(U | V_1, \ldots, V_n) \, dP = 0 \tag{1.1.3}$$

for each Borel set B of $\mathbb{R}^n$. On noting then that (1.1.1) is an obvious consequence of (1.1.3), we conclude that (1.1.2) implies (1.1.1). Hence, we have the corollary. ∎

1.1.4   Corollary (a restrictive version of the inversion theorem)

If ν is a real signed measure on $\mathbb{R}^n$ with Fourier transform ν^* such that it is integrable with respect to Lebesgue measure on $\mathbb{R}^n$, then ν is absolutely continuous with respect to Lebesgue measure on $\mathbb{R}^n$ with Random–Nikodym derivative

$$f_\nu(x) = \frac{1}{(2\pi)^n} \int_{\mathbb{R}^n} \exp\{-i\langle t, x \rangle\} \nu^*(t) \lambda(dt), \quad x \in \mathbb{R}^n, \tag{1.1.4}$$

where $\lambda(\bullet)$ denotes Lebesgue measure on $\mathbb{R}^n$.

Proof   Since ν^* is integrable with respect to Lebesgue measure, we have that, for the a and b as in Theorem 1.1.1,

$$\nu((a, b]) = \frac{1}{(2\pi)^n} \int_{\mathbb{R}^n} \left\{ \int_{(a,b]} e^{-i\langle t, x \rangle} \lambda(dx) \right\} \nu^*(t) \lambda(dt)$$

and hence, in view of Fubini's theorem and the Lebesgue dominated convergence theorem that for any $a, b \in \mathbb{R}^n$ with $a < b$

$$\nu((a, b]) = \frac{1}{(2\pi)^n} \int_{(a,b]} \left\{ \int_{\mathbb{R}^n} \exp\{-i\langle t, x \rangle\} \nu^*(t) \lambda(dt) \right\} \lambda \, dx). \tag{1.1.5}$$

(1.1.5) implies that ν is absolutely continuous with respect to Lebesgue measure on $\mathbb{R}^n$ with Randon–Nikodym derivative $f_\nu(\bullet)$. ∎

1.1.5   Theorem

If ν is a real signed measure on $\mathbb{R}^n$ with Fourier transform ν^*, then

$$\nu(\{x\}) = \lim_{\tau_r \to \infty, r=1,2,\ldots,n} \frac{1}{2^n \tau_1 \cdots \tau_n} \int_{-\tau_n}^{\tau_n} \cdots \int_{-\tau_1}^{\tau_1} \exp\{-i\langle t, x \rangle\} \nu^*(t) \, dt_1 \ldots dt_n.$$

$$\tag{1.1.6}$$

Proof On substituting $\int_{\mathbb{R}^n} \exp\{i\langle t, y\rangle\}\nu(dy)$ for $\nu^*(t)$ and using Fabini's theorem, we get (in obvious notation)

$$\frac{1}{2^n \prod\limits_{r=1}^{n} \tau_r} \int_{-\tau_n}^{\tau_n} \cdots \int_{-\tau_1}^{\tau_1} \exp\{-i\langle t, x\rangle\}\nu^*(t)\, dt_1 \ldots dt_n$$

$$= \int_{\mathbb{R}^n} \left(\prod_{r=1}^{n} \frac{\sin(\tau_r(y_r - x_r))}{\tau_r(y_r - x_r)}\right) \nu(dy)$$

$$= \nu(\{x\}) + \int_{\{x\}^c} \left(\prod_{r=1}^{n} \frac{\sin(\tau_r(y_r - x_r))}{\tau_r(y_r - x_r)}\right) \nu(dy), \tau_1, \ldots \tau_n > 0, \quad (1.1.7)$$

where we define for convenience $\sin(\tau_r(y_r - x_r))/\tau_r(y_r - x_r) = 1$ if $y_r = x_r$. In view of (1.1.7), the Lebesgue dominated convergence theorem then implies that (1.1.6) holds and hence we have the theorem. ∎

1.1.6 Remarks
(i) Corollary 1.1.2 follows also from Theorem 7.5 of Billingsley (1968) and hence implicitly via an argument based on the Weierstrass approximation theorem.

(ii) Corollary 1.1.3 implies immediately that (1.1.1) with 0 replaced by $E(U)$ is valid if and only if (1.1.2) with $U - E(U)$ in place of U is valid.

(iii) If ν is a real signed measure concentrated on the set

$$\prod_{r=1}^{n} \{m\lambda_r : m = 0, \pm1, \pm2, \ldots\}$$

(with $\lambda_r > 0$) (i.e. such that it gives value zero to each Borel set which is a subset of the complement of this set), then (1.1.7) implies immediately that for each point x of this set

$$\nu(\{x\}) = \frac{\prod\limits_{r=1}^{n} \lambda_r}{(2\pi)^n} \int_{-\pi/\lambda_n}^{\pi/\lambda_n} \cdots \int_{-\pi/\lambda_1}^{\pi/\lambda_1} \exp\{-i\langle t, x\rangle\}\nu^*(t)\, dt_1 \ldots dt_n;$$

this is obviously an extended form of the so-called inversion theorem for Fourier series.

1.1.7 Theorem
If for each of $r = 1, 2$, ν_r is a signed measure on $\mathbb{R}^n$ for which there exists a common open interval $\Theta \subset \mathbb{R}^n$ such that for every $\theta \in \Theta$ the function $\exp\{\langle\theta, \bullet\rangle\}$ is $|\nu_r|$-integrable, then the functions $\hat{\nu}_r : \Theta \to \mathbb{R}$ with values

$$\hat{\nu}_r(\theta) = \int_{\mathbb{R}^n} \exp\{\langle\theta, x\rangle\}\nu_r(dx) \quad (1.1.8)$$

are identical if and only if $\nu_1 = \nu_2$.

Proof The 'if' part of the assertion is straightforward. To prove the 'only if' part, assume now that $\hat{v}_1 = \hat{v}_2$ and fix a $\theta_0 \in \Theta$. It follows that there exists a neighborhood O of the origin in $\mathbb{R}^n$ such that for each point $y = (y_1, \ldots, y_n)$ of O, we have

$$\int_{\mathbb{R}^n} \exp \left\{ \langle \theta_0, x \rangle + \sum_{k=1}^{n} |y_k||x_k| \right\} |v_r|(dx) < \infty, \quad r = 1, 2.$$

Consequently, from Fubini's theorem, we get for each $t \in \mathbb{R}^n$, $y \in O$ and $r = 1, 2$

$$\int_{\mathbb{R}^n} \exp\{\langle \theta_0, x \rangle + i\langle t + y, x \rangle\} v_r(dx)$$

$$= \sum_{m_1=0}^{\infty} \cdots \sum_{m_n=0}^{\infty} \left(\prod_{k=1}^{n} \frac{(iy_k)^{m_k}}{m_k!} \right) \int_{\mathbb{R}^n} \exp\{\langle \theta_0, x \rangle + i\langle t, x \rangle\} \left(\prod_{k=1}^{n} x_k^{m_k} \right) v_r(dx),$$

which, in turn, implies inductively, in view of the relation $\hat{v}_1 = \hat{v}_2$ and the Lebesgue dominated convergence theorem, that for each $y \in O$ and $m = 0, 1, 2, \ldots$

$$\int_{\mathbb{R}^n} \exp\{\langle \theta_0, x \rangle + im\langle y, x \rangle\} v_1(dx) = \int_{\mathbb{R}^n} \exp\{\langle \theta_0, x \rangle + im\langle y, x \rangle\} v_2(dx)$$

(where the induction used is on m). From Corollary 1.1.2, we can then conclude that for each Borel set B of $\mathbb{R}^n$

$$\int_{B} \exp\{\langle \theta_0, x \rangle\} v_1(dx) = \int_{B} \exp\{\langle \theta_0, x \rangle\} v_2(dx),$$

implying that $v_1 = v_2$. This completes the proof of the theorem. ∎

For any nonnegative real measure μ on $\mathbb{R}^n$ for which

$$\left\{ \prod_{r=1}^{n} x_r^{k_r} : k_1, \ldots, k_n = 0, 1, \ldots \right\}$$

is a sequence of integrable Borel measurable functions, the sequence

$$\left\{ M_{k_1, \ldots, k_n} : k_1, \ldots, k_n = 0, 1, \ldots \right\},$$

where

$$M_{k_1, \ldots, k_n} = \int_{\mathbb{R}^n} \left(\prod_{r=1}^{n} x_r^{k_r} \right) \mu(dx)$$

is referred to as its moment sequence. Amongst the conditions that are sufficient for a moment sequence to determine the associated measure (uniquely), the most important one is due to Carleman; an elegant special case of this criterion was given earlier by Riesz. The conditions given by both Carleman and Riesz are implicitly linked with the uniqueness theorem for Fourier transforms (i.e. Corollary 1.1.2). In what follows we deal with these briefly; for a more detailed treatment of the 'moment problem' the reader is referred to Shohat and Tamarkin (1963) and Akhiezer (1965).

1.1.8   Theorem (the Riesz criterion)

Let $\{M_{k_1,\ldots,k_n} : k_1, \ldots, k_n = 0, 1, \ldots\}$ be a moment sequence and

$$\lambda_{2k} \equiv M_{2k,0,\ldots,0} + M_{0,2k,\ldots,0} + \cdots + M_{0,0,\ldots,2k}, \quad k = 1, 2, \ldots .$$

A sufficient condition for the moment sequence to determine the associated measure is that $\liminf_{k \geqslant 1} k^{-1}(\lambda_{2k})^{1/2k} < \infty$.

Proof   As in Feller (1966, p. 485), for each $h \in \mathbb{R}$ and $k = 1, 2, \ldots$

$$\left| e^{ih} - 1 - \frac{ih}{!!} - \cdots - \frac{(ih)^{2k-1}}{(2k-1)!} \right| \leqslant \frac{|h|^{2k}}{(2k)!}$$

implying that if μ is a nonnegative measure on $\mathbb{R}^n$ and $\zeta, t \in \mathbb{R}^n$, then

$$\int_{\mathbb{R}^n} \left| \exp\{i\langle \zeta, x \rangle\} \left(\exp\{i\langle t, x \rangle\} - 1 - \frac{i\langle t, x \rangle}{1!} - \cdots - \frac{(i\langle t, x \rangle)^{2k-1}}{(2k-1)!} \right) \right| \mu(dx)$$

$$\leqslant \int_{\mathbb{R}^n} \frac{|\langle t, x \rangle|^{2k}}{(2k)!} \mu(dx), \quad k = 1, 2, \ldots . \tag{1.1.9}$$

In view of the Cauchy–Schwarz and Hölder inequalities, we get

$$|\langle t, x \rangle|^{2k} \leqslant \|t\|^{2k} \|x\|^{2k}$$

$$= \|t\|^{2k} n^k \left(\sum_1^n x_r^2 / n \right)^k$$

$$\leqslant \|t\|^{2k} n^{k-1} \sum_1^n x_r^{2k}, \quad k = 1, 2, 3, \ldots ,$$

which implies that the right-hand side of (1.1.9) is less than or equal to $\|t\|^{2k} n^{k-1} \lambda_{2k}/(2k)!$ if μ is a measure of which $\{M_{k_1,\ldots,k_n}\}$ is the moment sequence. If $\liminf_{k \geqslant 1} k^{-1}(\lambda_{2k})^{1/2k} < \infty$, there exists a neighborhood O of the origin in $\mathbb{R}^n$ such that for every $t \in O$, $\liminf_{k \geqslant 1}\{\|t\|^{2k} n^{k-1} \lambda_{2k}/(2k)!\} = 0$. If μ and ν are measures on $\mathbb{R}^n$ having $\{M_{k_1,\ldots,k_n}\}$ as their moment sequence and μ^* and ν^* are the corresponding Fourier transforms, then for $k = 1, 2, \ldots$

$$|\mu^*(\zeta + t) - \nu^*(\zeta + t)| \leqslant 2\|t\|^{2k} n^{k-1} \lambda_{2k}/(2k)!, \quad t \in O, \quad \zeta \in [-\zeta^*, \zeta^*], \tag{1.1.10}$$

provided ζ^* satisfies the condition that the restrictions to $[-\zeta^*, \zeta^*]$ of μ^* and ν^* are identical. Assuming that $\liminf_{k \geq 1} k^{-1}(\lambda_{2k})^{1/2k} < \infty$, we get the left-hand side of (1.1.10) to be equal to zero. This, in turn, implies that, under the stated condition, $\mu^* = \nu^*$ and hence, in view of Corollary 1.1.2, $\mu = \nu$. This establishes the theorem. ∎

1.1.9 Theorem (the Carleman criterion)

Let $\{M_{k_1,\ldots,k_n}\}$ and $\{\lambda_{2k}\}$ be as defined in Theorem 1.1.8. A sufficient condition that the associated measure be determined by $\{M_{k_1,\ldots,k_n}\}$ is that

$$\sum_{k=1}^{\infty} (\lambda_{2k})^{-1/2k} = \infty.$$

(If $\lambda_{2k} \equiv 0$, we take $(\lambda_{2k})^{-1/2k} \equiv \infty$.)

Proof   See, for example, Shohat and Tamarkin (1963, pp. 14–21).

1.1.10 Remarks

(i) Theorems 1.1.8 and 1.1.9 as given above are indeed extended versions of the original results of Riesz and Carleman respectively given for the case of $n = 1$.

(ii) If the criterion of Theorem 1.1.8 is met, then it follows that, for some positive constant $c > 0$, there exists an increasing sequence $\{k_r : r = 1, 2, \ldots\}$ of positive integers such that

$$(k_{r+1} - k_r)\left(\lambda_{2k_{r+1}}\right)^{-1/2k_{r+1}} > c, \quad r = 1, 2, \ldots.$$

As $\{(\lambda_{2k})^{-1/2k} : k = 1, 2, \ldots\}$ is a decreasing sequence, we can hence claim that

$$\sum_{k=1}^{\infty} (\lambda_{2k})^{-1/2k} \geq k_1 \left(\lambda_{2k_1}\right)^{-1/2k} + \sum_{r=1}^{\infty} (k_{r+1} - k_r)\left(\lambda_{2k_{r+1}}\right)^{-1/2k_{r+1}} = \infty,$$

implying that the criterion of Theorem 1.1.9 holds. Thus, we have Theorem 1.1.8 as a corollary of Theorem 1.1.9.

(iii) As observed in Shohat and Tamarkin (1963, p. 22) (on correcting some minor misprints),

$$\int_0^{\infty} x^k e^{-\lambda x^\alpha} \sin(\lambda x^\alpha \tan \alpha \pi)\, dx = 0, \quad k = 0, 1, \ldots; \quad \lambda > 0, \quad 0 < \alpha < \tfrac{1}{2},$$

and

$$\int_{-\infty}^{\infty} x^k e^{-\lambda |x|^\alpha} \cos\left(\lambda |x|^\alpha \tan \frac{\pi \alpha}{2}\right) dx = 0, \quad k = 0, 1, \ldots; \quad \lambda > 0, \quad 0 < \alpha < 1.$$

In view of these identities, it follows that if $\{M_k : k = 0, 1, \ldots\}$ is the moment sequence of a measure on $\mathbb{R}$ that is absolutely continuous with respect to Lebesgue measure with Radon–Nikodym derivative f such that, for some $\lambda > 0$ and $0 < \alpha < \frac{1}{2}$,

$$f(x) \geqslant \exp\{-\lambda x^{\alpha}\} \quad \text{for sufficiently large positive } x \qquad (1.1.11a)$$

or

$$f(-x) \geqslant \exp\{-\lambda x^{\alpha}\} \quad \text{for sufficiently large positive } x, \qquad (1.1.11b)$$

or, for some $\lambda, \delta > 0$ and $0 < \alpha < 1$,

$$f(x) \geqslant \delta \exp\{-\lambda |x|^{\alpha}\} \quad \text{for all } x, \qquad (1.1.11c)$$

then it does not determine the associated measure. (Note that in the case of (1.1.11a), there exist $\delta, c > 0$ such that the function f^* defined by

$$f^*(x) = \begin{cases} f(x) - \delta \exp\{-\lambda(x-c)^{\alpha}\}\sin(\lambda(x-c)^{\alpha}\tan\alpha\pi) & \text{if } x \geqslant c \\ f(x) & \text{if } x < c, \end{cases}$$

with α and λ as in (1.1.11a), is nonnegative and satisfies

$$\int_{\mathbb{R}} x^k f^*(x)\,\mathrm{d}x = M_k, \quad k = 0, 1, \dots;$$

the functions f^* having the specified properties could be constructed analogously in the other cases.)

(iv) For $n > 1$, it is now easy to obtain a nonnegative real product measure $\Pi_{i=1}^{n}\mu_i$ with moment sequence such that it does not determine the associated measure. (Take here μ_i for which the moment sequence is well defined with at least one of them not determined by its moment sequence.)

(v) $\{M_k^{(\alpha)} : k = 0, 1, \dots\}$ defined, for $0 < \alpha \leqslant 2$, by

$$M_k^{(\alpha)} = \begin{cases} \mathrm{e}^{k^{\alpha}} & \text{if } 1 < \alpha \leqslant 2 \\ \mathrm{e}^{-k^{\alpha}} & \text{if } 0 < \alpha < 1 \\ k^k & \text{if } \alpha = 1 \end{cases} \qquad (1.1.12)$$

is the moment sequence corresponding to e^{X_α} where X_α is an extreme stable random variable with characteristic exponent α (see, for example, Eaton *et al*, 1971). What is observed in (iii) above and in Zolotarev (1964) implies that, for $1 < \alpha \leqslant 2$, the sequence $\{M_k^{(\alpha)}\}$ does not determine the corresponding distribution, while Theorem 1.1.8 implies that, for $0 < \alpha \leqslant 1$, the sequence in question determines the distribution. (One could obviously go further by noting, amongst other things, that the result remains valid even when k^{α} in (1.1.12) for $\alpha \neq 1$ is replaced by ck^{α} with $c > 0$.) Using a construction essentially due to Stieltjes, Heyde (1964) showed the validity of this interesting result for $\alpha = 2$.

(vi) The moment sequence of the Weibull distribution with distribution function

$$F(x) = \begin{cases} 1 - \mathrm{e}^{-\lambda x^{\alpha}}, & x \geqslant 0, \\ 0, & x < 0, \end{cases}$$

where $\lambda, \alpha > 0$, also has an inquisitive pattern. We have, in view of Theorem 1.1.8 and (iii) above, that the sequence determines the Weibull distribution in the class of all the distributions concentrated on $\mathbb{R}_+ (= [0, \infty))$ if and only if $\alpha \geq \frac{1}{2}$. Furthermore, Theorem 1.1.8 implies that it determines the distribution in question in the class of all distributions (on $\mathbb{R}$) if $\alpha \geq 1$.

For any signed measure ν on $\mathbb{R}^n$, if the interior of

$$\mathcal{D}_\nu = \{\theta \in \mathbb{R}^n : \exp\langle\theta, \bullet\rangle \text{ is } |\nu| - \text{integrable}\}$$

is nonempty, then we define the function $\hat{\nu} : \mathcal{D}_\nu \to \mathbb{R}$ such that

$$\hat{\nu}(\theta) = \int_{\mathbb{R}^n} \exp\{\langle\theta, x\rangle\}\nu(dx), \quad \theta \in \mathcal{D}_\nu$$

to be the Laplace transform of ν. Similarly, for any real sequence $\mathbf{a} = \{a_{j_1,\ldots,j_n} : j_1, \ldots, j_n = 0, 1, \ldots\}$, if the interior of

$$\mathcal{D}_\mathbf{a} = \left\{(z_1, \ldots, z_n) \in \mathbb{R}^n : \sum_{j_1=0}^{\infty} \cdots \sum_{j_n=0}^{\infty} \left| a_{j_1,\ldots,j_n} \prod_{r=1}^{n} z_r^{j_r} \right| < \infty \right\}$$

is nonempty, then the function $A : \mathcal{D}_\mathbf{a} \to \mathbb{R}$ given by

$$A(z_1, \ldots, z_n) = \sum_{j_1=0}^{\infty} \cdots \sum_{j_n=0}^{\infty} a_{j_1,\ldots,j_n} \prod_{r=1}^{n} z_r^{j_r}, \quad (z_1, \ldots, z_n) \in \mathcal{D}_\mathbf{a}$$

is defined to be the generating function of the sequence $\mathbf{a}$; call this a probability generating function whenever $\mathbf{a}$ is a probability distribution. (The latter definition could obviously be expressed in terms of a signed measure concentrated on $\mathbb{N}_0^n$, where $\mathbb{N}_0 = \{0, 1, \ldots\}$, i.e. in terms of a signed measure ν on $\mathbb{R}^n$ such that $|\nu|((\mathbb{N}_0^n)^c) = 0$.) Note, however, that we are not claiming here that the definitions given are universally followed. As $\hat{\nu}$ given above is a version of a multilateral Laplace transform, we have decided to call it a Laplace transform even though this is defined usually with $-\theta$ in place of θ; for the definition of a generating function, we have chosen the present version as it fits in well with the definitions of other integral transforms that we have encountered and we find it more convenient to use this. (One may observe in passing that the Laplace transform defined above reduces to the moment generating function of a probability measure when ν is a probability measure.) Theorem 1.1.7 implies that the restriction of a Laplace transform $\hat{\nu}$ to a nonempty open subset of the domain $\mathcal{D}_\nu$ of $\hat{\nu}$ determines ν; from the theorem or a simple power-series argument, it can further be seen that an analogous result holds for a generating function (and hence also for a probability generating function).

We conclude the discussion of the transforms in the present section by giving the relevant continuity theorems.

1.1.11   Theorem

Suppose that $P_m, m \geqslant 1$ and P are probability measures on $\mathbb{R}^n$ with the corresponding Fourier transforms (or, in other words, characteristic functions) as $P_m^*, m \geqslant 1$ and P^* respectively. Then $\{P_m\}$ converges weakly to P if and only if $\{P_m^*\}$ converges pointwise to P^*.

Proof   See, for instance, Billingsley (1968, pp. 46–47).

1.1.12   Theorem

Suppose that $\mu_m, m \geqslant 1$ are nonnegative measures on $\mathbb{R}^n$ for which the corresponding Laplace transforms $\hat{\mu}_m, m \geqslant 1$ are defined with domains of definition as supersets of some common nonempty open set $O \subset \mathbb{R}^n$. If, for each $\theta \in O, \{\hat{\mu}_m(\theta)\}$ converges to a real number $\hat{\mu}(\theta)$, then the function $\hat{\mu} : O \to \mathbb{R}$ is the restriction to O of the Laplace transform of some measure μ on $\mathbb{R}^n$. Moreover, in that case, the sequence $\{\mu_m\}$ satisfies the condition that it converges weakly to μ, i.e. $\{\mu_m((a, b])\}$ converges to $\mu((a, b])$ for each $a, b \in \mathbb{R}^n$ with $a < b$ such that $\mu((a, b)) = \mu([a, b])$.

Proof   Let $\theta_0 \in O$. The results follow trivially if $\hat{\mu}(\theta_0) = 0$. We can hence assume $\hat{\mu}(\theta_0) > 0$. There is also no loss of generality in assuming $\hat{\mu}_m(\theta_0) > 0$ for each $m \geqslant 1$. We can now define the sequence $\{P_m^{(\theta_0)}\}$ of probability measures on $\mathbb{R}^n$ such that for each $m \geqslant 1$ and each Borel set B of $\mathbb{R}^n$

$$P_m^{(\theta_0)}(B) = (\hat{\mu}_m(\theta_0))^{-1} \int_B \exp\{\langle \theta_0, x \rangle\} \mu_m(dx). \tag{1.1.13}$$

It is easily seen that this sequence is tight and relative to it for each $\theta \in O$ the function $\exp\{\langle \theta - \theta_0, \bullet \rangle\}$ is uniformly integrable. (We are slightly abusing here the notation of Billingsley (1968, p. 32) used for defining the uniform integrability; indeed, by the uniform integrability of the function $\exp\{\langle \theta - \theta_0, \bullet \rangle\}$, we mean that

$$\lim_{\alpha \to \infty} \sup_m \int_{\{x : \exp\{\langle \theta - \theta_0, x \rangle\} \geqslant \alpha\}} \exp\{\langle \theta - \theta_0, x \rangle\} P_m^{(\theta_0)}(dx) = 0.$$

By Prohorov's theorem (Billingsley, 1968, p. 37), $\{P_m^{(\theta_0)}\}$ is relatively compact and hence each of its subsequences $\{P_{m'}^{(\theta_0)}\}$ contains a further subsequence $\{P_{m''}^{(\theta_0)}\}$ converging weakly to some probability measure $P^{(\theta_0)}$ satisfying, in view of the continuity and the aforementioned uniform integrability relative to $\{P_m^{(\theta_0)}\}$ of $\exp\{\langle \theta - \theta_0, \bullet \rangle\}$,

$$\int_{\mathbb{R}^n} \exp\{\langle \theta - \theta_0, x \rangle\} P^{(\theta_0)}(dx) = \lim_{m'' \to \infty} \int_{\mathbb{R}^n} \exp\{\langle \theta - \theta_0, x \rangle\} P_{m''}^{(\theta_0)}(dx)$$

$$= \frac{\hat{\mu}(\theta)}{\hat{\mu}(\theta_0)}, \quad \theta \in O \tag{1.1.14}$$

(Billingsley, 1968, pp. 31–33). From (1.1.14), it is then immediate that the first part of the theorem is valid. As the uniqueness theorem for Laplace transforms implies because of (1.1.14) that $P^{(\theta_0)}$ is determined uniquely by the restriction to O of $\hat{\mu}$, Theorem 2.3 of Billingsley gives that $\{P_m^{(\theta_0)}\}$ converges weakly to $P^{(\theta_0)}$. By (1.1.13), for any interval $(a, b]$ of $\mathbb{R}^n$, we have

$$\mu_m((a, b]) = \hat{\mu}_m(\theta_0) \int_{(a,b]} \exp\{-\langle \theta_0, x \rangle\} P_m^{(\theta_0)}(\mathrm{d}x), \quad m = 1, 2, \ldots,$$

which implies in view of Part (iii) of Billingsley's Theorem 5.2 that the second part of our theorem is valid. ∎

1.1.13   Corollary

Let, for each $m \geqslant 1$, $\{a_{j_1,\ldots,j_n}^{(m)} : j_1, \ldots, j_n = 0, 1, \ldots\}$ be a nonnegative real sequence for which the corresponding generating function A_m is defined with domain of definition $\mathcal{D}^{(m)}$, and let there exist a nonempty open set O of $\mathbb{R}^n_+$ such that $O \subset \cap_{m=1}^{\infty} \mathcal{D}^{(m)}$. If, for each $z \in O$, $\{A_m(z)\}$ converges to a real number $A(z)$, then $A : O \to \mathbb{R}$ denotes the restriction to O of the generating function A^* of some nonnegative real sequence $\{a_{j_1,\ldots,j_n} : j_1, \ldots, j_n = 0, 1, \ldots\}$. Furthermore in that case,

$$a_{j_1,\ldots,j_n}^{(m)} \to a_{j_1,\ldots,j_n} \quad \text{as } m \to \infty \tag{1.1.15}$$

for each $j_1, \ldots, j_n \geqslant 0$; (1.1.15) implies that $\{A_m(z)\}$ converges to $A^*(z)$ for each $z \in [-z_0, z_0]$ for each $z_0 \in O$.

Proof   The corollary but for its last part follows from Theorem 1.1.12. To prove the last part of the result, it is sufficient to note that, under the relevant conditions, $\{a_{j_1,\ldots,j_n}^{(m)} \prod_{r=1}^{n} z_r^{j_r} : m = 1, 2, \ldots\}$ is uniformly integrable (with respect to the counting measure) on $\mathbb{N}_0^n$ for each $(z_1, \ldots, z_n) \in [-z_0, z_0]$ for each $z_0 \in O$. The assertion then follows trivially from Theorem 5.4 of Billingsley (1968). ∎

1.1.14   Remarks

(i) In view of the Lebesgue dominated convergence theorem, it follows that Theorem 1.1.7 remains valid if we take Θ as a set dense in an open interval of $\mathbb{R}^n$ instead of an open interval itself. This, in turn, implies that both Theorem 1.1.12 and Corollary 1.1.13 hold if we take O as a set dense in a nonempty open set instead of as an open set itself (of $\mathbb{R}^n$ and $\mathbb{R}^n_+$ respectively).

(ii) A specialized version of Theorem 1.1.12 relative to measures concentrated on $\mathbb{R}_+$ appears in Feller (1966, p. 410) as an extended continuity theorem. Another observation that is worth making at this stage is that although to prove the general Theorem 1.1.12 we have used an indirect approach involving Billingsley's (1968) results on convergence of probability measures, it is possible to prove this theorem directly using slightly modified versions of the relevant arguments implied in the given proof.

(iii) Under the assumptions in Corollary 1.1.13, the proof of Theorem 1.1.12 simplifies considerably and hence it goes without saying that for obtaining Corollary 1.1.13 we need only an argument that is substantially simpler than the one used to establish Theorem 1.1.12.

(iv) As shown earlier, if we are to go by the tradition, then a multilateral Laplace transform is to be defined with $\exp\{\langle\theta, x\rangle\}$ in our definition by $\exp\{-\langle\theta, x\rangle\}$. However, since we are not restricting ourselves to a measure ν on $\mathbb{R}_+$ or $\mathbb{R}_+^n$, the definition as given by us appears to us to be more natural for a transform than the one that is followed in the literature. However, the applications of the results on Laplace transforms in this monograph do not depend on what form for the definition we choose.

(v) For any nonnegative real measure for which the Laplace transform exists with O as an interior point of its domain of definition, the corresponding moment sequence satisfies the Riesz criterion of Theorem 1.1.8 and hence determines the measure.

(vi) Suppose $\hat{\nu}$ is the Laplace transform of a nonnegative measure ν on $\mathbb{R}^n$ with domain of definition $\mathcal{D}_\nu$, θ_0 is an interior point of $\mathcal{D}_\nu$, and $\{\theta_m : m = 1, 2, \ldots\}$ is a sequence of members of $\mathcal{D}_\nu$ such that all of the partial derivatives of $\hat{\nu}$ at θ_0 are determined by $\{\hat{\nu}(\theta_m) : m = 1, 2, \ldots\}$ (jointly with $\{\theta_m\}$). (The existence of $\{\theta_m\}$ with the stated property is easily seen.) Then as a corollary to (v) above, we have that, given θ_0 and $\{\theta_m\}$, $\{\hat{\nu}(\theta_m) : m = 1, 2, \ldots\}$ determines ν. (Defining the Mellin transform of a measure on $(0, \infty)^n$ appropriately, we can easily express this result in terms of a Mellin transform.) A specialized version of this result for $n = 1$ has recently been given by Lin (1992).

1.2   SOME AUXILIARY RESULTS ON PROBABILITY MEASURES AND STOCHASTIC PROCESSES

The following existence theorem for a product probability measure due to Tulcea (1949, 1950) plays a vital role in this monograph; the theorem has also appeared in Ash (1972, pp. 109–110) and Loéve (1963, pp. 137–138).

1.2.1   Theorem

Let $(\Omega_j, \mathcal{E}_j)$, $j = 1, 2, \ldots$ be measurable spaces, $\Omega = \Pi_{j=1}^\infty \Omega_j$ and $\mathcal{E} = \Pi_{j=1}^\infty \mathcal{E}_j$. Further, let P_1 be a probability measure on $\mathcal{E}_1$ and, for each $j = 1, 2, \ldots$ and each $(\omega_1, \ldots, \omega_j) \in \Pi_{i=1}^j \Omega_i$, $P_{j+1}(\omega_1, \ldots, \omega_j, \bullet)$ be a probability measure on $\mathcal{E}_{j+1}$. Assume that $P_{j+1}(\bullet, C)$ is $\Pi_{i=1}^j \mathcal{E}_i$ - measurable for each $C \in \mathcal{E}_{j+1}$ and $j = 1, 2, \ldots$. Then there exists a unique probability measure P on $\mathcal{E}$ such that

$$P\left(B_j \times \prod_{i=j+1}^{\infty} \Omega_i\right) = \int_{\Omega_1} P_1(d\omega_1) \int_{\Omega_2} P_2(\omega_1, d\omega_2) \cdots$$

$$\int_{\Omega_{j-1}} P_{j-1}(\omega_1, \ldots, \omega_{j-2}, d\omega_{j-1}) \int_{\Omega_j} I_{B_j}(\omega_1, \ldots, \omega_j) P_j(\omega_1 \ldots, \omega_{j-1}, d\omega_j),$$

$$B_j \in \prod_{i=1}^{j} \mathcal{E}_i, \quad j = 1, 2, \ldots . \tag{1.2.1}$$

(This theorem follows from the Caratheodory extension theorem on first noting that there exists a finitely additive set function $\hat{P}$ on the field $\mathcal{F}$ of measurable cylinders corresponding to the space $(\Omega, \mathcal{E})$ such that (1.2.1) is satisfied with $\hat{P}$ in place of P, and that $\hat{P}$ so obtained is continuous from above at the empty set. For the details, see Ash or Loéve.)

We also need some additional tools corresponding to stochastic processes in the study.

1.2.2 Theorem

(a) If $\{X_n, \mathcal{E}_n : n = 0, 1, \ldots\}$ is a submartingale satisfying $\sup_n E(X_n^+) < \infty$, then there exists an integrable random variable X_∞ that is measurable relative to $\sigma(\cup_{n=0}^{\infty} \mathcal{E}_n)$ such that (a.s.) $\lim_n X_n = X_\infty$ and $E(|X_\infty|) \leqslant \sup_n E(|X_n|)$.

(b) If the sequence in (a) is a martingale satisfying $\sup_n E(|X_n|^p) < \infty$ for some $p > 1$, then $\{X_n\}$ converges to X_∞ in pth mean (i.e. in L_p) and $\{X_n, \mathcal{E}_n : n = 0, 1, \ldots, \infty\}$ is a martingale, where X_∞ is as in (a) and $\mathcal{E}_\infty = \sigma(\cup_{n=0}^{\infty} \mathcal{E}_n)$.

Proof (a) follows from (i) of Theorem 4.15 on page 324 in Doob (1953) and on noting the validity of (a), (b) follows essentially from (ii) and (iii) on page 319 of Doob (1953). ■

1.2.3 Theorem

Let $(\Omega^j, \mathcal{E}^j, P_j)$, $j = 1, 2, \ldots$ be probability spaces with $((\Omega^j, \mathcal{E}^j)$ in standard notation and) measures P_j as symmetric (i.e. such that for every permutation $(i_1, \ldots, i_j)$ of $(1, \ldots, j)$ and members $E_1, \ldots, E_j$ of $\mathcal{E}$,

$$P_j(E_{i_1} \times \cdots \times E_{i_j}) = P_j(E_1 \times \cdots \times E_j))$$

and satisfying

$$P_j(E^{(j)}) = P_{j+1}(E^{(j)} \times \Omega), E^{(j)} \in \mathcal{E}^j.$$

Then

$$(P_{j+j'}(E_1 \times \cdots \times E_{j+j'}))^2 \leqslant P_{2j}(E_1 \times \cdots \times E_j \times E_1 \times \cdots \times E_j)$$

$$\bullet P_{2j'}(E_{j+1} \times \cdots \times E_{j+j'} \times E_{j+1} \times \cdots \times E_{j+j'}),$$

$$E_i \in \mathcal{E}, i = 1, 2, \ldots, j + j'; j, j' = 1, 2 \ldots . \tag{1.2.2}$$

Proof   In view of the properties of P_j's, it is sufficient if we prove the inequality of (1.2.2) for $j = j' = 1$ and hence that

$$(P_2(E_1 \times E_2))^2 \leqslant P_2(E_1 \times E_1) P_2(E_2 \times E_2), \quad E_1, E_2, \in \mathcal{E}. \tag{1.2.3}$$

(Note that with obvious notational alterations (1.2.2) can be reduced to the form (1.2.3).) For each $j \in \{1, 2, \ldots\}$, let $X_i^{(j)}, i = 1, 2, \ldots, j$ denote the projection mappings on $(\Omega^j, \mathcal{E}^j, P_j)$ and note that if $E_1, E_2 \in \mathcal{E}$, we have

$$E_j \left\{ \left(\frac{1}{j} \sum_{i=1}^{j} I_{\{X_i^{(j)} \in E_1\}} \right) \left(\frac{1}{j} \sum_{i=1}^{j} I_{\{X_i^{(j)} \in E_2\}} \right) \right\}$$

$$= \frac{1}{j} P_1(E_1 \cap E_2) + \frac{j(j-1)}{j^2} P_2(E_1 \times E_2), \quad j = 1, 2, \ldots . \tag{1.2.4}$$

Denoting the left-hand side of (1.2.4) by $M_j(E_1, E_2)$ and using the Cauchy–Schwarz inequality, we can then see that

$$(P_2(E_1 \times E_2))^2 = \left(\lim_{j \to \infty} M_j(E_1, E_2) \right)^2$$

$$\leqslant \left(\lim_{j \to \infty} M_j(E_1, E_1) \right) \left(\lim_{j \to \infty} M_j(E_2, E_2) \right)$$

$$= P_2(E_1 \times E_1) P_2(E_2 \times E_2).$$

Hence we have the result. ∎

1.2.4   Corollary

Let $(\Omega^j, \mathcal{E}^j, P_j), j = 1, 2, \ldots$ be as in Theorem 1.2.3. Then

$$P_{2j}(E_1 \times \cdots \times E_j \times E_1 \times \cdots \times E_j) \geqslant (P_j(E_1 \times \cdots \times E_j))^2,$$

$$E_i \in \mathcal{E}, \quad i = 1, 2, \ldots, j, \quad j = 1, 2, \ldots . \tag{1.2.5}$$

Proof   On taking $E_{j+1} = \cdots = E_{j+j'} = \Omega$, (1.2.2) reduces to (1.2.5). ∎

1.2.5   Theorem

Let

$$H(x) = \int_S H(x + y) \mu(dy) \tag{1.2.6}$$

be an integral equation with S as an Abelian topological semigroup with zero element, H as a nonnegative continuous function on S and μ as a σ-finite measure on (the Borel σ-field of) S. Let $S^*(\mu)$ be the smallest closed subsemigroup of S, with zero element and containing supp[μ]. (For any measure η on S, $\{x \in S :$

$\eta(U_x) > 0$ for each neighborhood U_x of $x\}$ is referred to as the support of η; we denote it by supp[η].) Then, for every $x \in S$ and $y, z \in S^*(\mu)$, we have

$$H(x + 2y)H(x + 2z) \geqslant (H(x + y + z))^2. \qquad (1.2.7)$$

Proof   It is sufficient if we prove that for every $n \geqslant 1$ and

$$y_1, \ldots, y_n, z_1, \ldots, z_n \in (\text{supp}[\mu]) \cup \{0\},$$

we have

$$H(x + 2(y_1 + \cdots + y_n))H(x + 2(z_1 + \cdots + z_n))$$
$$\geqslant (H(x + y_1 + \cdots + y_n + z_1 + \cdots + z_n))^2. \qquad (1.2.8)$$

There is no loss of generality in assuming $\{0\} \subset \text{supp}[\mu]$. We need a proof for (1.2.8) only when $x, y_1, \ldots, y_n, z_1, \ldots, z_n$ are such that

$$H(x + y_1 + \cdots + y_n + z_1 + \cdots + z_n) \neq 0.$$

In that case, we can take $H(x) \neq 0$ and, assuming x to be fixed, have probability spaces $(\Omega^j, \mathcal{E}^j, P_j)$, $j = 1, 2, \ldots$ meeting the requirements of Theorem 1.2.3 such that $\Omega = S$, $\mathcal{E} =$ the Borel σ-field of S and P_j are the probability measures on $\mathcal{E}^j$ for which for each $E_1, \ldots, E_j \in \mathcal{E}$:

$$P_j(E_1 \times \cdots \times E_j)$$
$$= (H(x))^{-1} \int_{E_j} \cdots \int_{E_1} H(x + y_1 + \cdots + y_j)\mu(dy_1) \cdots \mu(dy_j).$$

The inequality (1.2.8) now follows from (1.2.2) on noting that given

$$y_1, \ldots, y_n, z_1, \ldots, z_n \in \text{supp}[\mu],$$

in view of the continuity of H, there exists a sequence

$$\{O_{ij} : i = 1, 2, \ldots, 2n, \ j = 1, 2, \ldots\}$$

of open sets of S with positive μ values such that as $j \to \infty$,

$$P_{2n}\left(O_{1j} \times \cdots \times O_{2n,j}\right) \bigg/ \prod_{i=1}^{2n} \mu\left(O_{ij}\right)$$
$$\to H\left(x + y_1 + \cdots + y_n + z_1 + \cdots + z_n\right)/H(x),$$

$$P_{2n}\left(O_{1j} \times \cdots \times O_{nj} \times O_{1j} \times \cdots \times O_{nj}\right) \bigg/ \left(\prod_{i=1}^{n} \mu(O_{ij})\right)^2$$
$$\to H\left(x + 2\left(y_1 + \cdots + y_n\right)\right)/H(x)$$

and

$$P_{2n}(O_{n+1,j} \times \cdots \times O_{2n,j} \times O_{n+1,j} \times \cdots \times O_{2n,j}) \bigg/ \left(\prod_{i=n+1}^{2n} \mu(O_{ij}) \right)^2$$

$$\to H(x + 2(z_1 + \cdots + z_n))/H(x).$$

(In the case of a metric space S we can take $O_{ij}, i = 1, 2, \ldots, 2n$ to be open spheres with centers $y_1, \ldots, y_n, z_1, \ldots, z_n$ respectively and radii r_j converging to zero as $j \to \infty$.) Hence we have the theorem. ∎

1.2.6 Corollary

Under the assumptions in Theorem 1.2.5, we have, for every $x \in S$ and $y \in S^*(\mu)$,

$$H(x)H(x + 2y) \geqslant (H(x + y))^2.$$

Proof Take $z = 0$ in (1.2.7). ∎

1.2.7 Theorem

Let $(S^\infty, (\mathcal{B}(S))^\infty, P)$ be a probability space with S as a Polish space (i.e. as a complete separable metric space) and $\mathcal{B}(S)$ as the corresponding Borel σ-field, such that the coordinate mappings are exchangeable random elements (i.e. having P as symmetric in the sense of Hewitt and Savage (1955)). Then, for some sub σ-field $\mathcal{G}$ of $(\mathcal{B}(S))^\infty$, there exists a regular conditional probability $P(C|G)(\omega), C \in (\mathcal{B}(S))^\infty, \omega \in (\Omega =)S^\infty$ given G for which, for each ω, $P(\bullet|G)(\omega)$ yields the coordinate mappings to be independent and identically distributed random elements. Moreover, G can be taken to be either the invariant σ-field $\mathcal{I}$ or the tail σ-field $\mathcal{T}$ (relative to the coordinates of the space), with a common regular conditional probability having the stated property.

Proof Denote the coordinate mappings by $X_n, n \geqslant 1$ and define

$$\xi_n(B) = \frac{1}{n} \sum_{i=1}^{n} I_{\{X_i \in B\}}, \quad B \in \mathcal{B}(S), \quad n = 1, 2, \ldots .$$

Now, follow essentially the argument in Loéve (1963, p. 365) or Chow and Teicher (1979, pp. 220–221): For each $B \in \mathcal{B}(S)$,

$$E((\xi_m(B) - \xi_n(B))^2) = \frac{|m - n|}{mn}(P\{X_1 \in B\} - P\{X_1 \in B, X_2 \in B\})$$

$$\to 0 \quad \text{as } m, n \to \infty,$$

implying that $\{\xi_n(B) : n = 1, 2, \ldots\}$ converges in probability to some $\mathcal{T}$-measurable function $\xi(B)$ with values in [0,1]. Consequently, for each invariant set C of S^∞

and members $B_1, \ldots, B_m$ of $\mathcal{B}(S)$ and $m = 1, 2, \ldots$, we have

$$E\left(\left(\prod_{j=1}^{m} \xi_n(B_j)\right) I_C\right) \rightarrow E\left(\left(\prod_{j=1}^{m} \xi(B_j)\right) I_C\right) \quad \text{as } n \rightarrow \infty. \tag{1.2.9}$$

After using a minor manipulation involving an extended form of (1.2.4), we have that the left-hand side of (1.2.9) also tends as $n \rightarrow \infty$ to

$$P(\{X_1 \in B_1, \ldots, X_m \in B_m\} \cap C)$$

and hence that

$$P(\{X \in B_1, \ldots, X_m \in B_m\} \cap C) = E\left(\left(\prod_{j=1}^{m} \xi(B_j)\right) I_C\right)$$

for each $B_1, \ldots, B_m \in \mathcal{B}(S), m = 1, 2, \ldots$ and invariant set C. This in turn implies that

$$E\left(\prod_{j=1}^{m} \xi(B_j)|\mathcal{I}\right) = P\{X_1 \in B_1, \ldots, X_m \in B_m|\mathcal{I}\} \quad \text{a.s. }, B_1, \ldots, B_m \in \mathcal{B}(S),$$

$$m = 1, 2, \ldots . \tag{1.2.10}$$

As $\mathfrak{J} \subset \mathcal{T}$ and $\xi(B_j)$ are $\mathfrak{J}$-measurable, we have the left-hand side of (1.2.10) to be equal to $\prod_{j=1}^{m} \xi(B_j)$ a.s. implying that the right-hand side of the identity is equal to $P\{X_1 \in B_1, \ldots, X_m \in B_m|\mathcal{I}\}$ a.s. Noting in particular that the case $m = 1$ leads us to

$$P\{X_1 \in B|\mathcal{I}\} = \xi(B) = P\{X_1 \in B|\mathcal{T}\} \quad \text{a.s., } B \in \mathcal{B}(S),$$

we then get

$$P\{X_1 \in B_1, \ldots, X_m \in B_m|\mathcal{I}\} = \prod_{i=1}^{m} P\{X_i \in B_i|\mathcal{I}\}$$

$$= \prod_{i=1}^{m} P\{X_i \in B_i|\mathcal{T}\}$$

$$= P\{X_1 \in B_1, \ldots, X_m \in B_m|\mathcal{T}\} \text{ a.s.,}$$

$$B_1, \ldots, B_m \in \mathcal{B}(S), \quad m = 1, 2, \ldots . \tag{1.2.11}$$

For a further argument, we use the information provided in Theorem 6.6.6 on page 265 in Ash (1972) that a Polish space is Borel equivalent to a Borel subset of [0,1]. In view of this, on applying a standard technique, we get a regular conditional probability version corresponding to $P\{X_1 \in B|\mathcal{T}\}, B \in \mathcal{B}(S)$. (The result also follows as a corollary of Theorem 10.2.2 on page 270 in Dudley (1989).) For each $\omega \in S^\infty$, let Q_ω denote the corresponding member of the regular probability and $Q^*(\bullet, \omega)$ denote the product measure $(Q_\omega)^\infty(\bullet)$ on $(\mathcal{B}(S))^\infty$. Define

$$\mathfrak{C} = \{C \in (\mathcal{B}(S))^\infty : Q^*(C|\bullet) = P(C|\mathcal{I}) = P(C|\mathcal{T}) \quad \text{a.s.}$$

$$\text{and} \quad Q^*(C|\bullet) \text{ is measurable}\}.$$

In view of (1.2.11), it follows that $\mathfrak{C}$ contains all finite disjoint unions of measurable rectangles. (The unions in question form a field.) Also, it is clear that $\mathfrak{C}$ is a monotone class. Hence the monotone class theorem (see Ash, 1972, p. 191) implies that $\mathfrak{C} = (\mathcal{B}(S))^\infty$ and the validity of the theorem follows. ∎

1.2.8   Remarks

(i) Corollary 1.2.4 is essentially a result due to Hewitt and Savage (1955). Theorems 1.2.3 and 1.2.5 and Corollary 1.2.6 have been reported in a slightly simplified form in Davies and Shanbhag (1987).

(ii) The interim result (1.2.11) in the proof of Theorem 1.2.7 is valid for a more general topological space than a Polish space; for some further interesting observations on the theorem, see Dubins and Freedman (1979) and Aldous (1985).

(iii) Theorem 1.2.7 is an extended version of de Finetti's theorem and is essentially due to Olshen (1973). In the present monograph, we need only a specialized version of this result for $S = \mathbb{R}^n$. However, as the proof of the corollary is not vastly different from that of the general result, we decided to cover the general result above.

(iv) Corollary 1.2.6 and its specialized version with $S^*(\mu)$ replaced by $\text{supp}[\mu]$ are used in the following chapters in several places; the argument used in the proof of Theorem 1.2.5 simplifies slightly if we are seeking only the inequality in the corollary or its further special case referred to here.

1.3   SOME MISCELLANEOUS RESULTS

We give below some further tools that are needed (fully or partially) in the present study.

1.3.1   Lemma

Let F be a probability distribution function on the real line. Then, for every $0 < \delta < 1$,

$$\int_{-\infty}^\infty \frac{F(\mathrm{d}x)}{(1 - F(x-))^\delta} \begin{cases} = \dfrac{1}{1-\delta} & \text{if } F \text{ is continuous} \\[2mm] < \dfrac{1}{1-\delta} & \text{otherwise.} \end{cases} \tag{1.3.1}$$

Further, if the right extremity of F is not a discontinuity point of F, then

$$\int_{-\infty}^{\infty} \frac{F(dx)}{1 - F(x-)} = \infty. \tag{1.3.2}$$

Proof   By the monotone convergence theorem, we have for $0 < \delta < 1$

$$\int_{-\infty}^{\infty} \frac{F(dx)}{(1 - F(x-))^{\delta}} = \sum_{n=0}^{\infty} \binom{-\delta}{n} (-1)^n \int_{-\infty}^{\infty} (F(x-))^n F(dx).$$

Let $X_1, \ldots, X_{n+1}$ be a sample of $n + 1$ independent observations from F and

$$X_{(1)} \leqslant X_{(2)} \leqslant \cdots \leqslant X_{(n+1)}$$

be the corresponding ordered random variables. Then (reading $X_{(0)} = -\infty$) for all $n \geqslant 0$

$$(n + 1) \int_{-\infty}^{\infty} (F(x-))^n F(dx)$$

$$= P(X_{(n)} < X_{(n+1)}) \begin{cases} = 1 & \text{if } F \text{ is continuous or } n = 0 \\ < 1 & \text{otherwise.} \end{cases}$$

Hence it follows that for $0 < \delta < 1$

$$\int_{-\infty}^{\infty} \frac{F(dx)}{(1 - F(x-))^{\delta}} \begin{cases} = \sum_{n=0}^{\infty} \binom{-\delta}{n} (-1)^n \dfrac{1}{n+1} & \text{if } F \text{ is continuous} \\ < \sum_{n=0}^{\infty} \binom{-\delta}{n} (-1)^n \dfrac{1}{n+1} & \text{otherwise} \end{cases}$$

assuming $\sum_{n=0}^{\infty} \binom{-\delta}{n} (-1)^n 1/(n + 1) < \infty$. Since

$$\sum_{n=0}^{\infty} \binom{-\delta}{n} (-1)^n \frac{1}{n+1} = \int_0^1 \frac{dx}{(1 - x)^{\delta}} = \frac{1}{1 - \delta}, \quad 0 < \delta < 1,$$

the first part of the theorem follows. The second part of the theorem may be established as shown below.

Assume then that the right extremity of F is not a discontinuity point of F. It then follows that if (1.3.2) is not valid, we have a point x^* ($<$ the right extremity of F) such that $1 - F(x^*-) > 0$ and

$$\int_{[x^*, \infty)} (1 - F(x-))^{-1} F(dx) < 1,$$

which implies that

$$1 - F(x^*-) = \int_{[x^*, \infty)} (1 - F(x-)) \frac{F(dx)}{1 - F(x-)} < 1 - F(x^*-). \tag{1.3.3}$$

(To see the existence of x^*, note in particular that under the stated assumptions, for every sequence $\{x_n : n = 1, 2, \ldots\}$ of real numbers converging to the right extremity of F,

$$\int_{[x_n, \infty)} (1 - F(x-))^{-1} F(\mathrm{d}x) \to 0.)$$

Clearly (1.3.3) leads us to a contradiction and hence it follows that (1.3.2) does indeed hold. ∎

1.3.2   Lemma

Let $-\infty \leqslant \alpha < \beta \leqslant \infty$ and G be a Lebesgue–Stieltjes (measure) distribution function[†] (i.e. a distribution function determining a Lebesgue–Stieltjes measure) on $\mathbb{R}$. Further, let D be the set of discontinuity points of G in (α, β) and let f be a real- valued Borel measurable function on $\mathbb{R}$ such that

$$\sum_{x_i \in D} \int_{G(x_i-)}^{G(x_i+)} |f(y)| \, \mathrm{d}y < \infty. \qquad (1.3.4)$$

Denote by G_{c} the continuous part of G. If G^* is a real-valued function on the real line which can be different from G only at a countable number of points then

$$\int_\alpha^\beta f(G^*(x)) G_{\mathrm{c}}(\mathrm{d}x) = \int_{G(\alpha+)}^{G(\beta-)} f(y) \, \mathrm{d}y - \sum_{x_i \in D} \int_{G(x_i-)}^{G(x_i+)} f(y) \, \mathrm{d}y \qquad (1.3.5)$$

provided that either side of this equality exists.

Proof   Consider the measure space $(\mathbb{R}, \mathcal{B}, \mu_G)$, where $\mathcal{B}$ is the Borel σ-field of the real line and μ_G is the measure determined on this σ-field by G. Define

$$A = (\alpha, \beta) - D,$$

$$B = (G(\alpha+), G(\beta-)) - \bigcup_{x_i \in D} (G(x-), G(x_i)]$$

and

$$C = \{x : x \in \mathbb{R}, G(x) \in B\}.$$

Observe now that, in view of the condition (1.3.4) satisfied by f, the right-hand side of (1.3.5) reduces to $\int_B f(y) \, \mathrm{d}y$. Also, when G is viewed as a measurable function on the above measure space, then the induced measure corresponding to it agrees with the Lebesgue measure at least on B. Since the super set A of C satisfies $\mu_G(A - C) = 0$, Theorem 1.6.12 of Ash (1972) implies that the integral $\int_B f(y) \, \mathrm{d}y$ equals the left-hand side of the following obvious identity:

$$\int_A f(G(x)) \mu_G(\mathrm{d}x) = \int_{(\alpha, \beta)} f(G(x)) G_{\mathrm{c}}(\mathrm{d}x),$$

[†] Kingman and Taylor (1966) refer to this as a Stieltjies measure function.

The fact that $G^*(x)$ can be different from $G(x)$ only for countably many x's yields

$$\int_{(\alpha,\beta)} f(G(x))G_c(dx) = \int_{(\alpha,\beta)} f(G^*(x))G_c(dx)$$

and consequently we have the lemma. ∎

1.3.3 Remarks

(i) Lemma 1.3.1 is essentially due to Shanbhag (1980) and is an improved version of Lemma 1.1.8 in Kagan *et al*, (1973); the argument used to prove the lemma herein is substantially simpler than that used to prove its specialized version in the latter reference.

(ii) As implied by Remark 4 of Shanbhag and Kotz (1987), the second part of Lemma 1.3.1 can also be proved as a corollary to the Borel–Cantelli lemma.

(iii) It is natural to enquire whether there is a multivariate extension of Lemma 1.3.1 with the survivor function of F in place of $1 - F(x-)$. (If F is a distribution function on $\mathbb{R}^n$ and P_F is the probability measure determined by F, then by the survivor function S_F of F on $\mathbb{R}^n$, we mean the function defined by

$$S_F(x) = P_F\left(\prod_{i=1}^n [x_i, \infty)\right), \quad x \in \mathbb{R}^n,$$

where x_i are the coordinates of x.) Note that now we can have an F such that $P_F(\{x : S_F(x) = 0\}) > 0$ (or even $= 1$) and hence to have the integral in question to be well defined we need the restriction that $S_F(\bullet) > 0$ a.s. $[P_F]$. With the restriction, on using essentially the same argument as in the proof of Lemma 1.3.1, we get

$$\int_{\mathbb{R}^n} (S_F(x))^{-1} P_F(dx) = \infty$$

if F is continuous, implying that a partial extension of the lemma holds. However, in general, even under the restriction, that the result does not hold is shown by the following example.

Example   Consider a probability distribution function F on $\mathbb{R}^2$ such that

$$F = \sum_{i=1}^\infty a_i F_i,$$

where $a_i > 0$ for all i (with $\Sigma_{i=1}^\infty a_i = 1$) and, for each i, F is a probability distribution function concentrated on $(i-1, i] \times (-i-1, -i]$ such that $S_{F_i}(\bullet) > 0$ a.s. $[P_{F_i}]$. If we choose, in particular $a_i \propto i^{-1}(\log_e(i+1))^{-1}$, we get for each $0 < \delta \leqslant 1$

$$\int_{\mathbb{R}^2} (S_F(x))^{-\delta} P_F(dx) = \infty.$$

(iv) Lemma 1.3.2 is due to Kotz and Shanbhag (1980). If the condition (1.3.4) in this lemma is replaced by the stronger condition $\int_{G(\alpha+)}^{G(\beta-)} |f(y)| \, dy < \infty$, then both sides of (1.3.5) in the lemma exist and are finite. All the applications of the lemma to be met in this monograph satisfy this stronger condition and hence have the corresponding integrals well-defined and finite.

1.3.4   Lemma

Let $\psi_1, \ldots, \psi_k$, A and B be complex valued functions defined on a neighborhood of the origin such that they are locally integrable with respect to Lebesgue measure satisfying, for a sufficiently small δ,

$$\sum_{r=1}^{k} \psi_r(u + b_r v) = A(u) + B(v) + P_n(u, v), \quad |u|, |v| < \delta, \tag{1.3.5}$$

where b_r's are distinct nonzero constants and P_n is a polynomial of degree n. Then there exists a neighborhood of the origin such that the restrictions of ψ_r, A and B to it are all polynomials of degree $\leqslant \max(n, k)$.

Proof   Integrating (1.3.5) with respect to v over $[0, v']$ and then with respect to v' over $[0, \alpha]$ (with v' and α as sufficiently small but positive), we can see that A is twice differentiable on a neighborhood of the origin. The form of (1.3.5) then trivially implies that B and ψ_r are also twice differentiable on a neighborhood of the origin. Considering the equation (1.3.5) for $|u|, |v| < \delta^*$ for a sufficiently small δ^* and differentiating it with respect to u, we arrive at a new equation which implies (on taking, $v' = u + b_1 v$ for example, and $|v'|, |u| < \delta^{**}$ for sufficiently small δ^{**}) that an equation of the type of (1.3.5) with k replaced by $k - 1$ and n replaced by $n - 1$ and A replaced by its derivative is valid for $k \geqslant 1$. Hence, by induction on k, we see that A satisfies the requirement of the assertion. From the form of (1.3.5), it is then immediate that ψ_r and B are also as required. Hence we have the lemma. ∎

1.3.5   Lemma

Let F be a probability distribution function on $\mathbb{R}$ and g be a Borel measurable function that is integrable with respect to (the measure determined by) F. If, for some real $\rho \neq 0$,

$$\int_{\mathbb{R}} e^{itx} g(x) F(dx) = (\mu + i\rho t) \int_{\mathbb{R}} e^{itx} F(dx), t \in \mathbb{R}, \tag{1.3.6}$$

then F is absolutely continuous (with respect to Lebesgue measure) with density satisfying for almost all [Lebesgue] $x \in \mathbb{R}$ the differential equation

$$\rho f'(x) = (\mu - g(x)) f(x).$$

If g is continuous, then there is a differentiable version of f satisfying

$$f(x) \propto \exp\left\{-\frac{1}{\rho}\int_0^x (g(y) - \mu)\,dy\right\}, \quad x \in \mathbb{R},$$

where $\int_0^x$ is to be interpreted as $-\int_0^x$ if $x \leqslant 0$.

Proof There is no loss of generality in assuming that $\mu = 1$ and $\rho = -1$. (Note that (1.3.6) can be rewritten as

$$\int_{\mathbb{R}} e^{itx}\left(1 - \frac{g(x) - \mu}{\rho}\right) F(dx) = (1 - it)\int_{\mathbb{R}} e^{itx} F(dx), \quad t \in \mathbb{R}.)$$

We can then rewrite (1.3.6) as

$$(1 - it)^{-1}\int_{\mathbb{R}} e^{itx} g(x) F(dx) = \int_{\mathbb{R}} e^{itx} F(dx), \quad t \in \mathbb{R}. \tag{1.3.7}$$

Corollary 1.1.2 then yields that

$$\int_{-\infty}^x (1 - \exp\{-(x - y)\}) g(y) F(dy) = F(x), \quad x \in \mathbb{R},$$

which, in turn, implies that F is differentiable everywhere with derivative f satisfying

$$\int_{-\infty}^x \exp\{-(x - y)\} g(y) f(y)\,dy = f(x), \quad x \in \mathbb{R}. \tag{1.3.8}$$

From (1.3.8) it follows that f is continuous and almost everywhere differentiable with respect to Lebesgue measure such that

$$f'(x) = g(x)f(x) - f(x)$$
$$= (g(x) - 1)f(x) \quad \text{for a.a. } x \in \mathbb{R}.$$

Hence we have the first part of the lemma. The second part of the lemma follows on noting further that if g is continuous, then f is differentiable everywhere with $f'(x) = (g(x) - 1)f(x)$ for all $x \in \mathbb{R}$. ∎

1.3.6 Corollary

If in Lemma 1.3.5, $g(x) = \exp\{x\}$, $x \in \mathbb{R}$, then ρ and μ are both positive and

$$f(x) \propto \exp\left\{-\frac{1}{\rho}(e^x - \mu x)\right\}, \quad x \in \mathbb{R}.$$

(f here is clearly the density of the logarithm of a gamma random variable.)

1.3.7   Corollary

Let F be the distribution function of a random variable X such that $E(e^X) < \infty$. Suppose

$$\int_{\mathbb{R}} e^{(1+it)x} F(dx) = (\mu + \gamma t) \int_{\mathbb{R}} e^{itx} F(dx) \qquad (1.3.9)$$

for all $t \in [-\varepsilon, \varepsilon]$ for some $\varepsilon > 0$. Then $\gamma = i\rho$ with ρ as real, and the conclusions of Corollary 1.3.6 remain valid.

Proof   From (1.3.9) we have (subtracting the relation for $-t$ from the one for t)

$$2i \int_{\mathbb{R}} (e^x - \mu) \sin(tx) F(dx) = 2\gamma t \int_{\mathbb{R}} \cos(tx) F(dx)$$

with μ obviously as real. Hence it follows that $\gamma = i\rho$ for some real ρ. There is no loss of generality in taking $\mu = 1$. We have then the relation

$$\int_{\mathbb{R}} e^{itx} F^*(dx) = \int_{\mathbb{R}} e^{itx} F(dx), \qquad |t| \leqslant \varepsilon \qquad (1.3.10)$$

with F^* as the convolution of the distribution function $\int_{(-\infty, x]} e^y F(dy)$, $x \in \mathbb{R}$, and the distribution function corresponding to the characteristic function $(1 + i\rho t)^{-1}$, $t \in \mathbb{R}$. Let $c > 0$ be fixed; define

$$\phi(t) = \int_{(-\infty, -c]} e^{itx} F(dx) - \int_{[c, \infty)} e^{itx} F^*(dx), \qquad t \in \mathbb{R}$$

and

$$\psi(t) = \int_0^t \phi(u)\, du, \qquad t \in \mathbb{R},$$

(where the integral $\int_0^t$ is to be interpreted as $-\int_0^t$ if $t < 0$). From (1.3.10) it follows that $\psi(t)$ is a linear function of a characteristic function that has moments of all order with moments satisfying the Riesz criterion (see Theorem 1.1.8). This in turn implies easily that both F and F^* have moments of all order with moments satisfying the criterion in question. Consequently, it follows that the identity in (1.3.10) is valid for all $t \in \mathbb{R}$ and hence that (1.3.9) with $\gamma = i\rho$ holds for all $t \in \mathbb{R}$. The corollary now follows from Corollary 1.3.6. ∎

1.3.8   Remarks

(i) Lemmas 1.3.3 and 1.3.4 are slight improvements over Lemmas 1.5.1 and 6.1.1 respectively in Kagan *et al* (1973); it is also possible to give improved versions of several other results on functional equations given in this reference. (The statement of Lemma 1.5.9 in Kagan *et al* is not quite accurate.)

(ii) Corollaries 1.3.6 and 1.3.7 also appear in Kagan *et al* (1973). Our proof of Corollary 1.3.7, however, is different and corrects a minor error existing in the earlier proof. (The earlier proof assumes without any justification that the cases of $\int_{\mathbb{R}} \sin(tx) F(dx) \equiv 0$ do not arise.)

Yet another apparatus that we need is a version of Choquet's theorem.

1.3.9 Theorem

Let X be a metrizable compact convex subset of a locally convex topological vector space E, X_e the set of extreme points of X and x_0 an element of X. Then X_e is a G_δ set and there exists a probability measure P on X supported by X_e such that it represents x_0 (or, in other words, such that x_0 is its barycentre), i.e. it satisfies

$$f(x_0) = \int_{X_e} f(x) P_{x_0}(dx) \quad \text{for every } f \in E',$$

where E' is the topological dual of E.

For an account of Choquet's theorem and related results, see Phelps (1966). (The part of Theorem 1.3.8 that X_e is a G_δ set appears as Proposition 1.3 in the cited reference.) Choquet's theorem is an extension of the Krein–Milman theorem which states that if X is a compact convex subset of E, then it is equal to the closed convex hull of its extreme points.

1.3.10 Remarks

In addition to the tools we have mentioned in this chapter, there are some others to which we shall have to refer in our study. One of them which is also of independent interest is the Perron–Frobenius theorem which is stated as follows. For a proof, the reader is referred to Seneta (1973); see also Raghavan (1993).

We call an $n \times n$ matrix A nonnegative (positive) if all its elements are nonnegative (positive) and use the notation $A \geqslant 0$ ($A > 0$) to indicate such a property. Similarly we call a vector y nonnegative (positive) if all its elements are nonnegative (positive) and use the notation $y \geqslant 0$ ($y > 0$) to indicate such a property.

A matrix A of order $n \geqslant 2$ is said to be reducible if there exists a permutation matrix Π such that

$$\Pi A \Pi' = \begin{pmatrix} B & O \\ C & D \end{pmatrix}$$

where B and D are square matrices and O is a null matrix. Otherwise it is called irreducible; it is said to be primitive if $A^m > 0$ for some positive integer m.

The Perron-Frobenius theorem is as follows. Let $A \geqslant 0$ be an irreducible matrix of order n. Then:

(i) $Ay = \lambda_0 y$ for some $\lambda_0 > 0$, $y > 0$.

(ii) The eigenvalue λ_0 is geometrically simple.

(iii) The eigenvalue λ_0 is maximal in modulus among all the eigenvalues of A. That is, for any other eigenvalue μ of A, $|\mu| \leqslant \lambda_0$; if A is primitive, then the assertion holds with strict inequality.

(iv) The only nonnegative eigenvectors of A are just the positive scalar multiples of y.

(v) The eigenvalue λ_0 is algebraically simple.

(vi) Let $\lambda_0, \lambda_1, \ldots, \lambda_{(h-1)}$ be the distinct eigenvalues of A with $|\lambda_i| = \lambda_0, i = 1, 2, \ldots, h - 1$. Then they are precisely the solutions of the equation $\lambda^h - \lambda_0^h = 0$.

We also refer to Spitzer's integral representation theorem for stationary measures corresponding to a Bienaymé–Galton–Watson branching process (Athreya and Ney, 1972), certain tools in random walk such as the Wiener–Hopf factorization and important theorems on ladder variables (Feller, 1966). For details concerning these theorems, the reader is referred to the references given above.

CHAPTER 2

Simple Integral Equations: Versions of the Integrated Cauchy Functional Equation

2.1 INTRODUCTION

Consider the integral equation

$$H(x) = \int_S H(x + y)\mu(dy) \quad \text{for a.a. } [L]\, x \in S, \tag{2.1.1}$$

where S equals either $\mathbb{R}(= (-\infty, \infty))$ or $\mathbb{R}_+(= [0, \infty))$, μ is a σ-finite measure on S (i.e. on the Borel σ-field of S) satisfying $\mu(\{0\}) < 1$, $H : S \to \mathbb{R}_+$ is a Borel measurable function that is locally integrable (w.r.t. the Lebesgue measure), and a.a. $[L]$ refers to almost all with respect to the Lebesgue measure. This integral equation has been extensively studied under some constraints or otherwise by Deny (1961), Shanbhag (1977), Shimizu (1978), Lau and Rao (1982), and Alzaid, *et al* (1987) and several other authors. Its modified version, the partial convolution equation

$$H(x) = \int_{\mathbb{R}} H(x + y)\mu(dy) \quad \text{for a.a. } [L]\, x \in \mathbb{R}_+, \tag{2.1.2}$$

with $H : \mathbb{R} \to \mathbb{R}_+$ as a locally integrable Borel measurable function and μ as a σ-finite measure on $\mathbb{R}$ satisfying $\mu(\{0\}) < 1$, has been studied by Alzaid *et al* (1988) and several others. An important special case of (2.1.2) has appeared in Lindley (1952).

If $H : S \to \mathbb{R}_+$ is a locally integrable Borel measurable function on S and S_0 is a closed subset of S satisfying

$$H(x)H(y) = H(x + y) \quad \text{for a.a. } [L]\, x \in S \text{ for each } y \in S_0$$

then, under appropriate assumptions such as that S_0 is a finite set for which $S_0 \setminus \{0\}$ is nonempty and $H(y) \neq 0$ at least for one $y \in S_0 \setminus \{0\}$, we have a measure μ on S with support S_0 such that (2.1.1) holds. In view of this, one could view (2.1.1) as an extended version of the integrated Cauchy functional equation. Both (2.1.1) and its modified version (2.1.2) have various applications in characterization problems concerning probability distributions and stochastic processes; we discuss these later in this monograph. In the present chapter, we restrict ourselves to studying the problems of identifying the solutions to (2.1.1) and (2.1.2) by touching on, in particular, the contributions of Shanbhag (1977), Lau and Rao (1982), Alzaid *et al* (1987, 1988).

2.2   LAU–RAO–SHANBHAG THEOREMS

We now give the following theorems which are referred to as the Lau–Rao–Shanbhag theorems in the literature; individually, these results are also known as Shanbhag's lemma and the Lau-Rao theorem respectively.

2.2.1   Theorem

Let $\{(v_n, w_n): n = 0, 1, \ldots\}$ be a sequence of vectors with nonnegative real components such that $v_n \neq 0$ for some $n \geq 1$ and $w_1 \neq 0$. Then

$$v_m = \sum_{n=0}^{\infty} v_{m+n} w_n \quad m = 0, 1, \ldots \tag{2.2.1}$$

if and only if $\Sigma_{n=0}^{\infty} w_n b^n = 1$ and $v_n = v_0 b^n$, $n = 1, 2, \ldots$ for some $b > 0$.

Proof   The 'if' part is trivial. Let us now establish the 'only if' part. Assume that (2.2.1) is valid. Observe that $v_m(1 - w_0) \geq w_1 v_{m+1}$, $m \geq 0$. Since $w_1 v_{m+1} > 0$ for some m, it then follows that $w_0 < 1$. Consequently we have from (2.2.1) that $v_n = 0$ for some n if and only if $v_m = 0$ for all $m >$ this n. Since there is an n with $v_n \neq 0$ it follows that we should have $v_n \neq 0$ for all $n \geq 0$.

   Define $v_m^* = v_m/v_{m-1}$, $m = 1, 2, \ldots$ and $b = \sup\{v_m^* : m \geq 1\}$. Using the inequality $v_m(1 - w_0) \geq w_1 v_{m+1}$, $m \geq 0$, we see that $0 < b \leq (1 - w_0)/w_1$. If $v_m^* = b$ for all $m \geq 1$, then the result is immediate. On the other hand, if for some $m(\geq 0)$ we have $v_{m+1}^* < b$ then from (2.2.1) we get that

$$1 = \sum_{n=0}^{\infty} (v_{m+n}/v_m) w_n < \sum_{n=0}^{\infty} w_n b^n.$$

This implies that we can find $0 < \varepsilon \leq b/2$ and a positive integer n_0 such that $\Sigma_{n=0}^{n_0} w_n (b - \varepsilon)^n > 1$. Now, from (2.2.1)

$$v_m^* = v_{m+1}^* w_1 v_m^* + \left\{ \sum_{n(\neq 1)=0}^{\infty} w_n v_{m+n}^* v_{m+n-1} \right\} \Big/ v_{m-1}$$

$$\leq v_{m+1}^* w_1 v_m^* + b\{1 - w_1 v_m^*\}, \quad m \geq 1, \tag{2.2.2}$$

implying that given $0 < \varepsilon^* \leq b/2$ and m_0 such that $v_{m_0}^* \geq b - \varepsilon^*$, we have $v_{m_0+1}^* \geq b - (\varepsilon^*/w_1 v_{m_0}^*) \geq b - (2\varepsilon^*/w_1 b)$. From this it easily follows that given

$0 < \varepsilon \leqslant b/2$ and a positive integer n_0, we can find a positive integer m_0 such that $v_{m_0+n}^* \geqslant b - \varepsilon$ for $1 \leqslant n \leqslant n_0$. Consequently from (2.2.1) we have

$$1 \geqslant \sum_{n=0}^{n_0} \frac{v_{m_0+n}}{v_{m_0}} w_n \geqslant \sum_{n=0}^{n_0} w_n (b - \varepsilon)^n.$$

Hence we have a contradiction unless $v_m^* = b$ for all $m \geqslant 1$. This completes the proof. ∎

2.2.2   Theorem[†]

Let (2.1.1) be satisfied for $S = \mathbb{R}_+$ with H not as a function that is identically zero a.e. $[L]$ (i.e. almost everywhere w.r.t. the Lebesgue measure). Then either μ is arithmetic with some span λ and

$$H(x + n\lambda) = H(x)b^n, \quad n = 0, 1, \ldots \quad \text{for a.a. } [L]x \in \mathbb{R}_+$$

with b such that

$$\sum_{n=0}^{\infty} b^n \mu(\{n\lambda\}) = 1$$

or μ is nonarithmetic and

$$H(x) \propto \exp\{\eta x\} \quad \text{for a.a. } [L]x \in \mathbb{R}_+$$

with η such that

$$\int_{\mathbb{R}_+} \exp\{\eta x\} \mu(\mathrm{d}x) = 1.$$

Proof   There is no loss of generality in assuming that $\mu(\mathbb{R}_+) > 1$ since (2.1.1) is also satisfied by $H(x) \exp\{-\delta x\}$, $x \in S$ for any δ if $\mu(\mathrm{d}x)$ is replaced by $\exp\{\delta x\} \mu(\mathrm{d}x)$. In that case, we can find a $c \in (0, \infty)$ such that $\mu([0, c]) > 1$. In view of Fubini's theorem and the functional equation, it is easily seen that

$$\int_0^x H(y) \, \mathrm{d}y \geqslant \mu([0, c]) \int_c^x H(y) \, \mathrm{d}y, \quad x \geqslant c, \tag{2.2.3}$$

which implies, in view of the local integrability of H, $\int_0^x H(y) \, \mathrm{d}y$ for $x \geqslant c$ to be bounded and hence

$$\int_0^\infty H(y) \, \mathrm{d}y \left(= \lim_{x \to \infty} \int_0^x H(y) \, \mathrm{d}y \right) < \infty.$$

[†] Any measure on $\mathbb{R}$ or $\mathbb{R}_+$ that is not concentrated on $\{0\}$ is said to be arithmetic with span $\lambda (> 0, < \infty)$ if the (additive) subgroup of $\mathbb{R}$ generated by the support of the measure equals $\{n\lambda : n = 0, \pm1, \pm2, \ldots\}$; the measure is said to be nonarithmetic if it is not arithmetic.

Define

$$\hat{H}(x) = \int_x^\infty H(y)\,dy, \quad x \geqslant 0, \tag{2.2.4}$$

which is a nonnegative decreasing continuous function satisfying (because of Fubini's theorem) (2.1.1) for $S = \mathbb{R}_+$ (with a.a. [L] deleted) such that it is not identically zero. Noting that there exists an $s^* > 0$ such that $\mu([0, s^*]) < 1$, we can then conclude that $\hat{H}(x) > 0$ if and only if $\hat{H}(x + s^*) > 0$; consequently, we have $\hat{H}(x) > 0$ for all $x \in \mathbb{R}_+$.

Let $x_0 \in \mathbb{R}_+$ be fixed. From Corollary 1.2.6 or indeed its specialized version with $S^*(\mu)$ replaced by supp$[\mu]$, we have then

$$\hat{H}((n+1)s + x_0)\hat{H}((n-1)s + x_0) \geqslant (\hat{H}(ns + x_0))^2, \quad n = 1, 2, \ldots, \; s \in \text{supp}[\mu],$$

which implies in view of the arbitrariness of x_0 that

$$\left(\frac{\hat{H}((n+1)s + x)}{\hat{H}((n-1)s + x)} \right) \geqslant \left(\frac{\hat{H}(ns + x)}{\hat{H}((n-1)s + x)} \right)^2, \quad n \geqslant 1, \; s \in \text{supp}[\mu], \; x \in \mathbb{R}_+.$$

This implies that for each $s \in \text{supp}[\mu]$ and $x \in \mathbb{R}_+$, we have

$$\left\{ \frac{\hat{H}(x + ns)}{\hat{H}(x + (n-1)s)} : \quad n = 1, \ldots \right\}$$

to be an increasing sequence. Let $s_0 \in (0, \infty) \cap \text{supp}[\mu]$ be fixed. In view of the continuity of $\hat{H}(x + s_0)/\hat{H}(x)$, $x \in \mathbb{R}_+$, what we have seen immediately above implies that there exists an $x^* \in [0, s_0]$ such that

$$\frac{\hat{H}(x^* + s_0)}{\hat{H}(x^*)} \leqslant \frac{\hat{H}(x + s_0)}{\hat{H}(x)} \quad \text{for each } x \in \mathbb{R}_+. \tag{2.2.5}$$

If x is a point for which the equality in (2.2.5) is valid, we have from the integral equation on hand

$$\frac{\hat{H}(x^* + s_0)}{\hat{H}(x^*)} = \int_{\mathbb{R}_+} \frac{\hat{H}(x + s_0 + y)}{\hat{H}(x + y)} P_x(dy), \tag{2.2.6}$$

where P_x is the probability measure (on $\mathbb{R}_+$) for which for every Borel subset B of $\mathbb{R}_+$

$$P_x(B) = \int_B \frac{\hat{H}(x + y)}{\hat{H}(x)} \mu(dy).$$

The validity of (2.2.6) (with $\hat{H}$ obviously as positive continuous) implies in view of (2.2.5) then that for each $s \in \text{supp}[\mu]$

$$\frac{\hat{H}(x + s_0)}{\hat{H}(x)} = \frac{\hat{H}(x + s_0 + s)}{\hat{H}(x + s)},$$

which implies inductively on taking $x = x^* + (n-2)s_0$ and $s = s_0$ that

$$
\frac{\hat{H}(x^* + ns_0)}{\hat{H}(x^*)} \left(= \frac{\hat{H}(x^* + ns_0)}{\hat{H}(x^* + (n-1)s_0)} \left(\frac{\hat{H}(x^* + (n-1)s_0)}{\hat{H}(x^*)} \right) \right)
$$

$$
= \left(\frac{\hat{H}(x^* + s_0)}{\hat{H}(x^*)} \right)^n, \quad n = 2, 3, \dots . \tag{2.2.7}
$$

Observe now that for any given $x \in \mathbb{R}_+$, there exists a positive integer m such that $x + ms_0 \geqslant x^*$ and hence (2.2.7) and the fact that

$$
\{\hat{H}(x + ns_0)/\hat{H}(x + (n-1)s_0) : \quad n = 1, 2, \dots\}
$$

is increasing and $\hat{H}$ is decreasing imply for all $n \geqslant m$,

$$
\left(\frac{\hat{H}(x + s_0)}{\hat{H}(x)} \right)^n \leqslant \frac{\hat{H}(x + ns_0)}{\hat{H}(x)} \leqslant \frac{\hat{H}(x^*)}{\hat{H}(x)} \left\{ \frac{\hat{H}(x^* + (n-m)s_0)}{\hat{H}(x^*)} \right\}
$$

$$
= \left(\frac{\hat{H}(x^* + s_0)}{\hat{H}(x^*)} \right)^{n-m} \frac{\hat{H}(x^*)}{\hat{H}(x)},
$$

which yields trivially that $[\hat{H}(x + s_0)/\hat{H}(x)] \leqslant [\hat{H}(x^* + s_0)/\hat{H}(x^*)]$. In view of (2.2.5), we have then $[\hat{H}(x + s_0)/\hat{H}(x)]$ to be independent of x for $x \geqslant 0$. Since s_0 is arbitrary, this yields $\hat{H}(0)\hat{H}(x + y) = \hat{H}(x)\hat{H}(y)$ for all $x \in \mathbb{R}_+$ and $y \in \operatorname{supp}[\mu]$. Hence it follows that the result of the theorem is valid with $\hat{H}$ in place of H (see, for example, Marsaglia and Tubilla, 1975). From this, the validity of the required result is then immediate on appealing to the relation (2.2.4). ∎

2.2.3   Corollary

Let $\{(v_n, w_n) : n = 0, 1, \dots\}$ be a sequence of vectors with nonnegative real components such that $v_n \neq 0$ for at least one n, $w_0 < 1$, and the largest common divisor of the set $\{n : w_n > 0\}$ is unity. Then

$$
v_m = \sum_{n=0}^{\infty} v_{m+n} w_n, \quad m = 0, 1, \dots
$$

if and only if

$$
v_n = v_0 b^n, \quad n = 0, 1, 2, \dots \quad \text{and} \quad \sum_{n=0}^{\infty} w_n b^n = 1
$$

for some $b > 0$.

2.2.4 Remarks

(i) Theorem 2.2.1 is due to Shanbhag (1977), Theorem 2.2.2 is due to Lau and Rao (1982), and the present proof of Theorem 2.2.2 is taken essentially from Alzaid *et al* (1987); alternative proofs of Theorem 2.2.2 appear in Lau and Rao (1982), Ramachandran (1982), Rao and Shanbhag (1986) and elsewhere.

(ii) From the proof of Theorem 2.2.2, it is clear that it is based implicitly on exchangeable random variables.

(iii) But for the a priori condition that $w_0 < 1$, Corollary 2.2.3 is an extended version of Theorem 2.2.1; this is observed implicitly or explicitly by Lau and Rao (1982), Rao and Shanbhag (1986), and others.

(iv) Theorem 2.2.2 remains valid if the local integrability of H on $\mathbb{R}_+$ is replaced by the one on $(0, \infty)$. On the other hand it does not remain valid if the assumption of σ-finiteness of measure μ in (2.1.1) is dropped as is shown by the following example of Rao and Shanbhag (1986).

Example   Let μ be that measure on $\mathbb{R}_+$ for which the restriction to $(1, 2]$ agrees with the counting measure on $(1,2]$ and $\mu([0, 1] \cup (2, \infty)) = 0$. Define $H : \mathbb{R}_+ \to \mathbb{R}_+$ such that

$$H(x) = \begin{cases} 1 & \text{if } x \in [0, 1) \text{ or } x = 2, \\ 0 & \text{otherwise.} \end{cases}$$

Observe that we have here

$$H(x) = \int_{\mathbb{R}_+} H(x + y)\mu(\mathrm{d}y) \quad \text{for a.a. } [L]x \in \mathbb{R}_+$$

but H is not of the form stated in the theorem.

Should one consider instead of the multiplicative form the additive form of the Cauchy functional equation, then instead of (2.1.1) one would view the following equation as an extended version of the integrated Cauchy functional equation:

$$\int_{\mathbb{R}_+} H(x + y)\mu(\mathrm{d}y) = H(x) + c \quad \text{for a.a. } [L]x \in \mathbb{R}_+, \tag{2.2.8}$$

where $H : \mathbb{R}_+ \to \mathbb{R}$ is a locally integrable Borel measurable function and μ is a σ-finite measure on $\mathbb{R}_+$ with $\mu(\{0\}) < 1$. (The identity in (2.2.8) is understood as the one for which the left-hand side exists and equals the right-hand side.) The following example due to Rao and Shanbhag (1986) indicates that the general solution to (2.2.8) is not simple.

Example   Consider μ to be a probability measure on $\mathbb{R}_+$ such that it is determined by an infinitely divisible probability distribution with an entire characteristic function. From Picard's theorem (cf. Titchmarsh, 1978, p. 277) and the fact that the characteristic function involved here does not vanish, we can conclude that there exist infinitely many points (a_r, b_r) of $\mathbb{R}^2$ such that

$$\int_{\mathbb{R}_+} \exp(a_r x + i b_r x)\mu(\mathrm{d}x) = 1$$

or equivalently such that

$$\int_{\mathbb{R}_+} \exp(a_r x) \cos(b_r x) \mu(\mathrm{d}x) = 1 \tag{2.2.9}$$

and

$$\int_{\mathbb{R}_+} \exp(a_r x) \sin(b_r x) \mu(\mathrm{d}x) = 0. \tag{2.2.10}$$

If we now define

$$H_r(x) = \exp(a_r x) \cos(b_r x), \quad x \in \mathbb{R}_+,$$

it follows immediately, in view of (2.2.9) and (2.2.10), that

$$\int_{\mathbb{R}_+} H_r(x + y) \mu(\mathrm{d}y) = H_r(x), \quad x \in \mathbb{R}_+,$$

which in turn, illustrates that the solution to (2.2.8) is not in general as elegant as that of (2.1.1).

In spite of the observation made above, there do exist situations in which the solution to this latter version of the integrated Cauchy functional equation is far from complicated, as is shown by the theorem appearing below.

2.2.5   Theorem

If the function H in (2.2.8) is not a function which is identically equal to a constant a.e. [L] and the local integrability of H is replaced by H either increasing a.e. [L] or decreasing a.e. [L], then the equation cannot be valid unless either μ is nonarithmetic and H is of the form for which

$$H(x) = \begin{cases} \gamma + \alpha(1 - \exp\{\eta x\}) & \text{for a.a. } [L]x \text{ if } \eta \neq 0 \\ \gamma + \beta x & \text{for a.a. } [L]x \text{ if } \eta = 0 \end{cases}$$

or μ is arithmetic with span λ for some λ and H is of the form for which for each $n = 1, 2, \ldots$

$$H(x + n\lambda) = \begin{cases} H(x)\exp\{n\lambda\eta\} + \alpha'(1 - \exp\{n\lambda\eta\}) & \text{for a.a. } [L]x \text{ if } \eta \neq 0 \\ H(x) + \beta'n & \text{for a.a. } [L]x \text{ if } \eta = 0 \end{cases}$$

where $\alpha, \beta, \alpha', \beta', \gamma$ are all constants and η is such that

$$\int_{\mathbb{R}_+} \exp\{\eta x\} \mu(\mathrm{d}x) = 1.$$

Proof   There is no loss of generality in assuming that H is increasing a.e. [L]. Define, for each positive integer k, H_k such that

$$H_k(x) = k \int_{[0, 1/k]} H(x + y) \, \mathrm{d}y, \quad x \in \mathbb{R}_+.$$

Observe that on using the standard procedure as in the case of a distribution function on $\mathbb{R}$, H_k determines on $(0, \infty)$ (i.e. on the Borel σ-field of $(0, \infty)$) a measure that is absolutely continuous with respect to the Lebesgue measure. Consequently, in view of Fubini's theorem, it follows that for each k

$$H_k(x) = H_k(y) + \int_y^x h_k(z)\, dz, \quad 0 < y < x < \infty \qquad (2.2.11)$$

with h_k as a nonnegative Borel measurable function on $\mathbb{R}_+$ satisfying the integral equation of Theorem 2.2.2 such that its restriction to $(0, \infty)$ is locally integrable (with respect to a Lebesgue measure). Since $\{H_k\}$ converges to H a.e. $[L]$ on $\mathbb{R}_+$, (iv) of 2.2.4 implies in view of (2.2.11) the result of the present theorem; note that the first part of the theorem is now obvious and the second part follows on noting that if μ is arithmetic with span λ, then for $0 < y < x < \infty$ and $n, k \geqslant 1$

$$H_k(x + n\lambda) - H_k(x + (n-1)\lambda) \exp\{\lambda\eta\} = (1 - \exp\{\lambda\eta\}) H_k(y) + \int_y^{y+\lambda} h_k(z)\, dz$$

$$= H_k(y + \lambda) - \exp\{\lambda\eta\} H_k(y). \quad \blacksquare$$

2.2.6   Remarks

(i) Theorem 2.2.5 is due to Rao and Shanbhag (1986) and its proof given here is adapted from Shanbhag (1991).

(ii) Theorem 2.2.5 as well as its extended version given in the next chapter for more general spaces is useful in solving characterization problems concerning point processes such as Poisson and Yule processes.

(iii) From the statement of Theorem 2.2.5, it is implicit that under the assumptions in question, we can have neither $|c|\mu(\mathbb{R}_+) = \infty$ nor $\mu(\mathbb{R}_+) = 1$ and $\int_{\mathbb{R}_+} x\mu(dx) = \infty$.

(iv) If H in (2.2.8) is locally integrable and $\geqslant 0$ a.e. $[L]$ with $c \leqslant 0$, then it can easily be seen that the conclusions of Theorem 2.2.5 even without the a priori assumption that H is either increasing a.e. $[L]$ or decreasing a.e. $[L]$ remain valid. This result follows essentially from Theorem 2.2.2 on noting that we have here either $c = 0$ or $c \neq 0$, $\mu(\mathbb{R}_+) < 1$ and $H(x) + c(1 - \mu(\mathbb{R}_+))^{-1} \geqslant 0$ for a.a. $[L]$ $x \in \mathbb{R}$ with the function satisfying (2.1.1) with $S = \mathbb{R}_+$ but for a modification that it could be negative on a set of Lebesgue measure zero.

(v) If μ is a probability measure (i.e. if $\mu(\mathbb{R}_+) = 1$) and H is right continuous, then the integral equation of Theorem 2.2.2 is valid if and only if $H \equiv H(0)$ if μ is nonarithmetic and $H(x) = H(x + \lambda)$, $x \in \mathbb{R}_+$ if μ is arithmetic with span λ. Also, in that case, under the assumptions in Theorem 2.2.5, the integral equation (2.2.8) is satisfied if and only if H is of the form $H(x) = \alpha + \beta x$, $x \in \mathbb{R}_+$, with α as real and β as nonzero real satisfying $\int_{\mathbb{R}_+} x\mu(dx) = c/\beta$, if μ is nonarithmetic, and satisfies $H(x+\lambda) = H(x) + c(\sum_{n=0}^{\infty} n\mu(\{n\lambda\}))^{-1}$, $x \in \mathbb{R}_+$ if μ is arithmetic with span λ. The specialized results have applications in renewal theory and characterization problems relative to Poisson processes and have been studied respectively by Feller (1966, p. 351) and Isham *et al* (1975). (It may, however, be pointed out here that the result arrived at in Feller (1966) is slightly weaker than the corollary of Theorem 2.2.2 mentioned here especially since it assumes a priori that H is bounded.)

2.3 THE PARTIAL CONVOLUTION EQUATION AND THE WIENER-HOPF FACTORIZATION

We now study the problem of identifying the solution to the partial convolution equation (2.1.2) and its connection with the Wiener-Hopf factorization of a measure. To do this, we need some notation.

Let us consider the equation (2.1.2) and define, relative to μ, the measures μ_{1n} and μ_{2n} on $\mathbb{R}$ such that for every Borel set B of $\mathbb{R}$ and every integer $n \geqslant 1$,

$$\mu_{1n}(B) = \mu^n(\{(x_1, \ldots, x_n) \in \mathbb{R}^n : s_m \in \mathbb{R}_+, \quad m = 1, \ldots, n-1, s_n \in B\})$$

and

$$\mu_{2n}(B) = \mu^n(\{(x_1, \ldots, x_n) \in \mathbb{R}^n : s_m \in (-\infty, 0), \quad m = 1, \ldots, n-1, s_n \in B\})$$

$$(= \mu^n(\{(x_1, \ldots, x_n) \in \mathbb{R}^n : s_m > s_n, \quad m = 1, \ldots, n-1, s_n \in B\})),$$

where $s_m = x_1 + \cdots + x_m, m \geqslant 1$ and μ^n is the product measure $\Pi_{i=1}^n \mu_i$ with $\mu_i = \mu$ in the notation of Burrill (1972). Following the analogy with concepts in random walk in probability theory, we may refer to the measures ρ and τ defined below respectively as the descending ladder height measure and the (weak) ascending ladder height measure relative to μ,

$$\rho(\bullet) = \sum_{n=1}^{\infty} \mu_{1n}((-\infty, 0) \cap \bullet) \qquad (2.3.1)$$

and

$$\tau(\bullet) = \sum_{n=1}^{\infty} \mu_{2n}(\mathbb{R}_+ \cap \bullet). \qquad (2.3.2)$$

It is easy to check that if $\mu((-\infty, 0)) > 0$, then the closed subgroup of $\mathbb{R}$ generated by supp$[\rho]$ is the same as that generated by supp$[\mu]$. Also, if $\mu((0, \infty)) > 0$, then the closed subgroup of $\mathbb{R}$ generated by supp$[\tau]$ is the same as the one generated by supp$[\mu]$. These observations, in turn, imply that if $\mu((-\infty, 0)) > 0$, then either both ρ and μ are nonarithmetic or both are arithmetic with the same span, and if $\mu((0, \infty)) > 0$, then either both τ and μ are nonarithmetic or both are arithmetic with the same span.

The following theorem provides us with an important finding concerning the solution to the equation in question.

2.3.1 Theorem

Let the function H satisfying (2.1.2) not be a function that is equal to 0 a.e. [L] on (α, ∞), where $\alpha = \inf(\mathrm{supp}[\mu])$. Then ρ and τ as defined in (2.3.1) and (2.3.2) are Lebesgue–Stieltjes measures. (This is equivalent to the statement that they are both Radon measures and also to the statement that they are both regular measures.) Moreover,

$$H(x) = \int_{(-\infty,0)} H(x + y)\rho(\mathrm{d}y) + \xi(x)\exp(\eta x) \quad \text{for a.a. } [L]\, x \in \mathbb{R}_+, \quad (2.3.3)$$

where ξ is a nonnegative periodic function with $\xi(\bullet) = \xi(\bullet + s)$ for every $s \in \mathrm{supp}[\mu]$, and η is a real number such that

$$\int_{\mathbb{R}_+} \exp\{\eta x\}\tau(\mathrm{d}x) = 1. \quad (2.3.4)$$

(If η satisfying (2.3.4) does not exist, we take ξ in (2.3.3) to be identically equal to zero and η to be an arbitrary real number; also we may choose ξ to be equal to a constant everywhere if μ is nonarithmetic.)

Proof In view of the given assumptions, it easily follows that the assertions of the theorem are valid if $\mu((0, \infty)) = 0$. We may therefore assume that $\mu((0, \infty)) > 0$. As in the case of Theorem 2.2.2, the problem remains invariant if H and $\mu(\mathrm{d}y)$ are replaced respectively by $H(x)\exp\{-\delta x\}$, $x \in \mathbb{R}$, and $\exp\{\delta y\}\mu(\mathrm{d}y)$ for any $\delta \in \mathbb{R}$. Consequently, there is no loss of generality in assuming that $\mu((0, \infty)) > 1$ and hence that $0 < \int_x^\infty H(y)\,\mathrm{d}y < \infty$ for each $x \in \mathbb{R}$ following essentially the argument in the proof of Theorem 2.2.2. Since (2.1.2) is satisfied by this latter function, the theorem follows if we prove it by replacing H by this function provided in this case it is found additionally that (2.3.3) has $\xi(x)\exp\{\eta x\}$, $x \in \mathbb{R}_+$ to be decreasing. Define now $\{X_n : n = 0, 1, \ldots\}$ to be a temporally homogeneous Markov chain with state space $\mathbb{R}$ and transition function $P(\bullet, \bullet)$ such that

$$P(x, A) = \begin{cases} 1 & \text{if } x \in (-\infty, 0) \cap A \\ 0 & \text{if } x \in (-\infty, 0) \cap A^c \\ (\hat{H}(x))^{-1} \int_{A-x} \hat{H}(x + y)\mu(\mathrm{d}y) & \text{if } x \in \mathbb{R}_+ \text{ and } A \text{ is arbitrary,} \end{cases}$$

where $\hat{H}(x) = \int_x^\infty H(y)\,\mathrm{d}y$, $x \in \mathbb{R}$. For each $x \in \mathbb{R}_+$, define f_x to be the conditional probability that the chain enters $(-\infty, x)$ at least once given $X_0 = x$, and g_x to be the conditional probability that the process never enters $(-\infty, x)$ given $X_0 = x$. (Clearly these probabilities are to be calculated via the transition function.) We have then

$$f_x + g_x = 1, \quad x \in \mathbb{R}_+. \quad (2.3.5)$$

It is also easily seen that we have here

$$f_x = (\hat{H}(x))^{-1} \int_{(-\infty,0)} \hat{H}(x + y)\rho(\mathrm{d}y), \quad x \in \mathbb{R}_+, \quad (2.3.6)$$

which is clearly a continuous function yielding among other things that ρ is a Lebesgue–Stieltjes measure in view of the properties of $\hat{H}$. Also, from the properties of $\hat{H}$, it is obvious that τ is a Lebesgue–Stieltjes measure in the case of $\mu((-\infty, 0)) = 0$ since we have then $\tau = \mu$ and $\int_{\mathbb{R}_+} \hat{H}(y)\mu(dy) < \infty$. To have the first part of the theorem, we still have to show that τ is a Lebesgue–Stieltjes measure if $\mu((-\infty, 0)) > 0$. In this latter case, given any $s_0 \in (-\infty, 0) \cap \operatorname{supp}[\mu]$ and $y \in \mathbb{R}_+$, we have

$$\tau\left(\left(-\infty, y + \frac{ns_0}{2}\right)\right) \mu\left(\left(\frac{3s_0}{2}, \frac{s_0}{2}\right)\right)$$

$$\leqslant \tau\left(\left(-\infty, y + \frac{(n+1)s_0}{2}\right)\right) + \rho\left(\left(\frac{3s_0}{2}, 0\right)\right), \quad n = 0, 1, 2, \ldots, \quad (2.3.7)$$

which implies that

$$\tau\left(\left(-\infty, y + \frac{ns_0}{2}\right)\right) < \infty \quad \text{if} \quad \tau\left(\left(-\infty, y + \frac{(n+1)s_0}{2}\right)\right) < \infty.$$

Since $\tau((-\infty, 0)) = 0$, we have then by reverse induction that $\tau((-\infty, y)) < \infty$ for each y. This implies that τ is a Lebesgue–Stieltjes measure. (For obtaining (2.3.7), the identity given in brackets immediately after the definition of μ_{2n} may be used.) This establishes the first part of the theorem.

Define

$$\nu = \inf\{n \geqslant 1 : X_n = \inf_{m \geqslant 1}\{X_m\}\}$$

if it exists and $= \infty$ otherwise. Since τ is a Lebesgue–Stieltjes measure, it follows from the Borel–Cantelli lemma that irrespectively of the distribution of X_0, $P\{\nu < \infty\} = 1$. Consequently, we have

$$G_x = \hat{H}(x)P\{\nu < \infty, X_1 \geqslant x, X_2 \geqslant x, \ldots | X_0 = x\}$$

$$= \sum_{n=1}^{\infty} \hat{H}(x)P\{\nu = n, X_1 \geqslant x, X_2 \geqslant x, \ldots | X_0 = x\}$$

$$= \sum_{n=1}^{\infty} \int_{\mathbb{R}_+} \hat{H}(x + y)g_{x+y}\mu_{2n}(dy)$$

$$= \int_{\mathbb{R}_+} G_{x+y}\tau(dy), x \in \mathbb{R}_+, \quad (2.3.8)$$

where $G_x = \hat{H}(x)g_x$, $x \in \mathbb{R}_+$. Obviously G_x is decreasing and continuous in view of the properties of $\hat{H}$. Appealing to Theorem 2.2.2, we can then establish the second part of the theorem from (2.3.8) because of the validity of (2.3.5) and (2.3.6). ∎

2.3.2 Corollary

Let (2.1.1) be satisfied with $S = \mathbb{R}$ and H not as a function that is equal to zero a.e. $[L]$. Then either μ is nonarithmetic and H has the representation

$$H(x) = c_1 \exp\{\eta_1 x\} + c_2 \exp\{\eta_2 x\} \quad \text{for a.a. } [L] \, x \in \mathbb{R}, \qquad (2.3.9)$$

or μ is arithmetic with some span λ and H has the representation

$$H(x) = \xi_1(x) \exp\{\eta_1 x\} + \xi_2(x) \exp\{\eta_2 x\} \quad \text{for a.a. } [L] \, x \in \mathbb{R}, \qquad (2.3.10)$$

with c_1 and c_2 as nonnegative real numbers, ξ_1 and ξ_2 as periodic Borel measurable functions having period λ, and η_i, $i = 1, 2$ as real numbers such that $\int_{\mathbb{R}} \exp\{\eta_i x\}\mu(\mathrm{d}x) = 1$. (For the uniqueness of the representations but for the ordering of the terms, one may assume for example that $c_2 = 0$ and $\xi_2 \equiv 0$ if $\eta_1 = \eta_2$.)

Proof　In the case of arithmetic μ with span λ, there is no loss of generality in assuming that for each integer n, the function H is such that $H(x) \exp\{-\eta x\}$ is independent of x for $x \in [n\lambda, (n+1)\lambda)$, where η is as in Theorem 2.3.1. We have then from Theorem 2.3.1 for some $c \in \mathbb{R}_+$

$$H^*(x) = \int_{(-\infty, 0)} H^*(x + y)\rho^*(\mathrm{d}y) + c \quad \text{for a.a. } [L] \, x \in \mathbb{R},$$

where $H^*(x) = H(x) \exp\{-\eta x\}$ and $\rho^*(\mathrm{d}y) = \exp\{\eta y\}\rho(\mathrm{d}y)$. Consequently, in view of (iv) of Remarks 2.2.6, it follows that H is of the form (2.3.9) if μ is nonarithmetic and is of the form (2.3.10) if μ is arithmetic with span λ. (Note that if we find that H is of the form (2.3.9) or (2.3.10) without the information that $\int_{\mathbb{R}} \exp\{\eta_i x\}\mu(\mathrm{d}x) = 1$, then it is a trivial exercise to establish that unless respectively $c_i = 0$ or $\xi_i = 0$ a.e. $[L]$, it is necessary to have $\int_{\mathbb{R}} \exp\{\eta_i x\}\mu(\mathrm{d}x) = 1$ for the function H to satisfy (2.1.1) with $S = \mathbb{R}$.) ∎

2.3.3 Remarks

(i) Theorem 2.3.1 with trivial changes in its proof remains valid even when the local integrability of H on $\mathbb{R}$ is replaced by that on (α, ∞), where α is as defined in the statement of the theorem. Also both Theorem 2.3.1 and Corollary 2.3.2 hold even when we take H as a.e. $[L]$ nonnegative in place of nonnegative everywhere.

(ii) The proof of Theorem 2.3.1 simplifies considerably if we take in the place of (2.1.2) the equation (2.1.1) with $S = \mathbb{R}$. In this case, essentially by symmetry the fact that ρ is a Lebesgue–Stieltjes measure implies that τ is a Lebesgue–Stieltjes measure.

(iii) There obviously exist several other alternative arguments that are similar in spirit for arriving at Corollary 2.3.2. Also, it is worth noting here that Corollary 2.3.2 subsumes the following result that is essentially the one given in Ramachandran (1984):

Let $\{(v_n, w_n) : n = 0, \pm 1, \ldots\}$ be a sequence of two-component vectors with non-negative real components such that $w_0 < 1$ and at least one $v_n \neq 0$. Then

$$v_m = \sum_{n=-\infty}^{\infty} w_n v_{n+m}, \quad m = 0, \pm 1, \ldots$$

if and only if

$$v_m = B(m)b^m + C(m)c^m, \quad m = 0, \pm 1, \ldots$$

and $\Sigma_{m=-\infty}^{\infty} w_m b^m = \Sigma_{m=-\infty}^{\infty} w_m c^m = 1$ for some $b, c > 0$ and non-negative periodic functions B, C with the largest common divisor of $\{m : w_m > 0\}$ as their common period.

(iv) As observed by Alzaid *et al* (1988), the measure μ of Theorem 2.3.2 has the Wiener-Hopf factorization

$$\mu(B) = \rho(B) + \tau(B) - \tau^*\rho(B) \tag{2.3.11}$$

when B varies over the class of relatively compact Borel sets of $\mathbb{R}$. It is therefore tempting to attempt an alternative proof for Theorem 2.3.1 based on this factorization. In the proof of the theorem given above we have already observed that $\int_{(-\infty,0)} \hat{H}(x+y)\rho(dy) < \infty, x \in \mathbb{R}_+$. If we have additionally $\int_{\mathbb{R}_+} \hat{H}(x+y)\tau(dy) < \infty$, then it follows that (2.3.11) yields

$$G_x = \hat{H}(x) - \int_{(-\infty,0)} \hat{H}(x+y)\rho(dy)$$

$$= \int_{\mathbb{R}} \hat{H}(x+y)\mu(dy) - \int_{(-\infty,0)} \hat{H}(x+y)\rho(dy)$$

$$= \int_{\mathbb{R}} \{\hat{H}(x+z) - \int_{(-\infty,0)} \hat{H}(x+z+y)\rho(dy)\}\tau(dz)$$

$$= \int_{\mathbb{R}_+} G_{x+z}\tau(dz), \quad x \in \mathbb{R}_+,$$

and hence we have an alternative proof for (2.3.8). However, that one is not assured of $\int_{\mathbb{R}_+} \hat{H}(x+y)\tau(dy) < \infty$ is always shown by the following example adapted from Alzaid *et al* (1988).

Example   Let μ be the probability measure concentrated on $\{-1, 0, 1, 2, \ldots\}$ such that its moment generating function is given by

$$M(t) = 1 + \alpha\, e^{-t}(1 - e^t)^\beta, \quad t \leqslant 0,$$

where α and β are fixed positive numbers such that $1 < \beta < 2$ and $0 < \alpha\beta \leqslant 1$. (This moment generating function, but for a location change, was considered earlier by Seneta (1968) in connection with a certain problem in branching processes.) Let

H be such that

$$H(x) = \begin{cases} [x+1] & \text{if } x \geqslant 0 \\ 0 & \text{otherwise,} \end{cases}$$

where $[x+1]$ is the integer part of $x + 1$. Hence it follows that H satisfies the hypothesis of the theorem and

$$H(x) = \int_{\mathbb{R}} H(x+y)\mu(\mathrm{d}y), \quad x \in \mathbb{R}_+.$$

In this case ρ is the probability measure concentrated on $\{-1\}$ and in view of the Wiener–Hopf factorization of μ, τ is the probability measure with the moment generating function

$$M^*(t) = 1 - \alpha(1 - e^t)^{\beta-1}, \quad t \leqslant 0.$$

The expression for $M^*(t)$ implies that the τ in question has an infinite mean and hence we have here

$$\int_{\mathbb{R}_+} H(x+y)\tau(\mathrm{d}y) = \infty \quad \text{for each } x \in \mathbb{R}_+.$$

Now if we produce versions of H and μ giving $\mu(\mathbb{R}_+) > 1$ as in the proof of the cited theorem, then it easily follows that irrespectively of the value of δ taken

$$\int_{\mathbb{R}_+} \hat{H}(x+y)\tau(\mathrm{d}y) = \infty \quad \text{for each } x \in \mathbb{R}_+.$$

(v) The argument to arrive at $\int_{(-\infty,0)} \hat{H}(x+y)\rho(\mathrm{d}y) < \infty$, $x \in \mathbb{R}_+$ in the proof of Theorem 2.3.1 remains valid even if we consider any increasing continuous H satisfying (2.1.2) instead of $\hat{H}$. Then, essentially by symmetry, it follows that if we have, as in (ii) above, equation (2.1.1) with $S = \mathbb{R}$ instead of (2.1.2), then $\int_{\mathbb{R}_+} \hat{H}(x+y)\tau(\mathrm{d}y) < \infty$, $x \in \mathbb{R}$; also it is easily seen that now

$$\int_{(-\infty,0)} \hat{H}(x+y)\rho(\mathrm{d}y) < \infty, \quad x \in \mathbb{R}.$$

Consequently, it follows that in this latter case the approach based on the Wiener–Hopf factorization implied in (iv) could also be used to prove the second assertion of Theorem 2.3.1; indeed, we have now the validity of the assertion even when 'a.a. $[L]x \in \mathbb{R}_+$' in (2.3.3) is replaced by 'a.a. $[L]x \in \mathbb{R}$'.

(vi) If $\mathbb{R}_+$ in (2.2.8) is replaced by $\mathbb{R}$, then under the assumption that H is not a function that is equal to a constant a.e. $[L]$ (on $\mathbb{R}$) and H is either increasing or decreasing a.e. $[L]$, it follows, in view of Corollary 2.3.2, that every solution of the equation in question can be expressed as a convex combination of functions H_1 and H_2 of the form arrived at in Theorem 2.2.5 with (obviously the domain of definition as $\mathbb{R}$ instead of $\mathbb{R}_+$) and η replaced by η_1 and η_2 satisfying the conditions $\int_{\mathbb{R}} \exp\{\eta_i x\}\mu(\mathrm{d}x) = 1$, $i = 1, 2$. This also follows from a more general result to appear in Chapter 3.

It is now possible to identify the solution to (2.1.2) under a fairly general hypothesis.

2.3.4   Theorem

Let H be as in Theorem 2.3.1 except that the local integrability of H is now changed in that it is monotonic and right continuous and let $\rho^{(x)}$ denote the ρ measure on $\mathbb{R}$ with an alteration that the s_r's involved in its definition are replaced by $s_r + x$. Then, we have for each $x \in \mathbb{R}_+$

$$H(x) = \int_{(-\infty,0)} H(y)\rho^{(x)}(\mathrm{d}y) + \xi(x)\int_{[-x,0]} e^{\eta(x+y)}\left(\sum_{n=0}^{\infty}\rho^{*n}\right)(\mathrm{d}y) \qquad (2.3.12)$$

for some nonnegative real periodic function ξ such that it is constant if μ is nonarithmetic and a function with period λ if μ is arithmetic with span λ, and η is as defined in Theorem 2.3.1. (We define, as usual, ρ^{*n} for $n \geqslant 1$ to be the n-fold convolution of ρ with itself and ρ^{*0} to be the probability measure that is degenerate at zero.) Moreover, if f is any Borel measurable function on $(-\infty, 0)$ such that $\int_{(-\infty,0)} |f(y)|\rho^{(x)}(\mathrm{d}y) < \infty$ for each $x \in \mathbb{R}_+$ and ξ^* is any Borel measurable function of the form of ξ defined above with a modification that it is not necessarily nonnegative, then the function $\hat{H}$ given by

$$\hat{H}(x) = \int_{(-\infty,0)} f(y)\rho^{(x)}(\mathrm{d}y) + \xi^*(x)\int_{[-x,0]} e^{\eta(x+y)}\left(\sum_{n=0}^{\infty}\rho^{*n}\right)(\mathrm{d}y), \quad x \in \mathbb{R}_+$$

$$= f(x), \quad x \in (-\infty, 0)$$

satisfies (2.1.1) (even with 'a.a. [L]' replaced by 'all').

Proof   Under the more stringent assumption on H we have the validity of (2.3.3) with 'a.a. [L]' replaced by 'all'. Obviously the measure ρ is a Lebesgue–Stieltjes measure concentrated on $(-\infty, 0)$. For each $x \in \mathbb{R}_+$ and each Borel subset B of $\mathbb{R}$, we have then

$$\rho^{(x)}(B+x) = \sum_{n=1}^{\infty} \rho_n^{(x)}((-\infty, x) \cap B),$$

where $\rho_1^{(x)} = \rho$ and for each $n \geqslant 2$, $\rho_n^{(x)}$ is the convolution of measures $\rho_{n-1}^{(x)}$ $([-x, 0) \cap \bullet)$ and ρ (and hence is trivially a Lebesgue–Stieltjes measure concentrated on $(-\infty, 0)$). From the modified version of (2.3.3), we have also for each positive integer k and $x \in \mathbb{R}_+$

$$H(x) = \int_{(-\infty,-x)} H(x+y)\left(\sum_{n=1}^{\infty}\rho_n^{(x)}\right)(\mathrm{d}y) + \int_{[-x,0)} H(x+y)\rho^{*k}(\mathrm{d}y)$$

$$+ \xi(x)\int_{[-x,0]} e^{\eta(x+y)}\left(\sum_{n=0}^{k-1}\rho^{*n}\right)(\mathrm{d}y) \qquad (2.3.13)$$

on noting that the restrictions of ρ^{*n} and $\rho_n^{(x)}$ to $[-x, \infty)$ are identical for each $n \geqslant 1$. In view of the monotonicity (and hence the boundedness on $[0, x]$ for each $x \in \mathbb{R}_+$) of H and the fact that ρ is a Lebesgue–Stieltjes measure on $(-\infty, 0)$, it follows that the second term on the right-hand side of (2.3.13) tends to zero as $k \to \infty$. Consequently, on taking the limit as $k \to \infty$ of the right-hand side of (2.3.13), we arrive at the right-hand side of (2.3.12) and we get the validity of (2.3.12). It is easy to verify that the $\hat{H}$ of the theorem satisfies (2.1.2) with 'a.a. $[L]$' replaced by 'all' on noting in particular that taking without loss of generality $\eta = 0$ and following essentially the argument of Feller (1966) given to arrive at (3.9) of his Chapter XII, we get for each $x \in \mathbb{R}_+$

$$\tau(\mathbb{R}_+) + \sum_{n=0}^{\infty} \rho^{*n}([-x, \infty)) = 1 + \sum_{n=0}^{\infty} \rho^{*n}([-x, \infty))$$

$$= 1 + \left(\left(\sum_{n=0}^{\infty} \rho^{*n}\right) * \mu\right)([-x, \infty))$$

and hence

$$\left(\sum_{n=0}^{\infty} \rho^{*n}\right)([-x, \infty)) = \left(\left(\sum_{n=0}^{\infty} \rho^{*n}\right) * \mu\right)([-x, \infty))$$

whenever (2.3.4) is satisfied. ∎

2.3.5 Remark

If α as defined in the statement of Theorem 2.3.1 is finite, then Theorem 2.3.4 remains valid even when the assumption that H is monotonic is replaced by the one that it is locally bounded on $[\alpha, \infty)$ (i.e. bounded on every bounded subset of $[\alpha, \infty)$).

In order to discuss a further important theorem, we need the following lemma.

2.3.6 Lemma

Let μ be as in Theorem 2.3.1. Define for each $\theta \in \mathbb{R}$, $\mu^*(\theta) = \int_{\mathbb{R}} \exp(\theta x)\mu(dx)$, and $\tau^*(\theta)$ and $\rho^*(\theta)$ similarly corresponding to measures τ and ρ. Then, for any given θ, $\mu^*(\theta) = 1$ if and only if $\tau^*(\theta) = 1$ or $\rho^*(\theta) = 1$; moreover, $\min(\tau^*(\theta), \rho^*(\theta)) < 1$ unless $\tau^*(\theta) = \rho^*(\theta) = 1$.

Proof   The Wiener–Hopf factorization (2.3.11) in (iv) of Remarks 2.3.3 or its more general version $\mu + \tau^*\rho = \tau + \rho$ (which is easily seen to be valid on noting that $\mu([0, x]) + \tau^*\rho([0, x]) = \tau([0, x])$, $x \in \mathbb{R}_+$ and essentially by symmetry $\mu((-\infty, x)) + \tau^*\rho((-\infty, x)) = \rho((-\infty, x))$, $x \in (-\infty, 0))$ implies that

$$\tau([x, y])\rho([-x, 0]) \leqslant \tau([0, y]), \quad 0 \leqslant x < y < \infty.$$

If $\rho^*(\theta) > 1$, then, taking without loss of generality $\theta = 0$, we see from the above inequality that $\tau^*(\theta) < \infty$. (Note that if $\rho((-\infty, 0)) > 1$, then, for some $x \geqslant 0$,

$\rho([-x, 0]) > 1$, in which case we cannot have $\tau(\mathbb{R}_+) = \infty$ since in the case of $\tau(\mathbb{R}_+) = \infty$ for a sufficiently large y the inequality with '$\leqslant$' replaced by '$>$' is valid.) Essentially by symmetry, we have also that if $\tau^*(\theta) > 1$ then $\rho^*(\theta) < \infty$. If $\mu(\mathbb{R}) \leqslant 1$, then the definitions of τ and ρ yield that $\tau(\mathbb{R}) \leqslant 1$ and $\rho(\mathbb{R}) \leqslant 1$, and hence it follows that $\tau^*(\theta) \leqslant 1$ and $\rho^*(\theta) \leqslant 1$ whenever $\mu^*(\theta) \leqslant 1$ on taking without loss of generality $\theta = 0$. If we assume that $1 < \tau^*(\theta) < \infty$ and $1 < \rho^*(\theta) < \infty$, then the Wiener-Hopf factorization of μ yields that

$$1 - \mu^*(\theta) = (1 - \tau^*(\theta))(1 - \rho^*(\theta)) > 0 \qquad (2.3.14)$$

and hence that $\mu^*(\theta) < 1$, contradicting our earlier observation that $\tau^*(\theta) \leqslant 1$ and $\rho^*(\theta) \leqslant 1$ whenever $\mu^*(\theta) \leqslant 1$. Consequently, it follows that for each θ, either $\tau^*(\theta) \leqslant 1$ or $\rho^*(\theta) \leqslant 1$. Now, if θ_0 is such that $\tau^*(\theta_0) = 1$ (which implies that $\tau^*(\theta) > 1$ for each $\theta > \theta_0$ since $\mu(\{0\}) < 1$) or $\rho^*(\theta_0) = 1$ (which implies that $\rho^*(\theta) > 1$ for each $\theta < \theta_0$), then we can conclude, using the monotone convergence theorem, that respectively $\tau^*(\theta_0) = 1$ and $\rho^*(\theta_0) \leqslant 1$ or $\rho^*(\theta_0) = 1$ and $\tau^*(\theta_0) \leqslant 1$; in that case the identity

$$1 - \mu^*(\theta_0) = (1 - \tau^*(\theta_0))(1 - \rho^*(\theta_0)) \qquad (2.3.15)$$

arrived at via the Wiener-Hopf factorization yields that $\mu^*(\theta_0) = 1$. Furthermore, if $\mu^*(\theta_0) = 1$, then as observed above $\tau^*(\theta_0) \leqslant 1$ and $\rho^*(\theta_0) \leqslant 1$ and hence the identity (2.3.15) is valid. The identity in this latter case implies that we have either $\tau^*(\theta_0) = 1$ or $\rho^*(\theta_0) = 1$. From what we have found, it is now obvious that the lemma holds. ∎

2.3.7 Remarks

(i) Lemma 2.3.6 implies that if η satisfying (2.3.4) exists, then $\mu^*(\eta) = 1$, and that in Theorem 2.3.1 we have either $\mu(\mathbb{R}_+) \leqslant 1$ or $\mu((-\infty, 0)) \leqslant 1$.

(ii) One could also prove Lemma 2.3.6 slightly differently as follows. Since the result is trivially valid otherwise, we may assume $\mu((0, \infty)) > 0$ and $\mu((-\infty, 0)) > 0$; in that case the proof of Theorem 2.3.1 yields that there exists a real θ_0 such that

$$\theta_0 = \inf\{\theta : \rho^*(\theta) < 1\},$$

which implies in view of the monotone convergence theorem that $\rho^*(\theta_0) \leqslant 1$. For any $\varepsilon > 0$, if $\tau^*(\theta_0 - \varepsilon) > 1$, we are led to a contradiction essentially on using an argument in the existing proof of the lemma, as we get here $1 < \tau^*(\theta_0 - \varepsilon) < \infty$ and $1 < \rho^*(\theta_0 - \varepsilon) < \infty$. Consequently, it follows that $\tau^*(\theta) < 1$ for each $\theta < \theta_0$ and by the monotone convergence theorem $\tau^*(\theta_0) \leqslant 1$; if we now note in particular that $\mu^*(\theta) = 1$ implies $\rho^*(\theta) \leqslant 1$ and appeal to the identity

$$\mu^*(\theta) + \tau^*(\theta)\rho^*(\theta) = \tau^*(\theta) + \rho^*(\theta)$$

for $\theta \geqslant \theta_0$ we arrive at the required result.

   We are now ready to give the next theorem.

2.3.8   Theorem

Let μ and H be as in Theorem 2.3.1 with $\alpha > -\infty$ and $\rho^{(x)}$ for each $x \in \mathbb{R}_+$ be as defined in Theorem 2.3.4. Assume that the restriction of H to $[\alpha, \infty)$ is locally bounded and right-continuous. Define $\Theta = \{\theta : \mu^*(\theta) = 1\}$ with μ^* as defined in Lemma 2.3.6 and

$$m_\theta = \int_{[\alpha,\infty)} x \exp\{\theta x\} \mu(\mathrm{d}x), \quad \theta \in \mathbb{R}.$$

(Note that Theorem 2.3.1 trivially implies μ to be a Lebesgue–Stieltjes measure and hence m_θ is well defined for each $\theta \in \mathbb{R}$.) Then Θ is nonempty (without having more than two members), and if $\mu((-\infty, 0)) > 0$ and θ_0 is the only member in Θ or is the smaller of the two members in Θ, then we have $m_{\theta_0} \leqslant 0$. Moreover, in this case we have the assertions of Theorem 2.3.6 to be valid with:

(i) $\xi \equiv 0$ if Θ is a singleton and $m_{\theta_0} < 0$,

(ii) $\int_{[-x,0]} e^{\eta(x+y)} (\Sigma_{n=0}^\infty \rho^{*n})(\mathrm{d}y)$ replaced by $x\, e^{\theta_0 x} - \int_{(-\infty,0)} y\, e^{\theta_0 y} \rho^{(x)}(\mathrm{d}y)$ if Θ is a singleton and $m_{\theta_0} = 0$,

(iii) $\int_{[-x,0]} e^{\eta(x+y)} (\Sigma_{n=0}^\infty \rho^{*n})(\mathrm{d}y)$ replaced by $e^{\theta_1 x} - \int_{(-\infty,0)} e^{\theta_1 y} \rho^{(x)}(\mathrm{d}y)$ with $\theta_1 \in \Theta$ and $\theta_1 > \theta_0$ if Θ is a doubleton.

Proof   If $\mu((-\infty, 0)) > 0$, we have obviously in view of the fact that $\alpha > -\infty$ some real number δ such that $\rho^*(\delta) = 1$ and if $\mu((-\infty, 0)) = 0$, we have clearly $\xi \neq 0$ and, hence from the theorem, η exists such that $\tau^*(\eta) = 1$, where ρ^* and τ^* are as defined in Lemma 2.3.6. From what is mentioned in the lemma, we have $\mu^*(\delta) = 1$ if $\rho^*(\delta) = 1$ and also $\mu^*(\eta) = 1$ if $\tau^*(\eta) = 1$. Consequently it follows that Θ is nonempty. Clearly the assumption $\mu(\{0\}) < 1$ implies, in view of the lemma, that $\delta \leqslant \eta$ whenever we have $\rho^*(\delta) = 1$ and $\tau^*(\eta) = 1$. The lemma also assures us that for any θ, $\mu^*(\theta) = 1$ if and only if $\tau^*(\theta) = 1$ or $\rho^*(\theta) = 1$. It is therefore clear that Θ can have at most two points. (This also follows directly from the properties of moment generating functions of probability measures.) From Feller (1966, Theorem 1, p. 379 and Theorem 2, p. 380), it follows that if $\rho^*(\theta_0) = 1$, then $m_{\theta_0} \leqslant 0$. If we consider θ_0 as in the statement of the theorem and $\mu((-\infty, 0)) > 0$, then from the fact that $\delta \leqslant \eta$ whenever $\tau^*(\eta) = 1$, we have that $\delta = \theta_0$ and hence $m_{\theta_0} \leqslant 0$. Let us now establish the assertions (i), (ii), and (iii). As revealed in Remark 2.3.5, under the assumption of the present theorem, conclusions of Theorem 2.3.4 hold. Assertion (i) is then obvious since essentially due to the cited results in Feller (1966) we have that $\tau^*(\theta_0) < 1$ if $m_{\theta_0} < 0$ and, since $\Theta = \{\theta_0\}$, there is no η in this case such that $\tau^*(\eta) = 1$. It now remains to establish assertions (ii) and (iii). In the case of $m_{\theta_0} = 0$, Feller's Theorem 2 on page 380 implies the existence of $\eta = \theta_0$ and, and in the case of Θ containing two points, clearly we have $\eta = \theta_1 > \theta_0$ with $\theta_1 \in \Theta$ in view of what we have observed earlier. Let ξ be as in (2.3.12). We may here consider this to be a function defined on $\mathbb{R}$. Clearly the restriction of ξ to $[\alpha, \infty)$ is right-continuous and locally bounded. Define now in the case of assertion (ii)

$$H(x) = \begin{cases} \xi(x)(x-a)e^{\theta_0 x} & \text{if } x \geqslant \alpha \\ 0 & \text{otherwise} \end{cases}$$

and in the case of assertion (iii)

$$H(x) = \begin{cases} \xi(x)e^{\theta_1 x} & \text{if } x \geqslant \alpha \\ 0 & \text{otherwise.} \end{cases}$$

Observe that in both cases H satisfies (2.1.2) and we get

$$H(x) = \int_{(-\infty,0)} H(x+y)\rho(dy) + c\xi(x)e^{\eta x}, \quad x \in \mathbb{R}_+$$

for some positive constant c (which need not be the same in the two cases). In both cases, we have H satisfying the conditions required to arrive at (2.3.12) and hence the equation in question is valid with $c\xi$ replacing ξ. Since $c^{-1}\xi$ is of the form of ξ, assertions (ii) and (iii) easily follow.[†] ∎

The following corollaries of Theorems 2.3.1 and 2.3.8 respectively have some interesting applications to be discussed later in this monograph.

2.3.9   Corollary

Let $\{(v_n, w_n) : n = 0, 1, 2, \ldots\}$ be a sequence of vectors of nonnegative real components such that at least one $v_n \neq 0$ and $w_0 > 0$ and $\lambda > 0$. Further, let k and γ be positive integers such that the largest common divisor of k and those n for which $w_n > 0$ be γ. Then the sequence $\{(v_n, w_n)\}$ as defined satisfies the recurrence equations

$$v_{n+k} = \sum_{m=0}^{\infty} w_m v_{m+n}, \quad n = 0, 1, \ldots \qquad (2.3.16)$$

and for some integer k_1 that is an integral multiple of γ such that $k_1 > k$,

$$\lambda v_{n+k} = v_{n+k_1}, \quad n = 0, 1, \ldots, k + \gamma - 1 \qquad (2.3.17)$$

only if

$$v_n = q_n \lambda^{n/(k_1-k)}, \quad n = 0, 1, \ldots$$

for some nonnegative periodic sequence $\{q_n\}$ with period γ.

[†] The results from Feller (1966) that we have referred to herein are indeed the ones that, in a random walk with independent identically distributed jumps X_n for which $E(X_n)$ exists, the distribution of the first descending ladder height is proper only if $E(X_n) \leqslant 0$, and the distribution of the first weak ascending ladder height is improper if $E(X_n) < 0$ and proper if $E(X_n) = 0$.

Proof   It is sufficient if the result is established for $\gamma = 1$. Defining H and μ such that

$$H(x) = \begin{cases} v_{[x]+k}, & x > -k \\ 0, & \text{otherwise,} \end{cases}$$

and

$$\mu(\{n\}) = w_{n+k}, \quad n \geqslant -k, \quad \mu(\{-k, -k+1, \ldots\}^c) = 0,$$

where $[x]$ denotes the integral part of x, we see that the conditions for the validity of (2.3.2) with 'a.a. $[L]$' replaced by 'all' are met. The equation in this case implies in view of (2.3.17) and the fact that $\mu(\{-k\}) > 0$, that $H(x + k_1 - k) = \lambda H(x)$ for all $x \geqslant -k$. Since we can find an integer r such that $r(k_1 - k) \geqslant k$, we have subsequently

$$H(x) = \int_{[-k,\infty)} H(x + r(k_1 - k) + y)\lambda^{-r}\mu(dy), \quad x \in \mathbb{R}_+ \qquad (2.3.18)$$

with $r(k_1 - k) - k \geqslant 0$. There is no loss of generality in assuming that $\mu(\{0\}) > 0$ and hence that the measure ν defined such that

$$\nu(B) = \lambda^{-r}\mu(B - r(k_1 - k)) \quad \text{for every Borel set } B$$

is arithmetic with unit span. Consequently Theorem 2.2.2 (which is now a corollary of Theorem 2.3.1) implies, in view of (2.3.18) and the periodicity of $H(x)\lambda^{-x/(k_1-k)}$, $x \geqslant -k$, the required result. ∎

2.3.10   Corollary

Let $\{(v_n, w_n)\}$, k and γ be as defined in Corollary 2.3.9. Assume now that $\gamma = 1$. Define $D = \{b : b \geqslant 0, b^k = \Sigma_{m=0}^\infty w_n b^n\}$ and $\{f_{i,j} : i \geqslant k, j = 0, 1, \ldots, k - 1\}$ to be the sequence of absorption measures corresponding to the nonnegative matrix T (in the sense of Seneta (1973)) with state space $\{0, 1, 2, \ldots\}$ and the (i, j)th element as

$$T_{ij} = \begin{cases} \delta_{ij} & \text{if } i, j = 0, 1, \ldots, k - 1 \\ w_{j-i+k} & \text{if } i = k, k+1, \ldots \text{ and } j \geqslant i - k \\ 0 & \text{otherwise,} \end{cases}$$

where δ_{ij} is the Kronecker delta. Then the system of equations (2.3.16) is satisfied if and only if D is nonempty, $f_{n,0}, \ldots, f_{n,k-1}$ are finite for each $n \geqslant k$, and any one of the following conditions holds:

(i) D has only one point, $\Sigma_{n=0}^\infty (n - k)b^n w_n < 0$ for $b \in D$, and

$$v_n = f_{n,0}v_0 + \cdots + f_{n,k-1}v_{k-1}, \quad n = k, k+1, \ldots .$$

(ii) D has only one point, $\Sigma_{n=0}^\infty (n - k)b^n w_n = 0$ for $b \in D$ and for some $c \geqslant 0$,

$$v_n = f_{n,0}(v_0 - c_0 b^0) + \cdots + f_{n,k-1}(v_{k-1} - c(k - 1)b^{k-1}) + cnb^n,$$

with $b \in D$.

(iii) D contains two points and for some $c \geqslant 0$

$$v_n = f_{n,0}(v_0 - cb^0) + \cdots + f_{n,k-1}(v_{k-1} - cb^{k-1}) + cb^n, \quad n = k, k+1, \ldots,$$

with b as the larger of the two members of D.

2.3.11   Remarks

(i) The results and the proofs appearing in section 2.3 are either taken or adapted from Alzaid *et al* (1988). Alternative approaches for some of the results given here can be found in Lau and Rao (1984), Ramachandran (1982), Ramachandran and Prakasa Rao (1984), and Alzaid (1983) amongst others. In particular, a proof based on the Perron-Frobenius theorem (cf. Seneta, 1973, pp. 1-2) for Corollary 2.3.9 appears in Alzaid (1983).

(ii) If the integral equation of Corollary 2.3.2 is satisfied with 'a.a. [*L*]' replaced by 'all' and H as continuous, then the result of the corollary is obvious from Deny's (1961) general result. Indeed, from this latter result, the stronger version of Corollary 2.3.2 also follows, as illustrated by Rao and Shanbhag (1986), on noting that under the assumptions of the corollary, $\{n \int_0^{1/n} H(\bullet + y)\, dy : n = 1, 2, \ldots\}$ is a sequence of continuous functions on $\mathbb{R}$ satisfying the assumptions in Deny's theorem and converging a.e. [*L*] on $\mathbb{R}$ to H.

(iii) Rao and Shanbhag (1986) have given a proof for Theorem 2.2.2 based on the result of Deny referred to in (ii) above. (See also Remark 3 of Alzaid *et al* (1987) for a slight simplification in the proof.) This proof involves an application of the specialized version of Corollary 1.2.6 met in the proof of Theorem 2.2.2. If H is bounded and μ is finite, the Rao-Shanbhag proof simplifies considerably and goes through without any reference to the result in question, as observed by Fosam *et al* (1993).

(iv) The integral $\int_{(-\infty,0)} H(y)\rho^{(x)}(dy)$ in the statement of Theorem 2.3.4 has an interesting interpretation. If we define a Markov chain $\{X_n : n = 0, 1, \ldots\}$ as in the proof of Theorem 2.3.1 but with transition function $P(\bullet, \bullet)$ such that

$$P(x, A) = \begin{cases} 1 & \text{if } x \in (-\infty, 0) \cap A, \\ 0 & \text{if } x \in (-\infty, 0) \cap A^c, \\ (H(x))^{-1} \displaystyle\int_{A-x} H(x + y)\mu(dy) & \text{if } x \in \mathbb{R}_+ \text{ with } H(x) \neq 0 \\ & \qquad\qquad \text{and } A \text{ is arbitrary} \\ P(x_0, A) & \text{otherwise,} \end{cases}$$

where $x_0 \in \mathbb{R}_+$ is fixed satisfying $H(x_0) \neq 0$, then, for each $x \in \mathbb{R}_+$, the integral $\int_{(-\infty,0)} H(y)\rho^{(x)}(dy)$ can be expressed as $H(x)$ times the probability of absorption from x into $(-\infty, 0)$.

(v) Lindley (1952) obtains a corollary to Theorem 2.3.4 when $H(y) = 0$ for $y < 0$, H is bounded, and μ is a probability measure. In this case (2.3.12) reduces to

$$H(x) = C \sum_{n=0}^{\infty} \rho^{*n}([-x, 0]), \quad x \in \mathbb{R}_+$$

with C as a constant; the result has applications in queueing theory as we shall see later in this monograph.

We may conclude the present chapter by reproducing an observation of Fosam *et al* (1993) that is of relevance here. Suppose that we have (2.1.1) with H bounded and μ finite, subject to a modification that μ is now taken as a signed measure. Assume further that there exists a real sequence $\{a_n : n = 0, 1, \ldots\}$ such that $\Sigma_{n=0}^{\infty} |a_n| (|\mu|(S))^n < \infty$, $\Sigma_{n=0}^{\infty} a_n > 0 (\text{or} < 0)$ and

$$\nu = \left(\sum_{n=0}^{\infty} a_n \right)^{-1} \sum_{n=0}^{\infty} a_n \mu^{*n}$$

is a nonnegative measure with $\nu(\{0\}) < 1$, where $|\mu|$ is the total variation of μ. Then (2.1.1) holds with ν in place of μ and we get the results given in this chapter for (2.1.1) to be valid with ν in place of μ. (The remark, with obvious modifications, also applies to the more general equation (3.1.1) of the next chapter.) The results implied here are in the same spirit as those appearing in Krein (1962). The condition on μ that we have put to get ν as a nonnegative measure appears naturally in many situations; the following is an illustration.

Example   Consider a noninfinitely divisible characteristic function ϕ such that

$$\phi(t) = \exp \left\{ -\alpha + \int_{\mathbb{R}_+} e^{itx} \mu(\mathrm{d}x) \right\}, \quad -\infty < t < \infty$$

with $\alpha > 0$ and μ as a real signed measure on $\mathbb{R}_+$ with $\mu(\{0\}) = 0$. The existence of ϕ follows, see Lévy's example in Linnik (1964, p. 95). If we denote the measure corresponding to the Fourier transform $\exp\{\alpha - 1\}\phi(t)$, $t \in \mathbb{R}$ by ν, we see that

$$\nu = \left(\sum_{n=0}^{\infty} a_n \right)^{-1} \sum_{n=0}^{\infty} a_n \mu^{*n}$$

with $a_n = 1/n!$.

CHAPTER 3

A Version of Deny's Theorem and its Extensions: a Martingale Approach

3.1 INTRODUCTION

In the previous chapter, we studied some simple integral equations. We now consider more general equations that have received attention in the literature: we treated simple cases separately only because of their wider applicability and the interest they have generated amongst the audience. However, one could use the treatment as a platform for dealing with more general results, getting ideas about their structures and conditions under which they are achieved; this is so in spite of the need of more refined tools in the general case.

Unless stated otherwise, we assume the space S appearing in the present chapter to be a separable Abelian metric semigroup with zero element; in one or two places though, the results given do not require the space to be anything more than an Abelian topological semigroup. (Except in some specific examples, one could consider for simplicity $S = \Pi_{i=1}^{n} S_i$ with $n \geq 1$ and $S_i = \mathbb{R}$ or $\mathbb{R}_+$ or $-\mathbb{R}_+$ or $\mathbb{Z}(= \{0, \pm 1, \pm 2, \ldots\})$ or $\mathbb{N}_0(= \{0, 1, 2, \ldots\})$ or $-\mathbb{N}_0$ for $i = 1, 2, \ldots, n$.)

Consider now a generalized version of the integral equation (2.1.1) as

$$H(x) = \int_S H(x + y)\mu(dy), \quad x \in S, \tag{3.1.1}$$

where S is as mentioned above, H is a nonnegative real-valued continuous function on S, and μ is a measure on (the Borel σ-field of) S. (Note that the slight modification in (2.1.1) involving a locally integrable function in place of a continuous function and a.a. $[L]x$ in place of all x is not crucial as implied earlier since $\{k \int_0^{1/k} H(\bullet + y)\,dy : k = 1, 2, \ldots\}$ in that case converges a.e. $[L]$ to $H(\bullet)$ and its members are continuous satisfying (3.1.1).) Choquet and Deny (1960) and Deny (1961) have studied specialized versions of (3.1.1) under the additional assumptions that S is locally compact and is a group. Deny (1961) proved that if S is as assumed and the smallest closed subgroup of S containing supp$[\mu]$ equals S itself, then every nontrivial solution of (3.1.1) (i.e. $H \not\equiv 0$ satisfying (3.1.1)) can be expressed as a constant multiple of a weighted average of μ-harmonic exponential functions on S. (A function e : $S \to \mathbb{R}_+$ is called an exponential function on S if it is continuous and satisfies $e(x + y) = e(x)e(y)$ for all $x, y \in S$; an exponential function e on S is called a μ-harmonic exponential function on S if it satisfies $\int_S e(x)\mu(dx) = 1$.) As a corollary of Deny's result, it follows that if S and μ are as in Deny's result with μ satisfying additionally $\mu(S) = 1$, then a bounded real-valued continuous function H^* on S satisfies

$$H^*(x) = \int_S H^*(x + y)\mu(dy), \quad x \in S \qquad (3.1.2)$$

if and only if H^* is constant. This latter result was established by Choquet and Deny (1960).

Deny's proof of his result is based on potential theoretic arguments of Choquet theory. The set of functions H satisfying (3.1.1) with $H(0) = c$ fixed is convex and under Deny's assumptions the set of extreme points of this set is precisely the set of functions having each of its members to be equal to c times a μ-harmonic exponential function. Under the assumptions of Choquet and Deny (1960) the proof of Deny's theorem simplifies considerably. Doob et al (1960), Revuz (1975) and several others have given martingale proofs for various special cases of Deny's result.

As pointed out by Davies and Shanbhag (1987), unless some structural requirements are met, Deny's integral representation for the solution to (3.1.1) does not hold in the case of a semigroup; the following is an illustration.

Example   Let $S = \mathbb{R}_+^2$ and define μ by $\mu(\{0\} \times [0, b]) = b$, $\mu((0, 1] \times \mathbb{R}_+) = 0$ and $\mu((1, 1 + a] \times [0, b]) = ab, a, b \in \mathbb{R}_+$. If $q : \mathbb{R}_+ \to \mathbb{R}_+$ is continuous without being identically equal to zero but satisfying $q(a) = 0$ for $a \geqslant 1$, then the function H defined by $H(a, b) = q(a)\exp\{-b\}$, $(a, b) \in \mathbb{R}_+^2$ satisfies (3.1.1). However, H here cannot have the representation in terms of μ-harmonic exponential functions.

However, taking a hint from the example as well as from Corollary 1.2.6 used in the proof of Theorem 2.2.2, Davies and Shanbhag (1987) have shown via a martingale argument that, under some mild conditions, certain versions of Deny's theorem for semigroups hold. More refined and modified versions of these results have since appeared in Rao and Shanbhag (1989b) and Shanbhag (1991). The present chapter is devoted to studying these results and discussing their implications. Included amongst the corollaries of the results are the well known theorems of Bernstein and Bochner on completely monotonic functions.

3.2   AUXILIARY LEMMAS AND A SYMMETRIC MEASURE

In order to discuss the main results of the chapter, we require some preliminaries. As stated in Section 3.1, we assume S of (3.1.1) to be a separable Abelian metric semigroup with zero element. Define now $S^*(\mu)$ to be the smallest closed subsemigroup of S containing supp$[\mu]$ and the zero element. We begin by giving the following lemma.

3.2.1   Lemma

Let (3.1.1) be satisfied. Then there exist σ-finite measures μ_1 and μ_2 on S such that

$$H(x) = \int_S H(x + y)\mu_i(dy), \quad x \in S, \quad i = 1, 2, \tag{3.2.1}$$

the restriction μ_1 to $\{y : H(x + y) \neq 0 \text{ for some } x\}$ agrees with that of μ, $\text{supp}[\mu_1] = \text{supp}[\mu]$ and $\text{supp}[\mu_2] = S^*(\mu)$.

Proof   Let D be a countable dense subset of S. Because of the continuity of H we have

$$\{y : H(x + y) = 0 \quad \text{for all } x \in S\} = \{y : H(x + y) = 0 \quad \text{for all } x \in D\}.$$

Denoting this set by B_0, we see that it is a closed Borel set and that for $x \in S$

$$H(x) = \int_{S \setminus B_0} H(x + y)\mu(dy)$$

$$= \int_{S \setminus B_0} H(x + y)\mu(dy) + \int_{B_0} H(x + y)\nu(dy)$$

where ν is a finite measure on S with support $B_0 \cap \text{supp}[\mu]$. We define μ_1 by

$$\mu_1(C) = \mu(C \cap (S \setminus B_0)) + \nu(C)$$

for each Borel subset C of S. Clearly, then $\text{supp}[\mu_1] = \text{supp}[\mu]$ and the integral equation in (3.2.1) is satisfied for $i = 1$. Furthermore, we have

$$S \setminus B_0 = \{y : H(x + y) > 0 \quad \text{for some } x \in D\}$$

$$= \bigcup_{n=1}^{\infty} \left\{ y : H(x + y) > \frac{1}{n} \quad \text{for some } x \in D \right\}$$

and as

$$\mu\left(\left\{y : H(x + y) > \frac{1}{n}\right\}\right) \leqslant nH(x) < \infty$$

we see that μ_1 is σ-finite. This establishes the first part of the lemma.

This, in turn, implies that (3.1.1) with μ replaced by $\Sigma_{n=0}^{\infty} 2^{-n-1} \mu_1^{*n}$, where μ_1^{*n} denotes the n-fold convolution of μ_1 with itself, is valid. Appealing to the first part of lemma (i.e. the part concerning the existence of μ_1 satisfying the conditions specified), we can then see the existence of μ_2 meeting the requirements. ■

3.2.2   Remarks

(i) It now follows from Lemma 3.2.1 that the results of Theorem 1.2.5 and Corollary 1.2.6 remain valid even when μ is not assumed to be σ-finite if S is as in the present special case (or more generally a second countable Abelian topological semigroup). The applications of the results to be met henceforth take into account this information.

(ii) If μ_2 is used in place of μ in (2.2.6) in the proof of Theorem 2.2.2, then (2.2.7) in the proof follows via a slightly simpler argument.

(iii) It may be mentioned here that by the convolution of σ-finite measures $\nu_1, \ldots, \nu_n$ on S for $n \geq 1$, we mean the measure on S with values

$$\left(\prod_{i=1}^{n} \nu_i\right) (\{(x_1, \ldots, x_n) \in S^n : x_1 + \cdots + x_n \in B\}),$$

where B denotes any Borel set and $\prod_{i=1}^{n} \nu_i$ denotes the product measure relative to $\nu_1, \ldots, \nu_n$. The definition, with the usual convention that ν_i^{*0} denotes the probability measure on S that is concentrated at $\{0\}$, is used in the proof of Lemma 3.2.1 above.

If we have (3.1.1) satisfied with H such that $H(0) > 0$, then it is easily seen on using the product measure theorem of Tulcea (Theorem 1.2.1) that there exists a unique probability measure P_H defined on $\mathcal{B}(S^{\mathbb{N}})$, the Borel σ-field of $S^{\mathbb{N}}$, where $\mathbb{N} = \{1, 2, \ldots\}$, such that for each $n \geq 1$ and Borel sets $B_1, B_2, \ldots, B_n$ of S

$$P_H\left(B_1 \times \cdots \times B_n \times \prod_{n+1}^{\infty} S\right) = \frac{1}{H(0)} \int_{B_n} \cdots \int_{B_1} H\left(\sum_{i=1}^{n} x_i\right) \mu_2(dx_1) \ldots \mu_2(dx_n),$$

where μ_2 is as in Lemma 3.2.1.

The probability measure P_H so obtained is symmetric in the sense of Hewitt and Savage (1955). Consider now the probability space $(S^{\mathbb{N}}, \mathcal{B}(S^{\mathbb{N}}), P_H)$ and define on it $X_n, n \geq 1$ to be the projection (or the coordinate) mappings and $Y_n = \Sigma_{j=1}^{n} X_j$, $n \geq 0$ following the usual convention that $Y_0 \equiv 0$. Define for each $n \geq 0$ and $\omega \in S^{\mathbb{N}}$

$$\xi_n(x, \omega) = \begin{cases} H(x + Y_n(\omega))/H(Y_n(\omega)) & \text{if } H(Y_n(\omega)) > 0 \quad \text{and} \quad H(x) > 0 \\ 0 & \text{otherwise;} \end{cases}$$

also define

$$\xi(x, \omega) = \begin{cases} \lim_{n \to \infty} \xi_n(x, \omega) & \text{if the limit exists and is finite} \\ 0 & \text{otherwise.} \end{cases}$$

From the integral equation, it easily follows that for each $x \in S$, $\{\xi_n(x, \bullet) : n = 0, 1, 2, \ldots\}$ is a nonnegative supermartingale[†] and hence by 1.2.2(a) converges a.s. $[P_H]$ to $\xi(x, \bullet)$. The following lemma makes two important observations concerning $\{\xi_n(x, \bullet)\}$.

[†] We can take this to be so relative to sub σ-fields $\mathcal{F}_n$, $n \geq 0$ of $\mathcal{B}(S^{\mathbb{N}})$ such that $\mathcal{F}_0 = \{\phi, S^{\mathbb{N}}\}$ and $\mathcal{F}_n = \sigma(X_1, \ldots, X_n)$, $n \geq 1$; other choices of $\mathcal{F}_n$ are also obviously possible.

3.2.3   Lemma

If for any $x \in S$ and $y \in S^*(\mu)$, $\{\xi_n(x, \bullet)\}$ or $\{\xi_n(x + 2y, \bullet)\}$ is a martingale, then $\{\xi_n(x + y, \bullet)\}$ is a martingale. Also $\{\xi_n(x + y, \bullet)\}$ is a martingale converging in L_p for some $p > 1$ whenever $\{\xi_n(x, \bullet)\}$ or $\{\xi_n(x + 2y, \bullet)\}$ is a martingale converging in L_p for some $p > 1$. (The two p's need not be equal.)

Proof   Let $x \in S$ and $y \in S^*(\mu)$ be given. Clearly $\{\xi_n(x, \bullet)\}$ is a martingale if and only if

$$\mu_2(\{z : H(x + Y_n(\omega) + z) > 0\} \cap \{z : H(Y_n(\omega) + z) = 0\}) = 0$$

$$\text{for a.a. } [P_H]\omega \quad \text{and} \quad n \geqslant 0. \qquad (3.2.2)$$

Since the inequality in Corollary 1.2.6 is valid, it follows that (3.2.2) with x replaced by $x + y$ is valid whenever (3.2.2) is valid. This, in turn, implies that $\{\xi_n(x + y, \bullet)\}$ is a martingale if $\{\xi_n(x, \bullet)\}$ is. Furthermore, if $\{\xi_n(x, \bullet)\}$ is a martingale converging in L_p for some $p > 1$, then appealing once more to the inequality referred to above and using Hölder's inequality, we have for the p in question

$$E((\xi_n(x + y, \bullet))^{2p/(1+p)}) \leqslant E((\xi_n(x, \bullet)\xi_n(x + 2y, \bullet))^{p/(1+p)})$$
$$\leqslant (E((\xi_n(x, \bullet))^p))^{1/(1+p)}(E(\xi_n(x + 2y, \bullet)))^{p/(1+p)}$$
$$\leqslant G_p(x)(H(x + 2y)/H(0))^{p/(1+p)}, \quad n \geqslant 0, \quad (3.2.3)$$

where

$$G_p(x) = (\sup_{n \geqslant 0} E((\xi_n(x, \bullet))^p))^{1/(1+p)} < \infty.$$

In view of (3.2.2) and our observation that we have here $\{\xi_n(x + y, \bullet)\}$ to be a martingale, 1.2.2(b) then implies that $\{\xi_n(x + y, \bullet)\}$ is a martingale converging in L_p for some $p > 1$. The present conclusions remain valid with obvious alterations in the arguments if we take $\{\xi_n(x + 2y, \bullet)\}$ in place of $\{\xi_n(x, \bullet)\}$. Hence, we have the lemma. ∎

3.2.4   Remarks

(i) Since $\{\xi_n(0, \bullet)\}$ is a bounded martingale, Lemma 3.2.3 implies that, for each $y \in S^*(\mu)$, $\{\xi_n(y, \bullet)\}$ is a martingale converging in L_p for some $p > 1$. Lemma 4 of Davies and Shanbhag (1987) corresponding to the behavior of $\{\xi_n(y, \bullet)\}$ yields a stronger result that the sequence is a martingale converging in L_p for each $p > 0$; however, the weaker result is sufficient for our purposes.

(ii) It is of relevance to our study to give an example illustrating that there exist cases in which for some x, $\{\xi_n(x, \bullet)\}$ is not a martingale. The following example is constructed for this purpose.

Example   Let $S = \mathbb{N}_0^2$ and μ be a measure on S such that

$$\mu(\{(0, 2)\}) = \mu(\{(1, 0)\}) = 1$$

and $\mu(\{(x_1, x_2)\}) = 0$ if $(x_1, x_2) \neq (0, 2)$ or $(1,0)$. Define H on S such that

$$H(x_1, x_2) = \begin{cases} c & \text{if } x_1 = 0, \ x_2 \neq 1 \quad \text{or} \quad x_1 = 1, 2, \ldots, \ x_2 = 1 \\ 2c & \text{if } x_1 = 0, \ x_2 = 1 \\ 0 & \text{otherwise,} \end{cases}$$

where c is a positive constant. We have then (3.1.1) satisfied with $H(0, 0) > 0$. However, for $x = (1, 1)$, we cannot have $\{\xi_n(x, \bullet)\}$ to be a martingale since (3.2.2) is not satisfied.

(iii) If $H(x) > 0$ for all $x \in S$, then we have trivially that (3.2.2) is satisfied for each $x \in S$ and hence that $\{\xi_n(x, \bullet)\}$ is a martingale for each $x \in S$.

(iv) If S is a group and the smallest closed subgroup of S containing supp$[\mu]$ equals S itself, then the inequality

$$H(2(x - y) + z)H(z) \geqslant (H(x - y + z))^2, \quad x, y \in S^*(\mu), \quad z \in S$$

(i.e. the inequality

$$H(2x + (z - 2y))H(2y + (z - 2y)) \geqslant (H(x + y + (z - 2y)))^2,$$
$$x, y \in S^*(\mu), \quad z \in S),$$

which is a corollary of Theorem 1.2.5, yields immediately that

$$H(2x + z)H(z) \geqslant (H(x + z))^2, \quad x, z \in S.$$

Consequently, we have here, assuming $H(0) > 0$, $\{\xi_n(x, \bullet)\}$ for each $x \in S$, to be a well defined martingale (on noting that $H(x) > 0$ for each $x \in S$) satisfying

$$\xi_n(2x, \bullet) \geqslant (\xi_n(x, \bullet))^2, \quad n \geqslant 0$$

and hence

$$\sup_{n \geqslant 0} E(\xi_n(x, \bullet))^2 \leqslant H(2x)/H(0) < \infty,$$

which, in turn, implies in view of Theorem 1.2.2, that $\{\xi_n(x, \bullet)\}$ is a martingale converging in L_2 for each $x \in S$. (If we take $H(0) = 0$, then it turns out that $H \equiv 0$.)

(v) If $H(x) \geqslant H(x + y)$ for each $x, y \in S$ (as in the case of $\hat{H}$ in the proof of Theorem 2.2.2) in addition to $H(0) > 0$, we have $\{\xi_n(x, \bullet)\}$ to be a bounded martingale and hence a martingale converging in L_p for each $p > 0$ for each $x \in S$.

3.3   GENERAL THEOREMS ON THE INTEGRAL EQUATIONS

We give now some results on the integral equation extending, amongst other things, the Choquet–Deny theorem.

3.3.1   Theorem

Let (3.1.1) be satisfied with $H(0) > 0$ so that the probability space $(S^{\mathbb{N}}, \mathcal{B}(S^{\mathbb{N}}), P_H)$ is well defined. Define $\hat{S}$ to be the set of points x for which $\{\xi_n(x, \bullet) : n = 0, 1, 2, \ldots\}$ is a martingale converging in L_p for some $p > 1$. Then $S^*(\mu) \subset \hat{S}$,

$$H(x) = H(0)E(\xi(x, \bullet)), x \in \hat{S}, \quad H(x) \geqslant H(0)E(\xi(x, \bullet)), x \in \hat{S}^c, \quad (3.3.1)$$

$$\xi(x + y, \bullet) = \xi(x, \bullet)\xi(y, \bullet) \quad \text{a.s. } [P_H] \quad \text{for all } x \in \hat{S} \quad \text{and} \quad y \in S^*(\mu),$$
$$(3.3.2)$$

$$\int_S \xi(x, \bullet)\mu(dx) = 1 \quad \text{a.s. } [P_H], \quad (3.3.3)$$

$$\xi(x, \bullet) \equiv 0 \quad \text{if } H(x) = 0. \quad (3.3.4)$$

Proof   The result of 3.2.4 (i) implies that $S^*(\mu) \subset \hat{S}$. Since for each $x \in \hat{S}$, $\{\xi_n(x, \bullet)\}$ is a regular martingale (i.e. a martingale converging in L_1), and, in general, it is a supermartingale, (3.3.1) follows. If $\{\xi_n(x, \bullet)\}$ is a regular martingale, we have, after a minor manipulation

$$E(|\xi_{n+1}(x, \bullet) - \xi_n(x, \bullet)|)$$

$$= (H(0))^{-1} \int_S \cdots \int_S \left| \frac{H\left(x + \sum_{i=1}^n y_i + y\right)}{H\left(\sum_{i=1}^n y_i\right)} - \frac{H\left(x + \sum_{i=1}^n y_i\right)}{H\left(\sum_{i=1}^n y_i\right)} \frac{H\left(y + \sum_{i=1}^n y_i\right)}{H\left(\sum_{i=1}^n y_i\right)} \right|$$

$$\times H\left(\sum_{i=1}^n y_i\right) \mu_2(dy_1)\ldots\mu_2(dy_n)\mu_2(dy)$$

$$= \int_S E(|\xi_n(x + y, \bullet) - \xi_n(x, \bullet)\xi_n(y, \bullet)|)\mu_2(dy), \quad n \geqslant 0, \quad (3.3.5)$$

where the ratios with $H(\Sigma_{i=1}^n y_i)$ in the denominator are to be understood as equal to zero whenever $H(\Sigma_{i=1}^n y_i) = 0$. (To see the validity of (3.3.5), note, in particular, that if $H(\Sigma_{i=1}^n y_i) = 0$, then $H(\Sigma_{i=1}^n y_i + y) = 0$ for each $y \in \text{supp}[\mu_2]$.) In that case we have in view of Fatou's lemma

$$0 = \lim_{n \to \infty} E(|\xi_{n+1}(x, \bullet) - \xi_n(x, \bullet)|)$$

$$\geqslant \int_S E(|\xi(x + y, \bullet) - \xi(x, \bullet)\xi(y, \bullet)|)\mu_2(dy)$$

implying that

$$\xi(x + y, \bullet) = \xi(x, \bullet)\xi(y, \bullet) \quad \text{a.s. } [P_H] \quad \text{for a.a. } [\mu_2]y. \quad (3.3.6)$$

Assume that $x \in \hat{S}$ and $x' \in S^*(\mu)$. Then, by Lemma 3.2.3, $x + x' \in S$; we have also trivially $x' \in \hat{S}$. From (3.3.6), we have then

$$\xi(z + y, \bullet) = \xi(z, \bullet)\xi(y, \bullet), \quad z = x, x', \quad \xi(x + x' + y, \bullet) = \xi(x + x', \bullet)\xi(y, \bullet)$$

$$\text{a.s. } [P_H] \quad \text{for a.a. } [\mu_2]y,$$

which implies, in turn, that there exists a sequence $\{y_n : n = 1, 2, \ldots\}$ of members of $S^*(\mu)$ converging to x' such that

$$\xi(z + y_n, \bullet) = \xi(z, \bullet)\xi(y_n, \bullet), \quad z = x, x',$$

$$\xi(x + x' + y_n, \bullet) = \xi(x + x', \bullet)\xi(y_n, \bullet) \quad \text{a.s. } [P_H], \quad n \geqslant 1. \quad (3.3.7)$$

Consequently, we have, in view of Corollary 1.2.6 and Lemma 3.2.3 (together with the trivial fact that $0 \in \hat{S}$ or that $S^*(\mu) \subset \hat{S}$),

$$0 = \lim_{n \to \infty} E\{\xi(z + 2x', \bullet) - 2\xi(z + x' + y_n, \bullet) + \xi(z + 2y_n, \bullet)\}$$

$$\geqslant \limsup_{n \to \infty} E\left\{\frac{(\xi(z + x', \bullet))^2}{\xi(z, \bullet)} - 2\xi(z + x', \bullet)\xi(y_n, \bullet) + \frac{(\xi(z + y_n + \bullet))^2}{\xi(z, \bullet)}\right\}$$

$$= \limsup_{n \to \infty} E\left\{\left(\frac{\xi(z + x', \bullet)}{(\xi(z, \bullet))^{1/2}} - \xi(y_n, \bullet)(\xi(z, \bullet))^{1/2}\right)^2\right\}$$

$$\geqslant 0, \quad z = 0, x, \quad (3.3.8)$$

where each of the ratios with $\xi(z, \bullet)$ or $(\xi(z, \bullet))^{1/2}$ in the denominator is defined to be equal to zero on $\{\xi(z, \bullet) = 0\}$; to arrive at (3.3.8), we have noted, amongst other things, that $\xi(z + y, \bullet) = 0$ a.s. $[P_H]$ on $\{\xi(z, \bullet) = 0\}$ if $y \in S^*(\mu)$. (3.3.8) implies that $\{\xi(y_n, \bullet)(\xi(z, \bullet))^{1/2}\}$ converges in L_2 to $\xi(z+x', \bullet)/(\xi(z, \bullet))^{1/2}$, which subsumes the result that $\{\xi(y_n, \bullet)\}$ converges in L_2 to $\xi(x', \bullet)$ (on noting that $\xi(0, \bullet) = 1$ a.s. $[P_H]$). Since $\xi(x + x', \bullet) = 0$ a.s. $[P_H]$ on $\{\xi(x, \bullet) = 0\}$, we have then (3.3.2) to be valid. (3.3.4) follows trivially by the definition of ξ. Since $S^*(\mu) \subset \hat{S}$, (3.3.1), (3.3.2) and (3.3.4) imply on using Fubini's theorem that

$$E\left(\left(1 - \int_S \xi(x, \bullet)\mu(\mathrm{d}x)\right)^2\right) = 0. \quad (3.3.9)$$

(Observe that the restriction of μ to $\{x : H(x) > 0\}$ is σ-finite or

$$\int_S \xi(x, \bullet)\mu(\mathrm{d}x) = \int_S \xi(x, \bullet)\mu_1(\mathrm{d}x)$$

with μ_1 as in Lemma 3.2.1, and hence we are justified in claiming the validity of (3.3.9) via Fubini's theorem.) (3.3.9) implies (3.3.3) immediately; one could also arrive at (3.3.3) using the argument of Davies and Shanbhag (1987). ∎

3.3.2   Corollary

Let S be as defined before, $H : S \to \mathbb{R}_+$ be a bounded continuous function, and μ be a subprobability measure on S (i.e. a measure such that $\mu(S) \leqslant 1$). Then (3.1.1) is satisfied if and only if either $H \equiv 0$ or $\mu(S) = 1$ and

$$H(x + y) = H(x) \quad \text{for each } x \in S \quad \text{and} \quad y \in \text{supp}[\mu]. \tag{3.3.10}$$

Proof   It is sufficient if we show that whenever $H \not\equiv 0$, (3.3.1) is valid only if $\mu(S) = 1$ and (3.3.10) holds. Consider $H(0) > 0$. The boundedness of H and Theorem 3.3.1 imply that the martingale limit $\xi(y, \bullet)$ for each $y \in \text{supp}[\mu]$ is such that

$$E((\xi(y, \bullet))^n) = H(ny)/H(0) < K < \infty \tag{3.3.11}$$

implying that $\xi(y, \bullet) \leqslant 1$ a.s. $[P_H]$ and hence that $H(y) \leqslant H(0)$. The integral equation then yields that $\mu(S) = 1$ and $H(y) = H(0)$ for each $y \in \text{supp}[\mu]$. Applying the result to each $H(x + \bullet)$ with $H(x) \neq 0$ and using the result that $H(x + y) = 0$ whenever $H(x) = 0$ and $y \in \text{supp}[\mu]$, we get then that if $H \not\equiv 0$, we have $\mu(S) = 1$ and

$$H(x + y) = H(x), \quad x \in S, \quad y \in \text{supp}[\mu].$$

Hence we have the corollary. ∎

Although we have given above the result 3.3.2 as a corollary to Theorem 3.3.1, the theorem to follow with its proof shows that there exists an elementary technique for arriving at a more general result. We also give below some interesting corollaries of the theorem in question.

3.3.3   Theorem

Let S be an Abelian topological semigroup, h a bounded nonnegative Borel measurable function on S and μ a subprobability measure on S. Then

$$h(x) = \int_S h(x + y)\mu(dy), \quad x \in S \tag{3.3.12}$$

if and only if either $h \equiv 0$ or $h(x) \neq 0$ for some x,

$$h(x + y) = h(x) \quad \text{for a.a. } [\mu]y \in S \quad \text{for each } x \in S \tag{3.3.13}$$

and $\mu(S) = 1$. Moreover, if h is continuous, (3.3.13) implies that $h(x + y) = h(x)$ for each $y \in \text{supp}[\mu]$ for each $x \in S$.

Proof   The 'if' part of the first assertion is trivial. Let us now establish the 'only if' part of the assertion; to do this, it is sufficient if we assume the existence of a point $x_0 \in S$ and a Borel set B_0 such that

$$h(x_0)\mu(B_0) < \int_{B_0} h(x_0 + y)\mu(dy) \tag{3.3.14}$$

and arrive at a contradiction. Assume then the existence of x_0 and B_0 meeting the requirements and note also that we have in that case $h(x_0) > 0$ and $\mu(B_0) > 0$. In view of Fubini's theorem and a trivial special case of the Kolmogorov consistency theorem, it follows from (3.3.12) that there exists an infinite sequence $\{Z_n : n = 1, 2, \ldots\}$ of 0–1-valued exchangeable random variables (defined on a probability space) such that for each $n \geqslant 1$

$$P\{Z_1 = 1, \ldots, Z_n = 1\}$$

$$= (h(x_0))^{-1} \int_{B_0} \cdots \int_{B_0} h(x_0 + y_1 + \cdots + y_n)\mu(dy_1)\ldots\mu(dy_n).$$

From Corollary 1.2.4, we have immediately that

$$P\{Z_1 = 1, \ldots, Z_{2^n} = 1\} \geqslant (P\{Z_1 = 1, \ldots, Z_{2^{n-1}} = 1\})^2$$

$$\vdots$$

$$\geqslant (P\{Z_1 = 1\})^{2^n}, \quad n \geqslant 1.$$

This, in turn, implies that

$$P\{Z_1 = 1, \ldots, Z_{2^n} = 1\}/(\mu(B_0))^{2^n} \geqslant (P\{Z_1 = 1\}/\mu(B_0))^{2^n}, \quad n \geqslant 1,$$

leading to a contradiction since, due to (3.3.14) and the boundedness of h, the left-hand side of the inequality is bounded relative to n while the right-hand side tends to ∞ as $n \to \infty$. Hence we have the required result. The second assertion of the theorem is now obvious. ■

3.3.4   Corollary

Let S be as in the theorem, $g : S \to \mathbb{R}$ be a bounded Borel measurable function and $\mu^{(1)}$ and $\mu^{(2)}$ be subprobability measures on S such that $\mu^{(1)} + \mu^{(2)}$ is a probability measure on S. Then

$$g(x) = \int_S g(x + y)(\mu^{(1)} - \mu^{(2)})(dy), \quad x \in S \qquad (3.3.15)$$

if and only if for each $x \in S$,

$$g(x + y) = \begin{cases} g(x) & \text{for a.a. } [\mu^{(1)}] \ y \in S \\ -g(x) & \text{for a.a. } [\mu^{(2)}] \ y \in S; \end{cases} \qquad (3.3.16)$$

moreover, if g is continuous, (3.3.15) implies that for each $x \in S$ (3.3.16) is valid with 'for a.a. $[\mu^{(i)}] \ y \in S$' replaced by 'for $y \in \text{supp}[\mu^{(i)}]$' respectively.

Proof   Let c be a constant such that $c - |g(x)|$ is nonnegative for all $x \in S$. Define for each $n \geqslant 0$

$$h(n, x) = c + (-1)^n g(x), \quad x \in S.$$

Note that

$$h(n, x) = \int_S h(n, x+y)\mu^{(1)}(dy) + \int_S h(n+1, x+y)\mu^{(2)}(dy), \quad n = 0, 1, \ldots, x \in S,$$

which obviously is of the form of (3.3.12) with $\mathbb{N}_0 \times S$ in place of S (where $\mathbb{N}_0 = \{0, 1, 2, \ldots\}$) and μ as a probability measure. Appealing to the theorem, we can then arrive at the corollary immediately. ∎

3.3.5 Corollary

Corollary 3.3.4 holds if the assumption that g is bounded is replaced by the weaker assumption that $g(\bullet + y) - g(\bullet)$ is bounded for a.a. $[\mu^{(1)} + \mu^{(2)}]y \in S$ (with the implicit requirement that for each $x \in S$, $g(x + \bullet)$ is $(\mu^{(1)} + \mu^{(2)})$-integrable) provided $\mu^{(2)}(S) > 0$.

Proof   It is sufficient if we establish the 'only if' part of the first assertion. Let N be a $(\mu^{(1)} + \mu^{(2)})$-null set such that $g(\bullet + y) - g(\bullet)$ is bounded for each $y \in N^c$. Now, if $y \in N^c$, on applying Corollary 3.3.4 to $g(\bullet + y) - g(\bullet)$, we get from (3.3.15) for each $x \in S$

$$g(x+y+z) - g(x+z) = \begin{cases} g(x+y) - g(x) & \text{for a.a. } [\mu^{(1)}]z \in S. \\ -(g(x+y) - g(x)) & \text{for a.a. } [\mu^{(2)}]z \in S. \end{cases} \quad (3.3.17)$$

On integrating out y from the second equation of (3.3.17) with respect to measure $\mu^{(2)}$ and simplifying, we get

$$\int_S (g(x+y+z) + g(x+y))\mu^{(1)}(dy) - (g(x+z) + g(x))$$

$$= \mu^{(2)}(S)\{g(x+z) + g(x)\} \quad \text{for a.a. } [\mu^{(2)}]z \in S. \quad (3.3.18)$$

(Note, in particular, Fubini's theorem implies here that for each $x \in S$

$$g(x+y+z) + g(x+y) = g(x+z) + g(x) \quad \text{for a.a. } [\mu^{(2)}]y \quad \text{for a.a. } [\mu^{(2)}]z.)$$

From the first equation of (3.3.17), we get for each $x \in S$

$$g(x+y+z) - g(x+y) = g(x+z) - g(x) \quad \text{for a.a. } [\mu^{(2)}]y \quad \text{for a.a. } [\mu^{(1)}]z,$$

and, from the second equation (writing y for z and vice versa), we get for each $x \in S$

$$g(x+y+z) - g(x+y) = -(g(x+z) - g(x)) \quad \text{for a.a. } [\mu^{(2)}]y \quad \text{for a.a. } [\mu^{(1)}]z.$$

Since $\mu^{(2)}(S) > 0$, these two relations in turn imply that for each $x \in S$

$$g(x+z) = g(x) \quad \text{for a.a. } [\mu^{(1)}]z. \quad (3.3.19)$$

Using (3.3.19) in (3.3.18), as $\mu^{(2)}(S) > 0$, it is then easily seen that

$$g(x + z) = -g(x) \quad \text{for a.a. } [\mu^{(2)}]z. \tag{3.3.20}$$

Since it now follows that (3.3.19) and (3.3.20) are implied by (3.3.15) (for the modified version of g), we have the required result. ∎

3.3.6   Corollary

Let S, $\mu^{(1)}$, $\mu^{(2)}$ and g be as in Corollary 3.3.4 but for a modification that g is now continuous (possibly satisfying the weaker version appearing in Corollary 3.3.5 of the boundedness condition if $\mu^{(2)}(S) > 0$). If there exists a dense subset A of S such that for every $x \in A$, we have $y \in S^*(\mu^{(1)}, \mu^{(2)})$ with $x + y \in S^*(\mu^{(1)}, \mu^{(2)})$, where $S^*(\mu^{(1)}, \mu^{(2)})$ is the smallest closed subsemigroup of S containing $(\mathrm{supp}[\mu^{(1)}]) \cup (\mathrm{supp}[\mu^{(2)}])$, then (3.3.15) is met if and only if

$$g(x) = g(0), \quad x \in S$$

and either $g(0) = 0$ or $\mu^{(2)}(S) = 0$.

Proof   The 'if' part is trivial. To prove the 'only if' part, note that (3.3.15) implies in view of Corollary 3.3.4 that

$$g(x + y) = g(x), \quad x \in S, \quad y \in S^*(\mu^{(1)}, \mu^{(2)}),$$

which, in turn, implies that

$$g(x) = g(x + y) = g(0), \quad x \in A, \quad y \in S^*(\mu^{(1)}, \mu^{(2)})$$

and hence, in view of the denseness of A in S, that $g(x) = g(0)$, $x \in S$. If (3.3.15) is met with $g \equiv g(0)$, then it is immediate that we have either $g(0) = 0$ or $\mu^{(2)}(S) = 0$. ∎

3.3.7   Remarks

(i) Theorem 3.3.1 is essentially adapted from Davies and Shanbhag (1987) and but for minor alterations appears in Shanbhag (1991). Indeed, assuming additionally S to be complete, Shanbhag (1991) uses a somewhat shorter approach based on de Finetti's theorem corresponding to random variables taking values in a Polish space (see, for example, Olshen (1973) for the relevant information) to prove the theorem.

(ii) Corollary 3.3.6 yields directly the Choquet–Deny theorem as a special case and hence this result, as well as Theorem 3.3.3 and Corollaries 3.3.4 and 3.3.5, could be considered as extensions of the theorem in question. Corollaries 3.3.4 and 3.3.5 with h continuous and $S = \mathbb{R}_+$ or $S = \mathbb{R}$, have been dealt with either by Shimizu (1978) or by Ramachandran et al (1988); the general results are either adapted or taken from Rao and Shanbhag (1989a, 1989b). A version of Corollary 3.3.6 in the case of $\mu^{(2)}(S) = 0$ also appears in Davies and Shanbhag (1987).

(iii) A somewhat indirect proof for Theorem 3.3.3 follows via the result on page 225 in Feller (1966) that there exists a distribution on [0,1] such that for each $n \geqslant 1$, $P\{Z_1 = 1, \ldots, Z_n = 1\}$ appearing in the proof of the theorem denotes its nth moment giving

$$P\{Z_1 = 1, \ldots, Z_n = 1\} \geqslant (P\{Z_1 = 1\})^n.$$

Also, a closer scrutiny of the existing proof of the theorem reveals that, with very minor alterations, one can produce from the proof a new proof which does not require a reference to the Kolmogorov consistency theorem; note that it is sufficient to argue here in terms of the specialized versions of probability measures P_j's of Theorem 1.2.3 without bringing into consideration an infinite dimensional probability space.

(iv) If we choose $\{y_n : n = 1, 2, \ldots\}$ to be a sequence of members of $S^*(\mu)$ for which the first equation in (3.3.8) is valid instead of it to be a sequence converging to x', then the proof of Theorem 3.3.1 remains valid even when we take S as a second countable Abelian topological semigroup in place of a separable Abelian metric semigroup. Furthermore, if we assume additionally $\mu((\text{supp}[\mu])^c) = 0$, then if we define the measure μ_1 on S such that it is concentrated on $\{x : H(x) > 0\}$ with its restriction to this set agreeing with that of μ and define the measure μ_2 on S such that it is also concentrated on $\{x : H(x) > 0\}$ with its restriction to the set in question agreeing with that of $\Sigma_{n=0}^{\infty} 2^{-(n+1)} \mu_1^{*n}$ (without insisting that all the requirements of Lemma 3.2.1 be met), then Theorem 3.3.1 holds for an Abelian topological semigroup (with zero element) even without the requirement of second countability. (This can be seen on using in (3.3.6) and subsequently, the additional information that $\{\xi_n(x, \bullet)\}$ is a martingale only if, for each y lying in $S^*(\mu) \setminus S^*(\mu_1)$, $H(x+y) = 0 = H(y)$ and hence $\xi(x+y, \bullet) = 0 = \xi(x, \bullet)\xi(y, \bullet)$; note also that $S^*(\mu) \setminus S^*(\mu_1) \subset \hat{S}$.)

(v) A special case of Corollary 3.3.4 with $\mu^{(2)}(S) = 0$ has been established recently by Székely and Zeng (1990) following essentially an argument, which involves the martingale convergence theorem and the Hewitt–Savage zero–one law, in Meyer (1966, pp. 151–152). (Incidentally, the argument referred to here is due to Doob *et al* (1960).) A martingale argument substantially simpler than the one used in the proof of Theorem 3.3.1 without involving any zero–one law could also be produced to prove this latter result. To see this, take a modified version of (3.3.12) with h not necessarily nonnegative (but, as in the previous case, as bounded) and $\mu(S) = 1$. In that case, we can consider the probability space $(S^{\infty}, (\mathcal{B}(S))^{\infty}, \mu^{\infty})$ (in the usual notation) and essentially as in Meyer (1966), define on it the bounded martingale $\{h(x + \Sigma_{m=1}^{n} X_m) : n = 1, 2, \ldots\}$ for each $x \in S$, where X_n, $n \geqslant 1$ as before are projection mappings. If we denote the a.s. limit of $\{h(x + \Sigma_{m=1}^{n} X_m)\}$ by h_x for each $x \in S$ the argument leading to (3.3.6) in the proof of Theorem 3.3.1 yields that, for each $x \in S$,

$$h_{x+y}(\bullet) = h_x(\bullet) \quad \text{a.s. } [\mu^{\infty}] \quad \text{for a.a. } [\mu] \quad y \in S. \tag{3.3.21}$$

(Indeed, the argument simplifies in the present case.) As $E(h_x(\bullet)) = h(x)$, $x \in S$, we can then immediately conclude that (3.3.21) implies (3.3.13). Hence we have that (3.3.12) for the modified case implies (3.3.13); this is sufficient to have the result sought as its other parts are obvious.

3.4   A VERSION OF DENY'S THEOREM AND ITS EXTENSIONS

At this point, it is natural to ask whether Theorem 3.3.1 holds with $\hat{S}$ as S or $S^*(\mu)$ in (3.3.2) replaced by S (assuming $\mu(\{0\}) < 1$). From the example given in (ii) of Remarks 3.2.4, we have that there exist cases in which $\{\xi_n(x, \bullet)\}$ is not a martingale for a particular x. The following example of Rao and Shanbhag (1989a) (which is adapted from the example in Section 3.1) illustrates further that even when $\{\xi_n(x, \bullet)\}$ is a martingale for each $x \in S$ and $\hat{S} = S$ it is not necessary that the modified result corresponding to (3.3.2) be valid.

Example 1   Let $S = \mathbb{N}_0^2$ and define on S, μ to be the measure satisfying $\mu(\{0\} \times \{n\}) = 2^{-1}$, $n = 0, 1, \ldots, \mu(\{1, \ldots, m\} \times \mathbb{N}_0) = 0$ and $\mu(\{n_1\} \times \{n_2\}) = 1$, $n_1 = m + 1, m + 2, \ldots, n_2 = 0, 1, \ldots$, where m is a fixed positive integer $\geqslant 2$. Define $q : \mathbb{N}_0 \to \mathbb{R}_+$ such that $q(n) = 0$ for all $n > m$ and $q(n) \neq 0$ otherwise. The H on S such that $H(n_1, n_2) = 9(n_1)(2)^{-n_2}$, $n_1, n_2 = 0, 1, \ldots$ satisfies (3.1.1). In this case, $\{\xi_n(x, \bullet)\}$ turns out to be a bounded martingale for each $x \in S$ and we have $\hat{S} = S$. However, we cannot have here (3.3.2) with $S^*(\mu)$ replaced by S satisfied since the H has $H(1, 1) > 0$ and for some x, $H(x) = 0$.

Yet another example illustrates that even in cases in which for each x, $\{\xi_n(x, \bullet)\}$ is a martingale, we could still have $\hat{S} \neq S$.

Example 2   Let $S = \mathbb{Z}^2$ and μ_2 be a probability measure on S such that it is concentrated on $\mathbb{Z} \times \{0\}$ such that there are two roots 0 and α say (with $\alpha \neq 0$) to the equation

$$\sum_{m=-\infty}^{\infty} e^{tm} \mu_2(\{(m, 0)\}) = 1.$$

Define

$$H(m, n) = \begin{cases} 1 & \text{if } (m, n) \in \mathbb{Z} \times \{0\} \\ e^{\alpha m} & \text{otherwise.} \end{cases}$$

In this case, since H satisfies (3.1.1) and does not vanish, we have for each $x \in S$, $\{\xi_n(x, \bullet)\}$ to be a martingale. Also we have here for each $x = (m, n)$ with $n \neq 0$, $\xi(x, \bullet) = 0$ a.s. $[P_H]$ and hence $\{\xi_n(x, \bullet)\}$ is not even a regular martingale; consequently we have in this example $\hat{S} \neq S$.

In view of what we have revealed, it is now clear that, unless some further assumptions are made, the modified version of Theorem 3.3.1 mentioned above does not hold. Introduce then the following condition.

Condition I   Let $\mu(\{0\}) < 1$ and there be a subset D of S that is dense in $S + (B \setminus \{0\})$ for some Borel set B with $\mu(B^c) = 0$ such that for every $x \in D$, we have $k \geqslant 1$ and $y_1, \ldots, y_k \in S^*(\mu)$ satisfying $x + \Sigma_{i=1}^{i-1} y_i \in S + y_r, r = 1, 2, \ldots, k$ and $x + \Sigma_{i=1}^{k} y_i \in S^*(\mu)$.

(There are several equivalent ways of expressing the condition; we have chosen the present form only for convenience.)

Note that in the case when S is a group mentioned in (iv) of Remarks 3.2.4, for which we have $\hat{S} = S$, Condition I is trivially met. The following theorem establishes that under the validity of Condition I the more general extended version of Theorem 3.3.1 sought above indeed holds.

3.4.1   Theorem

Let Condition I be valid in (3.1.1). Then either the H is identically equal to zero or $H(0) > 0$ and the assertions of Theorem 3.3.1 hold with $\hat{S} = S$ and $S^*(\mu)$ in (3.3.2) replaced by S.

Proof   There is no loss of generality in assuming D to be a semigroup and hence we shall do so. If $H(x) = 0$, $x' \in D$, then, in view of the inequality in Corollary 1.2.6, we have $H(x + x' + \Sigma_1^k y_i) = 0$ and successively

$$H(x + x' + \sum_1^{k-1} y_i), \ldots, H(x + x')$$

to be all equal to zero, where $y_1, \ldots, y_k$ are as in Condition I when the point considered in D is x'; (3.1.1) then yields $H(x + x') = 0$ and by the continuity of H we get $H \equiv 0$ on $x + S$. This in turn implies that either $H \equiv 0$, or $H(0) > 0$ and $\{\xi_n(x, \bullet)\}$ is a martingale for each $x \in S$. We now restrict ourselves to the latter situation. An argument analogous to the one used for concluding (obviously in a somewhat different notation) that $H(x + x') = 0$ whenever $H(x) = 0$ and $x' \in D$ but with Lemma 3.2.3 instead of the inequality of Corollary 1.2.6 implies that $x + x' \in \hat{S}$ whenever $x \in \hat{S}$ and $x' \in D$, which, in turn, implies that $D \subset \hat{S}$ since we have trivially $0 \in \hat{S}$. Consider now $x' \in D$ and $y_1, \ldots, y_k$ to be the corresponding y_i's as in Condition I; define $y^* = \Sigma_1^k y_i$. Appealing to (3.3.2), we can then see that if $x \in \hat{S}$

$$0 = |\xi(x + x' + y^*, \bullet) - \xi(x, \bullet)\xi(x' + y^*, \bullet)|$$

$$= |\xi(x + x', \bullet) - \xi(x, \bullet)\xi(x', \bullet)||\xi(y^*, \bullet) \quad \text{a.s. } [P_H] \qquad (3.4.1)$$

with $\xi(x' + y^*, \bullet) = 0$ a.s. $[P_H]$ on $\{\xi(y^*, \bullet) = 0\}$. Since essentially the argument involved earlier in concluding that $H(x+x') = 0$ when $H(x+x'+y^*) = 0$ implies that $\xi(x', \bullet) = 0$ a.s. $[P_H]$ on $\{\xi(x' + y^*, \bullet) = 0\}$ and $\xi(x + x', \bullet) = 0$ a.s. $[P_H]$ on $\{\xi(x + x' + y^*, \bullet) = 0\}$, (3.4.1) yields that

$$\xi(x + x', \bullet) = \xi(x, \bullet)\xi(x', \bullet). \qquad (3.4.2)$$

This implies amongst other things (in view of the fact that $H(0)E(\xi(x', \bullet)) = H(x')$, $H(0)E(\xi(2x', \bullet)) = H(2x'))$ that

$$
\begin{aligned}
H(0)H(2x') &= (H(0))^2 E(\xi(2x', \bullet)) \\
&= (H(0))^2 E((\xi(x', \bullet))^2) \\
&\geqslant (H(0)E(\xi(x', \bullet)))^2 = (H(x'))^2
\end{aligned}
$$

and hence by the continuity of H, we have

$$
H(0)H(2x) \geqslant (H(x))^2, \quad x \in S + (B \setminus \{0\}), \tag{3.4.3}
$$

where B is as in Condition I. Applying the result (3.4.3) to each $H(y + \bullet)$ where y is such that $H(y) > 0$ and also recalling the earlier observation that $H(y) = 0$ implies $H(y + x) = 0$ for each $x \in S$, we then get

$$
H(y)H(2x + y) \geqslant (H(x + y))^2, \quad x \in S + (B \setminus \{0\}), \quad y \in S, \tag{3.4.4}
$$

which implies in view of Theorem 1.2.2 essentially via the relevant argument in (iv) of Remarks 3.2.4 that $\{\xi_n(x, \bullet)\}$ is a martingale converging in L_2 for each $x \in S + (B \setminus \{0\})$ and hence that $S + (B \setminus \{0\}) \subset \hat{S}$. In view of Lemma 3.3.1 amongst other things, we have then

$$
\xi(x, \bullet) = (1 - \mu_1(\{0\}))^{-1} \int_{B \setminus \{0\}} \xi(x + z, \bullet)\mu_1(dz) \quad \text{a.s. } [P_H], \quad x \in S \tag{3.4.5}
$$

(with μ_1 as defined in the lemma). (3.4.5) follows since Fatou's lemma yields the result with '$\geqslant$' instead of '$=$' in view of the integral equation (3.1.1) with μ_1 in the place of μ, and we have due to the integral equation referred to and Fubini's theorem that the expectations of its two sides are equal. Using Fubini's theorem, (3.4.2) hence yields

$$
\xi(x + y, \bullet) = \xi(x, \bullet)\xi(y, \bullet) \quad \text{a.s. } [P_H] \quad \text{for each} \quad x \in S, \quad y \in D \tag{3.4.6}
$$

and we also get $E(\xi(x, \bullet)) = H(x)/H(0)$, $x \in S$. If $x \in S + (B \setminus \{0\})$; then there exists a sequence $\{x_n : n = 1, 2, \ldots\}$ of members of D converging to x and consequently yielding in view of (3.4.4) and (3.4.6) that

$$
\begin{aligned}
0 &= \lim_{n \to \infty} E\{\xi(2x_n, \bullet) - 2\xi(x + x_n, \bullet) + \xi(2x, \bullet)\} \\
&\geqslant \limsup_{n \to \infty} E\{(\xi(x_n, \bullet) - \xi(x, \bullet))^2\} \geqslant 0
\end{aligned} \tag{3.4.7}
$$

(in which we use, amongst others, the facts that $\xi(2x, \bullet) \geqslant (\xi(x, \bullet))^2$ a.s. $[P_H]$ and $\xi(2x_n, \bullet) = (\xi(x_n, \bullet))^2$ a.s. $[P_H]$ for each $n \geqslant 1$), which implies that $\{\xi(x_n, \bullet) : n = 1, 2, \ldots\}$ converges in L_2 to $\xi(x, \bullet)$. Consequently, we have that if $x, x' \in S + (B \setminus \{0\})$, then there exist sequences $\{x_n : n = 1, 2, \ldots\}$ and $\{x'_n : n = 1, 2, \ldots\}$ of members of D such that $\{\xi(x_n, \bullet)\}$, $\{\xi(x'_n, \bullet)\}$ and $\{\xi(x_n + x'_n, \bullet)\}$ converge in

L_2 respectively to $\xi(x, \bullet)$, $\xi(x', \bullet)$ and $\xi(x + x', \bullet)$. (To see the convergence in L_2 in the third case, we have assumed D to be a semigroup and hence $x_n + x_n' \in D$ for each $n \geqslant 1$.) We have also from (3.4.2) or (3.4.6)

$$\xi(x_n + x_n', \bullet) = \xi(x_n, \bullet)\xi(x_n', \bullet), n \geqslant 1 \quad \text{a.s. } [P_H],$$

which implies that

$$\xi(x + x', \bullet) = \xi(x, \bullet)\xi(x', \bullet) \quad \text{a.s. } [P_H].$$

Using Fubini's theorem and (3.4.5) we can hence conclude that (3.3.2) with $S^*(\mu)$ replaced by S is valid. Repeating the argument given above to observe via (3.4.2) that $S + (B \setminus \{0\}) \subset \hat{S}$, we can then conclude, in view of the relation $E(\xi(x, \bullet)) = H(x)/H(0)$, $x \in S$ that $\hat{S} = S$. Hence we have the theorem. ■

3.4.2   Corollary

If, in Theorem 3.3.1, S is a group with its smallest closed subgroup containing supp$[\mu]$ as itself, then the assertions in the theorem with $\hat{S}$ as S and $S^*(\mu)$ in (3.3.2) replaced by S hold.

Proof   The hypothesis of the corollary implies that $H(0) > 0$ and supp$[\mu]$ has at least one point $x \neq 0$. If we denote the point by x_0, then the validity of the integral equation at the point $x = -x_0$ implies that $\mu(\{0\}) < 1$ and hence it easily follows that Condition I is met. Consequently, we have the result in question. ■

3.4.3   Remarks

(i) With minor alterations in its proof (i.e. by taking $\{x_n : n = 1, 2, \ldots\}$ and $\{x_n' : n = 1, 2, \ldots\}$ to be sequences of members of D for which the first equation in (3.4.7) and the corresponding equations with x, $\{x_n\}$ replaced by x', $\{x_n' : n = 1, 2, \ldots\}$, and $x + x'$, $\{x_n + x_n' : n = 1, 2, \ldots\}$ respectively hold, when x, x' are given points of $S + (B \setminus \{0\})$, directly without involving any pointwise convergence), Theorem 3.4.1 can be seen to be valid even when, as in the case of Theorem 3.3.1, we take S as a second countable Abelian topological semigroup in place of a separable Abelian metric semigroup. Consequently, with the alteration in question, Corollary 3.4.2 also holds. Moreover, if it is assumed that $\mu((\text{supp}[\mu])^c) = 0$, then modifying the definition of μ_2 as in (iv) of Remarks 3.3.7, we can see that both Theorem 3.4.1 and Corollary 3.4.2 in the modified form hold even without the requirement of second countability.

(ii) In view of what we have pointed out in (iv) of Remarks 3.2.4, it is clear that the proof of Theorem 3.4.1 simplifies slightly under the hypothesis of Corollary 3.4.2.

(iii) If $\hat{\mu}$ is any other measure relative to which the integral equation (3.1.1) with $\hat{\mu}$ in the place of μ is valid, then Theorem 3.4.1, Corollary 3.4.2 and their modified versions in (i) above also hold with $\mu(\{0\}) < 1$ and $\mu(B^c) = 0$ in Condition I replaced respectively by $\hat{\mu}(\{0\}) < 1$ and $\hat{\mu}(B^c) = 0$.

A natural question arising now is the following: does there exist under the assumptions in Theorem 3.4.1, possibly with some additions, an integral representation in terms of μ-harmonic exponential functions of the type in Deny's theorem to the solution to (3.1.1)? However, before we make any attempt to answer this question, we may give the following corollaries that are of relevance to it.

3.4.4 Corollary

If S is countable and Condition I is met, then, under discrete topology, every solution H with $H(0) > 0$ to (3.1.1) can be expressed as a constant multiple of a weighted average of μ-harmonic exponential functions.

Proof   We can find a member A of $\mathcal{B}(S^{\mathbb{N}})$ with $P_H(A^c) = 0$ such that for each $\omega \in A$, $\xi(\bullet, \omega)$ is a μ-harmonic exponential function. We have also $H(x) = H(0)E(\xi(x, \bullet))$, $x \in S$. Consequently, taking $\hat{\xi}(\bullet, \omega) = \xi(\bullet, \omega)$ if $\omega \in A$ and $= \xi(\bullet, \omega_0)$ if $\omega \in A^c$, where ω_0 is some fixed point of A, we can see that here $H(x) = H(0)E(\hat{\xi}(x, \bullet))$, $x \in S$ with $\hat{\xi}(\bullet, \omega)$ as a μ-harmonic exponential function for each $\omega \in \Omega$. ∎

3.4.5 Corollary

Let $n \geqslant 1$ and $S = \Pi_{i=1}^n S_i$ with $S_i = \mathbb{Z}$ or $\mathbb{N}_0$ or $-\mathbb{N}_0$ or $\mathbb{R}$ or $\mathbb{R}_+$ or $-\mathbb{R}_+$. Let λ be the restriction to S of a Haar measure on the smallest subgroup of $\mathbb{R}^n$ containing S. Let $h : S \to \mathbb{R}_+$ be a Borel measurable function that is locally integrable with respect to λ, and μ be a σ-finite measure on S such that Condition I is met. Then

$$h(\bullet) = \int_S h(\bullet + y)\mu(dy) \quad \text{a.e. } [\lambda]$$

implies that

$$h(\bullet) = \int_{[-\infty,\infty]^n} e^{\langle \bullet, x\rangle}\alpha(dx) \quad \text{a.e. } [\lambda], \tag{3.4.8}$$

where α is a measure on $[-\infty, \infty]^n$ such that $\alpha(\{x \in [-\infty, \infty]^n : \int_S e^{\langle x, y\rangle}\mu(dy) \neq 1$ or $\langle x, y\rangle$ is undefined or equal to ∞ for some $y \in S\}) = 0$ (and we define $e^{-\infty} = 0$, $e^\infty = \infty$ and $0 \cdot (\pm\infty) = 0$ with obviously the product as commutative).

Proof   Without loss of generality, we can assume $S = \Pi_{i=1}^n S_i$ with the first m of the S_i's to be either equal to $\mathbb{Z}$ or $\mathbb{N}_0$ and the remaining S_i's to be equal to $\mathbb{R}$ or $\mathbb{R}_+$ where m is a fixed integer such that $0 \leqslant m \leqslant n$. For every positive integer k, take a_k to be the point in $\mathbb{R}_+^n$ that has first m coordinates to be equal to zero and the remaining coordinates equal to k^{-1}. Observe that for each k, $H_k(\bullet) = \int_{Q_k} h(\bullet + y)\lambda(dy)/\lambda(Q_k)$, with $Q_k = [0, a_k]$, is continuous satisfying the requirements of Theorem 3.4.1. Since $H_k(\bullet) \to h(\bullet)$ a.e. $[\lambda]$, in view of the continuity Theorem 1.1.12 for Laplace transforms of measures subject to the modification of Remark 1.1.14 (i), to have the result it is sufficient if we prove that if H

is as in Theorem 3.4.1 then it has in the present case the integral representation of (3.4.8). We require a proof in the case of only the latter assertion of the theorem. If $x \in S$ and $\{x_n : n = 1, 2, \ldots\}$ is a sequence of members of S with rational components converging to x, then it follows that $\{\xi(x_n, \bullet)\}$ converges in L_2 to x and hence

$$\xi(x, \bullet) = \prod_{i=1}^{n} (a_i(\bullet))^{x(i)} \quad \text{a.s. } [P_H]$$

for some nonnegative real a_i's with these to be positive if S_i is uncountable or a group, where $x(i)$ is the ith component of x (and $0^0 = 1$). The result is now immediate. ∎

3.4.6   Corollary

Let S be as in Corollary 3.4.4, F be the restriction to S of a distribution function on $\mathbb{R}^n$ and μ be a σ-finite measure on S satisfying Condition I such that

$$F(\bullet) = \int_S F(\bullet + y)\mu(dy) + G(\bullet) \quad \text{a.e. } [\lambda] \tag{3.4.9}$$

for some G possessing the property that $\Delta_{y-x}G(x) = 0$ for every $y \geqslant x$ (where Δ is the delta operator as used in the definition of a distribution function on $\mathbb{R}^n$) and λ as in Corollary 3.4.5. Then, for some measure α concentrated on $[-\infty, \infty]^n$ of the type in Corollary 3.4.5,

$$\Delta_{y-x}F(x) = \int_{[-\infty,\infty]^n} \left(\prod_{i=1}^{n} \frac{e^{z_i y_i} - e^{z_i x_i}}{z_i^*} \right) \alpha(dz) \quad \text{for } x, y \in S^* \quad \text{with} \quad y \geqslant x,$$

$$\tag{3.4.10}$$

where $S^* = \Pi_{i=1}^n S_i^*$ with $S_i^* = S_i \setminus \{0\}$ if $S_i = -\mathbb{R}_+$ and $= S_i$ otherwise, x_i, y_i and z_i denote the ith components of x, y and z respectively, $z_i^* = z_i$ if $z_i \in \mathbb{R}$ and $= 1$ or -1 according to whether $z_i = \infty$ or $-\infty$, and $(e^{z_i y_i} - e^{z_i x_i})/z_i$ is defined, for convenience, to be equal to $y_i - x_i$ for $z_i = 0$.

Proof   Consider without loss of generality S as in the proof of Corollary 3.4.4 and F as λ-locally integrable; define F_k as in the proof referred to taking F in the place of H. Observe that for every k, F_k determines on S^*, in an obvious way, a measure that is absolutely continuous with respect to the restriction of λ. (The measure in question is the one which gives the value $\Delta_{y-x}F_k(x)$ to the set $(x, y] \cap S^*$ for each $x, y \in S$ with $x \leqslant y$ and the value zero to the complement in S^* of the union of all such sets.) It follows that the Radon–Nikodym derivative f_k arrived at in that case satisfies (3.4.8) with a modification that S is replaced by S^{**}, where S^{**} is defined analogously to S^* but with $\mathbb{N}_0$ in place of $-\mathbb{R}_+$. This in turn implies that F_k satisfies (3.4.10) with an appropriate α_k. Since $\Delta_{y-x}F_k(x) \to \Delta_{y-x}F(x)$ for $x, y \in S^*$ with $y \geqslant x$ essentially due to the continuity theorem mentioned in the proof of Corollary 3.4.5, we arrive at the required result. ∎

Let us now attend to the question raised above concerning the integral representation for H. Corollaries 3.4.4 and 3.4.5 indicate that one could have the representation sought under a fairly general hypothesis. In what follows, we give a theorem
confirming this. However, in order to do this, we require a further condition.

Condition II   Let S be a closed subsemigroup (with zero element) of a locally
compact second countable Abelian metric group and there be a compact subset
K of S, which is the closure of a nonempty open subset of the group, such that
$\mathrm{supp}[\mu] \setminus \{0\} \subset S + K$.

(Once again, there are several equivalent ways of expressing the condition and
we have chosen the present form only because we felt it to be appealing; note also
that if S is the closure of an open subset of a locally compact second countable
Abelian metric group, then Condition II is trivially met.)

We now give the theorem in question.

3.4.7   Theorem

Let Conditions I and II be met in (3.1.1) and $H(0) > 0$. Then we have a probability
space $(\Omega, \mathcal{F}, P)$ and a product measurable function $\eta : S \times \Omega \to \mathbb{R}_+$ such that
$H(x) = H(0)E(\eta(x, \bullet))$, $x \in S$, $\eta(x, \bullet) \equiv 0$ if $H(x) = 0$, and $\eta(\bullet, \omega)$ is a
μ-harmonic exponential function for each $\omega \in \Omega$.

Proof   Consider the probability space and ξ as in Theorem 3.4.1 and K as in
Condition II. Define λ to be the restriction to S of a Haar measure on G and

$$N_1 = \left\{ \omega : \int_K \xi(y, \omega)\lambda(\mathrm{d}y) = 0 \right\}.$$

Observe that in view of Fubini's theorem

$$0 = E\left\{ \left(\int_K \xi(y, \bullet)\lambda(\mathrm{d}y) \right) I_{N_1} \right\}$$

$$= \int_K E(\xi(y, \bullet)I_{N_1})\lambda(\mathrm{d}y). \qquad (3.4.11)$$

The conclusions of Theorem 3.4.1 easily imply that if $\{x_n : n = 1, 2, \ldots\}$ is a
sequence of members of S and x a member of S such that $\{x_n\}$ converges to x,
then $\{\xi(x_n, \bullet)\}$ converges in L_2 to $\xi(x, \bullet)$. Since (3.4.11) implies $E(\xi(x, \bullet)I_{N_1}) = 0$
for a.a. $[\lambda]x$ in K, we can then immediately conclude that $E(\xi(x, \bullet)I_{N_1}) = 0$ for
all $x \in K$. The modified version of (3.3.2) with $\hat{S}$ and $S^*(\mu)$ replaced by S (given
by Theorem 3.4.1) then implies that $E(\xi(x, \bullet)I_{N_1}) = 0$ for all $x \in \mathrm{supp}[\mu] \setminus \{0\}$.
Consequently, from (3.3.3), we get once again on appealing to Fubini's theorem
that

$$0 = E\left(\left(\int_{\mathrm{supp}[\mu]\setminus\{0\}} \xi(x, \bullet)\mu_1(\mathrm{d}x) \right) I_{N_1} \right)$$

$$= (1 - \mu(\{0\})) P_H(N_1), \qquad (3.4.12)$$

where μ_1 is a σ-finite measure on S with its restriction to $\{x : H(x) > 0\}$ agreeing with that of μ (as given by Lemma 3.2.1) or specifically the measure on S given by

$$\mu_1(\bullet) = \mu(\bullet \cap \{x : H(x) > 0\}).$$

(3.4.12) implies that $P_H(N_1) = 0$. We now define for $x \in S$

$$\xi^*(x, \bullet) = \int_K \xi(x + y, \bullet)\lambda(dy). \tag{3.4.13}$$

The modified version of (3.3.2) with $\hat{S}$ and $S^*(\mu)$ replaced by S referred to above also implies due to Fubini's theorem that

$$\xi^*(x + y, \bullet) = \xi^*(x, \bullet)\xi(y, \bullet) \quad \text{a.s. } [P_H] \quad \text{for each } x, y \in S. \tag{3.4.14}$$

Let $\{O_n : n = 1, 2, \ldots\}$ be a countable open covering of S such that each O_n is a relatively compact subset of S. Since

$$H(0)E\left(\int_{K+O_n} \xi(y, \bullet)\lambda(dy)\right) = \int_{K+O_n} H(y)\lambda(dy) < \infty, \quad n = 1, 2, \ldots$$

it follows that we can find a P_H-null set N_2 such that for each $\omega \in \Omega \setminus N_2$, $\int_{K+O_n} \xi(y, \omega)\lambda(dy) < \infty$ for all $n \geq 1$. An application of Theorem 1.1.5 on page 3 or (b) of the theorem on page 4 of Rudin (1960) implies that the restriction of $\xi^*(\bullet, \omega)$ to O_n is continuous (or, to be more specific, uniformly continuous) for each $n \geq 1$ and $\omega \in \Omega \setminus N_2$. Consequently, we have $\xi^*(\bullet, \omega)$ to be a continuous function on S for each $\omega \in \Omega \setminus N_2$. Denote now by N_3 the P_H-null set of ω for which $\int_S \xi^*(x, \omega)\mu(dx) \neq \xi^*(0, \omega)$ and, considering S' to be a fixed countable dense subset of S, denote by N_4 the null set of ω for which $\xi^*(x, \omega) \neq \xi^*(0, \omega)\xi(x, \omega)$ for some x or $\xi^*(x + y, \omega) \neq \xi^*(x, \omega)\xi(y, \omega)$ for some x, y with $x, y \in S'$.

If we now take $N = \cup_{i=1}^4 N_i$, then N is a P_H-null set and for each $\omega \in \Omega \setminus N$, $\xi^*(\bullet, \omega)$ is a nonnegative continuous function on S satisfying $\xi^*(0, \omega) > 0$,

$$\frac{\xi^*(x, \omega)}{\xi^*(0, \omega)} = \xi(x, \omega) \quad \text{a.s. } [P_H], \qquad \int_S \frac{\xi^*(x, \omega)}{\xi^*(0, \omega)}\mu(dx) = 1 \quad \text{for each } x \in S$$

and

$$\frac{\xi^*(x + y, \omega)}{\xi^*(0, \omega)} = \frac{\xi^*(x, \omega)}{\xi^*(0, \omega)}\frac{\xi^*(y, \omega)}{\xi^*(0, \omega)} \quad \text{for each } x, y \in S.$$

Defining then $\eta : S \times \Omega \to \mathbb{R}_+$ by

$$\eta(x, \omega) = \begin{cases} \xi^*(x, \omega)/\xi^*(0, \omega), & x \in S, \ \omega \in \Omega \setminus N \\ \eta(x, \omega_0), & x \in S, \ \omega \in N, \end{cases}$$

where $\omega_0 \in \Omega \setminus N$, we see that η satisfies the claims of Theorem 3.4.7 and the proof of the theorem is hence complete. ∎

3.4.8   Corollary

Let (3.1.1) be satisfied with S as a locally compact second countable Abelian metric group and the smallest closed subgroup of S containing supp$[\mu]$ as S itself. Then either $H(x) = 0$ for all $x \in S$ or we have a probability space $(\Omega, \mathcal{F}, P)$ and a product measurable function $\xi : S \times \Omega \to \mathbb{R}_+$ such that $H(x) = H(0)E(\xi(x, \bullet))$, $x \in S$, and $\xi(\bullet, \omega)$ is a μ-harmonic exponential function for each $\omega \in \Omega$.

Proof   If $\mu(\{0\}) < 1$, then the requirements of Conditions I and II are clearly satisfied. In the case of $\mu(\{0\}) \geqslant 1$, it follows at once that $H(x + y) = 0$ for every $x \in S$ and nonzero support point y of μ from which it is easily seen (on appealing to the assumption on supp$[\mu]$ together with the continuity of H) that $H(x) = 0$ for all $x \in S$. ∎

There is a certain uniqueness property associated with the representations appearing in Theorems 3.4.1 and 3.4.7 (and also the one implied in Remark 3.4.3(i)). This is given by the following theorem.

3.4.9   Theorem

Suppose we are given a topological semigroup S with zero element, a continuous nonnegative real function H on S with $H(0) > 0$, and a measure μ on S satisfying Condition I such that the restriction of the measure to $\{x : H(x) > 0\}$ is σ-finite. If $(\Omega, \mathcal{E}, P)$ is a probability space and $\zeta : S \times \Omega \to \mathbb{R}_+$ is a product measurable function satisfying

(i) $H(x) = H(0)E(\zeta(x, \bullet))$, $x \in S$,

(ii) $\zeta(x, \bullet)\zeta(y, \bullet) = \zeta(x + y, \bullet)$ a.s. $[P]$, $x, y \in S$,

(iii) $\int_S \zeta(x, \bullet)\mu(dx) = 1$ a.s. $[P]$,

(iv) $\zeta(x, \bullet) \equiv 0$ if $H(x) = 0$,

then the distribution of $\{\zeta(x, \bullet) : x \in S\}$ (i.e. the probability measure induced by the stochastic process on $\mathbb{R}_+^S$) is unique. (Note that the function H of the theorem satisfies (3.1.1).)

Proof   On using Theorem 1.1.8 it follows in view of the constraints to be met by ζ that the distribution of $\left\{ \int_B \zeta(x, \bullet)\mu(dx) : B \in \mathcal{B}(S) \right\}$ (with $\mathcal{B}(S)$ as the Borel σ-field of S) is uniquely determined (irrespective of what probability space and ζ we choose). Also it is easily seen that if $s \in$ supp$[\mu]$, then, irrespective of the choice of the probability space and ζ, there exists a sequence $\{O_n(s) : n = 1, 2, \ldots\}$ of open sets of S such that

$$\left\{ \int_{O_n(s)} (\zeta(x, \bullet)/\mu(O_n(s)))\mu(dx) : n = 1, 2, \ldots \right\}$$

converges in L_2 (or indeed in L_p for any $p > 0$) to $\zeta(s, \bullet)$. Consequently it follows that if $\{s_1, \ldots, s_r\}$ is any finite set of support points of μ, then the distribution of

$(\zeta(s_1, \bullet), \ldots, \zeta(s_r, \bullet))$ is unique. Evoking L_2-convergence once more, we can then see that the distribution of $\{\zeta(s, \bullet) : s \in S^*(\mu)\}$ is unique. Consider $x_0 \in D$ and $y_1, \ldots, y_k$ as in Condition I. Then there exist $x_1, \ldots, x_k \in S$ such that

$$\zeta(x_r, \bullet)\zeta\left(x_0 + \sum_{i=1}^{r} y_i, \bullet\right) = \left(\zeta\left(x_0 + \sum_{i=1}^{r-1} y_i, \bullet\right)\right)^2 \quad \text{a.s. } [P], \quad r = 1, \ldots, k,$$

implying that $\zeta(x_0, \bullet) = 0$ a.s. $[P]$ on $\{\zeta(x_0 + \Sigma_{i=1}^{k} y_i, \bullet) = 0\}$. It then easily follows (from (ii)) that

$$\zeta(x_0, \bullet) = \begin{cases} \zeta\left(x_0 + \sum_{1}^{k} y_i, \bullet\right)\Big/\zeta\left(\sum_{1}^{k} y_i, \bullet\right) & \text{a.s. } [P] \text{ on } \left\{\zeta\left(\sum_{1}^{k} y_i, \bullet\right) > 0\right\} \\ 0 & \text{a.s. } [P] \text{ on } \left\{\zeta\left(\sum_{1}^{k} y_i, \bullet\right) > 0\right\} \end{cases}$$

giving that the distribution of $\{\zeta(x, \bullet) : x \in D\}$ is uniquely determined. Using L_2-convergence, we can then see that the distribution of $\{\zeta(x, \bullet) : x \in S + (B \setminus \{0\})\}$ is unique. For $r \geqslant 1$, if $x_1, \ldots, x_r \in S$ and $z \in B \setminus \{0\}$, then, for $\theta_1, \ldots, \theta_r > 0$ such that $\Sigma_1^r \theta_i < 1$,

$$E\left(\prod_{1}^{r}(\zeta(x_i + z, \bullet))^{\theta_i}\right)(\zeta(z, \bullet)^{1-\Sigma_1^r \theta_i})$$

or its simpler version

$$E\left(\left(\prod_{1}^{r}(\zeta(x_i, \bullet))^{\theta_i}\right)\zeta(z, \bullet)\right) \quad \text{(given by (ii))}$$

is unique. Since Fubini's theorem gives in view of the properties of ζ

$$(1 - \mu(\{0\}))E\left(\prod_{1}^{r}(\zeta(x_i, \bullet))^{\theta_i}\right) = \int_{B\setminus\{0\}} E\left(\left(\prod_{1}^{r}(\zeta(x_i, \bullet))^{\theta_i}\right)\zeta(\dot{z}, \bullet)\right)\mu(dz),$$

the uniqueness theorem for Mellin transforms establishes the theorem. (It is a simple exercise to see that the version of the uniqueness theorem for Mellin transforms implied here is a corollary to Theorem 1.1.7.) ∎

3.4.10 Remarks

(i) Theorem 3.4.7 remains valid even when we replace in Condition II 'supp[μ]' by 'supp[$\hat{\mu}$]', where $\hat{\mu}$ is any other measure relative to which the integral equation with μ replaced by $\hat{\mu}$ holds.

(ii) A close scrutiny of the results of Corollaries 3.4.5 and 3.4.6 implies that the measure α involved in (3.4.8) and (3.4.10) can be taken to be concentrated on the following subset of $[-\infty, \infty]^n$:

$$\left\{ x : x \in \prod_{i=1}^{n} \tilde{S}_i \quad \text{with} \quad \tilde{S}_i = \mathbb{R} \quad \text{if } S_i \text{ is a group or uncountable and} \right.$$

$$= (-\infty, \infty] \quad \text{or} \quad [-\infty, \infty) \quad \text{according to whether}$$

$$\left. S_i = -\mathbb{N}_0 \quad \text{or} \quad S_i = \mathbb{N}_0, \quad \text{and} \quad \int_S \exp\{\langle x, y \rangle\} \mu(\mathrm{d}y) = 1 \right\}.$$

Also if μ in these results is concentrated on $\Pi_{i=1}^{n} S_i^{**}$, where $S_i^{**} = S_i \setminus \{0\}$ if $S_i = \pm\mathbb{N}_0$ and $= S_i$ otherwise, then (3.4.8) and (3.4.10) hold with $[-\infty, \infty]^n$ replaced by $\mathbb{R}^n$ (and with z_i in place of z_i^* in the latter case).

(iii) The results of the present section are either taken or adapted from Davies and Shanbhag (1987), Rao and Shanbhag (1989a) and Shanbhag (1991). Corollary 3.4.4 in the special case $S^*(\mu) = S$ was established by Ressel (1985); this specialized result also follows as an immediate corollary of Theorem 3.3.1. Corollary 3.4.8 is indeed a version of Deny's theorem.

(iv) If S is as in Corollaries 3.4.5 and 3.4.6, then Condition I is met if the smallest closed subgroup of $\mathbb{R}^n$ containing supp$[\mu]$ equals the smallest subgroup of $\mathbb{R}^n$ containing S, and μ is concentrated on

$$\{x = (x_1, \ldots, x_n) \in S : x_i \neq 0 \quad \text{if } S_i \text{ is not a group}\} \cup \{0\}$$

with obviously $\mu(\{0\}) < 1$.

(v) Using Theorem 3.4.9 it can easily be verified that if we choose it as in (ii) above, then the α of Corollaries 3.4.5 or 3.4.6 is unique.

3.5   BERNSTEIN'S AND BOCHNER'S THEOREMS ON ABSOLUTELY MONOTONIC FUNCTIONS

Several of the problems in characterizations and infinite divisibility involve two of the famous theorems due to Bernstein and Bochner respectively (see for example Rao and Rubin, 1964; Steutel, 1970; Talwalker, 1970; Puri and Rubin, 1974). These results are known to be connected with potential theory as observed in Phelps (1966). We now reveal that it is possible to have these as results on (3.1.1) from Corollary 3.4.5.

3.5.1   Bernstein's theorem

Let f be a completely monotonic function on $(0, \infty)$. Then f has the integral representation

$$f(x) = \int_{\mathbb{R}_+} \exp\{-yx\} \nu(\mathrm{d}y), \quad x \in (0, \infty)$$

with ν as a uniquely determined measure on $\mathbb{R}_+$.

Proof   Define

$$H(n, x) = \begin{cases} f(x) - c, & n = 0 \\ (-1)^n f^{(n)}(x), & n = 1, 2, \ldots, \end{cases} \tag{3.5.1}$$

where $f^{(n)}(x)$ is the nth derivative of f and $c = \lim_{x \to \infty} f(x)$. (Note that $\lim_{x \to \infty} f(x)$ exists because, by assumption, f is a nonnegative decreasing real function.) We have then

$$H(n, x) = \int_0^\infty H(n + 1, x + y) \, dy, \quad x > 0,$$

with $H(n, \bullet)$ as a nonnegative continuous function having $\lim_{x \to \infty} H(n, x) = 0$. Consequently Corollary 3.4.5 implies, in view of what is revealed in Remark 3.4.10(v) (or Theorem 1.1.7), that

$$H(0, x) = \int_{(0,\infty)} \exp\{-yx\}\alpha(dy), \quad x \in (0, \infty)$$

for some uniquely determined measure α on $\mathbb{R}_+$. Since $f(x) = H(0, x) + c$ with c obviously nonnegative, we have then the required result. ∎

3.5.2   Bochner's theorem

Let f be a completely monotonic function on $(0, \infty)^n$. Then f has the integral representation

$$f(x) = \int_{\mathbb{R}_+^n} \exp\{-\langle y, x \rangle\}\nu(dy), \quad x \in (0, \infty)^n$$

with ν as a uniquely determined measure on $\mathbb{R}_+^n$.

Proof   In view of Bernstein's theorem given above, it follows that the result is valid for $n = 1$. We can therefore evoke an inductive argument to prove the result. Assume that $n > 1$ and that the result is valid for $n - 1$. Consequently, for each $r = 1, 2, \ldots, n$ and $c \in (0, \infty)$,

$$f(x_1, \ldots x_{r-1}, c, x_{r+1}, \ldots, x_n), (x_1, \ldots, x_{r-1}, x_{r+1}, \ldots, x_n) \in (0, \infty)^{n-1}$$

has the representation in question; appealing to the decreasing nature, for each r and $(x_1, \ldots, x_{r-1}, x_{r+1}, \ldots, x_n)$, of $f(x_1, \ldots, x_{r-1}, c, x_{r+1}, \ldots, x_n)$ as a function of c and the continuity Theorem 1.2.3, we can then see that

$$\lim_{c \to \infty} f(x_1, \ldots, x_{r-1}, c, x_{r+1}, \ldots, x_n)$$

exists possessing the representation. This, in turn, implies that there is no loss of generality in assuming that

$$\lim_{x_r \to \infty} f(x_1, x_2, \ldots, x_n) = 0, \quad (x_1, \ldots, x_{r-1}, x_{r+1}, \ldots, x_n) \in (0, \infty)^{n-1},$$

$$r = 1, 2, \ldots, n.$$

In that case, we have the function H defined by

$$H(m, x_1, \ldots, x_n) = (-1)^{mn} \frac{\partial^{mn}}{\partial x_1^m \ldots \partial x_n^m} f(x_1, \ldots, x_n),$$

$$m = 0, 1, \ldots, (x_1, \ldots, x_n) \in (0, \infty)^n$$

satisfying the following equation with $H(m, \bullet)$ as a nonnegative (componentwise decreasing) continuous function for each $m = 0, 1, \ldots$:

$$H(m, x_1, \ldots, x_n) = \int_0^\infty \cdots \int_0^\infty H(m + 1, x_1 + y_1, \ldots, x_n + y_n)\, dy_1 \ldots dy_n,$$

$$(x_1, \ldots, x_n) \in (0, \infty)^n, \quad m = 0, 1, \ldots \; .$$

On appealing to Corollary 3.4.5 together with the result implied in Remark 3.4.10(v) (or Theorem 1.1.7), we then see that $f(\bullet)(= H(0, \bullet))$ has the representation in the assertion. Hence we have the theorem. ∎

3.5.3   Remarks

(i) Even though Bochner's theorem is more general than Bernstein's, we have dealt with Bernstein's theorem separately here because of its intrinsic interest and usefulness.

(ii) It is a simple exercise to show that if $k \geqslant 2$ and f_k is a completely monotonic function on $(0, \infty)^k$, then the function f_{k-1} on $(0, \infty)^{k-1}$ given by

$$f_{k-1}(x_1, \ldots, x_{k-1}) = \lim_{x_k \to \infty} f_k(x_1, \ldots, x_{k-1}, x_k), (x_1, \ldots, x_{k-1}) \in (0, \infty)^{k-1}$$

is (well defined and) completely monotonic. (The coordinatewise decreasing nature of f_k implies that f_{k-1} is well defined, and, appealing to the same phenomenon of the moduli of the partial derivatives of f_k, one gets in view of the Lebesgue dominated convergence theorem and an inductive argument that

$$\frac{\partial^{n_1 + \cdots + n_{k-1}}}{\partial x_1^{n_1} \ldots \partial x_{k-1}^{n_{k-1}}} f_{k-1}(x_1, \ldots, x_{k-1})$$

$$= \lim_{x_k \to \infty} \frac{\partial^{n_1 + \cdots + n_{k-1}}}{\partial x_1^{n_1} \ldots \partial x_{k-1}^{n_{k-1}}} f_k(x_1, \ldots, x_{k-1}, x_k), (x_1, \ldots, x_{k-1}) \in (0, \infty)^{k-1},$$

$$n_1, \ldots, n_{k-1} \geqslant 0,$$

with the identity as meaningful.) If this latter result is used instead of the continuity Theorem 1.2.3 in the proof of Bochner's theorem above, we get an alternative argument for arriving at the theorem in question.

(iii) Somewhat simpler versions of the arguments appearing in the proofs of Theorems 3.5.1 and 3.5.2 can be used to obtain well known theorems on the Hausdorff moment problem. In particular, these yield the following theorem.

3.5.4   Theorem

A sequence $\{\mu_{n_1,\ldots,n_k} : n_1,\ldots,n_k = 0, 1, \ldots\}$ of real numbers represents the moment sequence of some probability distribution concentrated on $[0, 1]^k$ if and only if $\mu_{0,\ldots,0} = 1$ and

$$(-1)^{m_1+\cdots+m_k} \Delta_1^{m_1} \ldots \Delta_k^{m_k} \mu_{n_1,\ldots,n_k} \geqslant 0, m_1,\ldots,m_k, n_1,\ldots,n_k = 0, 1, \ldots,$$

where Δ_i is the usual difference operator acting on the ith coordinate; the moment sequence in question obviously determines the distribution.

CHAPTER 4

Multiple Integral Equations and Stability Theorems

4.1 INTRODUCTION

Let S be a separable metric Abelian semigroup (unless stated otherwise) as in Chapter 3, $k \geqslant 2$ be a positive integer, $H_i : S \to \mathbb{R}_+$, $i = 1, 2, \ldots, k$ be (nonnegative) continuous functions, and μ_i, $i = 1, 2, \ldots, k$ be measures on S such that

$$H_i(x) = \sum_{j=1}^{k} \int_S H_{i+j-1}(x + y)\mu_j(\mathrm{d}y), \quad x \in S, \quad i = 1, 2, \ldots, k \qquad (4.1.1)$$

where $H_{k+1}, \ldots, H_{2k-1}$ are defined for convenience, and are respectively equal to $H_1, \ldots, H_{k-1}$. Shimizu (1978) and, more recently, Ramachandran *et al* (1988) have dealt with the problem of identifying the solution to (4.1.1) either for $S = \mathbb{R}$ or $S = \mathbb{R}_+$. A more general case of the problem when S is a Polish Abelian semigroup satisfying some mild conditions has been studied by Rao and Shanbhag (1989b). The multiple integral equations (4.1.1) for special cases of S and k appear in characterization problems of probability distributions dealing, especially, with stable distributions.

We devote the present chapter to studying the problem of solving (4.1.1) for general S. In the process of doing this, we show that the problem in question can be solved by reducing it to one of solving equation (3.1.1). Shimizu (1980), Gu and Lau (1984), and Rao and Shanbhag (1989b) have considered specialized versions of (3.1.1) with an error term; these latter equations are met while dealing with stability versions of characterization theorems of probability distributions and stochastic processes. These equations as well as the corresponding versions relative to (4.1.1) are also studied in this chapter. In addition, we discuss briefly in this chapter alternative potential theoretic approaches such as the ones based on the Choquet theorem for solving the integral equations.

Some of the results appearing in this chapter are taken from Rao and Shanbhag (1990, 1992) and Rao *et al* (1992).

4.2 MULTIPLE INTEGRAL EQUATIONS

We study in this section the problem of solving (4.1.1) by reducing it to that of solving (3.1.1). Observe that $\{0, 1, \ldots, k - 1\}$ is an Abelian semigroup with zero element under the operation '$+$' defined by

$$x + y = \begin{cases} x + y & \text{if } x + y < k \\ x + y - k & \text{if } x + y \geqslant k, \end{cases}$$

where '$+$' and '$-$' appearing on the right-hand side are taken in the usual sense for real numbers; the operation considered here is obviously 'the addition modulo k'. Denoting this latter semigroup by $\mathbb{N}_0^{(k)}$, and $\mathbb{N}_0^{(k)} \times S$ by S^*, let $H^* : S^* \to \mathbb{R}_+$ be such that

$$H^*(n, x) = H_{n+1}(x), \quad n \in \mathbb{N}_0^{(k)}, \quad x \in S \tag{4.2.1}$$

and μ^* be the measure on S^* for which

$$\mu^*(\{n\} \times B) = \mu_{n+1}(B), \quad n \in \mathbb{N}_0^{(k)}, \quad B \in \mathcal{B}(S),$$

$\mathcal{B}(S)$ is, as defined earlier, the Borel σ-field of S. We can then rewrite (4.1.1) as

$$H^*(x^*) = \int_{S^*} H^*(x^* + y^*)\mu^*(dy^*), \quad x^* \in S^*. \tag{4.2.2}$$

Clearly (4.2.2) is a specialized version of the integral equation (3.1.1) covered in Chapter 3. We have then one of the main results of the section as follows.

4.2.1 Theorem

Let Condition I introduced in Section 3.4 be met with μ and S replaced respectively by μ^* and S^* (with obvious notational alterations). Then H_i appearing in (4.1.1) is independent of i with H_1 to be either identically equal to zero or such that $H_1(0) > 0$ and we have a probability space $(\Omega, \mathcal{F}, P)$ and a product measurable function $\xi : S \times \Omega \to \mathbb{R}_+$ satisfying

  (i) $H_1(x) = H_1(0)E(\xi(x, \bullet))$, $x \in S$,

  (ii) $\xi(x + y, \bullet) = \xi(x, \bullet)\xi(y, \bullet)$ a.s., $x, y \in S$,

  (iii) $\int_S \xi(x, \bullet) \left(\Sigma_{i=1}^k \mu_i \right)(dx) = 1$ a.s., $\xi(x, \bullet) = 0$ whenever $H_1(x) = 0$.

Proof   Theorem 3.4.1 yields that either $H^* \equiv 0$ or $H^*(0^*) > 0$, where 0^* is the zero element of S^*, and there exists a probability space $(\Omega, \mathcal{F}, P)$ and a product measurable function $\xi^* : S^* \times \Omega \to \mathbb{R}_+$ such that

  (i)' $H^*(x^*) = H^*(0^*)E(\xi^*(x^*, \bullet))$, $x^* \in S^*$,

  (ii)' $\xi^*(x^* + y^*, \bullet) = \xi^*(x^*, \bullet)\xi^*(y^*, \bullet)$ a.s.

  (iii)' $\int_{S^*} \xi^*(x^*, \bullet)\mu^*(dx^*) = 1$ a.s., $\xi^*(x^*, \bullet) \equiv 0$ whenever $H^*(x^*) = 0$.

If $H^* \equiv 0$, clearly all H_i are identically equal to zero. On the other hand, if $H^*(0^*) > 0$ and the assertion of the existence of $(\Omega, \mathcal{F}, P)$ and ξ^* holds, then (ii)' implies that

$$\xi^*((n, 0), \bullet) = (\xi^*((1, 0), \bullet))^n \quad \text{a.s. for } n \in \mathbb{N}_0^{(k)}$$

and

$$(1 =)\xi^*(0^*, \bullet) = (\xi^*((1, 0), \bullet))^k \quad \text{a.s.}$$

yielding, in particular, that $\xi^*((1, 0), \bullet) = 1$ a.s. That the theorem holds is now obvious. ∎

Rao and Shanbhag (1989b) used a somewhat different formulation for solving (4.1.1) for $k = 2$. Their result is contained in the following theorem. However, to state the theorem, we first need some notation.

Abusing the notation used in Chapter 3 slightly, define $S^*(\mu^*)$ to be the smallest closed subsemigroup of S^* containing the zero element and $\text{supp}[\mu^*]$. Define S_0^* to be the section $\{x \in S : (0, x) \in S^*(\mu^*)\}$ of $S^*(\mu^*)$.

Here is then the theorem just mentioned.

4.2.2 Theorem

Let $\text{supp}[\Sigma_{i=1}^k \mu_i] \subset S_0^*$ (or equivalently, in obvious notation, $S^*(\Sigma_{i=1}^k \mu_i) = S_0^*$) and Condition I of Section 3.4 be met with $\Sigma_{i=1}^k \mu_i$ in the place μ. Then each H_i in (4.1.1) satisfies (3.1.1) with μ replaced by $\Sigma_{i=1}^k \mu_i$. Moreover, in this case either all H_i are identically equal to zero, or we have a probability space $(\Omega, \mathcal{F}, P)$ and for $i = 1, 2, \ldots, k$, product measurable functions $\xi_i : S \times \Omega \to \mathbb{R}_+$ satisfying

(i) $H_i(x) = H_i(0)E(\xi_i(x, \bullet)), E(\xi_i(x, \bullet)) < \infty, x \in S$,

(ii) $\xi_i(x + y, \bullet) = \xi_i(x, \bullet)\xi_i(y, \bullet)$ a.s., $x, y \in S$,

(iii) $\int_S \xi_i(x, \bullet)\left(\Sigma_{j=1}^k \mu_j\right)(dx) = 1$ a.s., $\xi_i(x, \bullet) \equiv 0$ whenever $H_i(x) = 0$ and $H_i(0) \neq 0$.

Proof Let $H_1(0) > 0$. (4.2.2) then satisfies the requirements of Theorem 3.3.1 (with obvious notational modifications). Denoting, in the present case, the probability space and the function ξ of the theorem by $(\Omega^*, \mathcal{B}(S^{*\mathbb{N}}), P_{H^*}^*)$ and ξ^* respectively, we see that, for each $i = 0, 1, \ldots, k - 1$,

$$H_{i+1}(y) = H_1(0)E^*(\xi^*((i, y), \bullet)) \quad \text{for each } y \in \text{supp}[\mu_{i+1}],$$

where E^* denotes the expectation relative to the probability space in question. In view of the assumption $\text{supp}[\Sigma_{i=1}^k \mu_i] \subset S_0^*$, we have, for each $i = 0, 1, \ldots, k - 1$, that $\xi^*((i, y), \bullet) = \xi^*((0, y), \bullet)$ a.s. $[P_{H^*}^*]$ for each $y \in \text{supp}[\mu_{i+1}]$, which follows as a consequence of the corresponding identity with the kth power on both sides, and hence

$$H_{i+1}(y) = H_1(y), \quad y \in \text{supp}[\mu_{i+1}].$$

If we have an x such that $H_1(x) \neq 0$, then applying the above result to $H_i(x + \bullet)$, we see that

$$H_{i+1}(x + y) = H_1(x + y), \quad y \in \text{supp}[\mu_{i+1}]$$

and hence, from (4.1.1), we get

$$H_1(x) = \int_S H_1(x + y) \left(\sum_{j=1}^{k} \mu_j \right) (dy). \tag{4.2.3}$$

By symmetry, (4.2.3) also remains valid with H_1 replaced by H_i whenever x is such that $H_i(x) \neq 0$. Since, for each $x \in S$, the equation (4.2.3) with H_1 replaced by $\Sigma_{i=1}^{k} H_i$ remains valid, it follows then trivially that, irrespectively of whether or not $H_i(x) \neq 0$, the one with H_1 replaced by H_i remains valid. This establishes the first assertion of the theorem. Theorem 3.4.1 then implies that either all H_i are identically equal to zero or we have probability spaces $(\Omega_i, \mathcal{F}_i, P_i)$ and product measurable functions $\xi_{(i)} : S \times \Omega_i \to \mathbb{R}_+$ such that (i), (ii) and (iii) in the statement of the theorem with ξ_i and a.s. replaced respectively by $\xi_{(i)}$ and a.s. $[P_i]$ hold for $i = 1, 2, \ldots, k$. In the latter case, considering $(\Omega, \mathcal{F}, P)$ to be the product probability space corresponding to $(\Omega_i, \mathcal{F}_i, P_i)$, $i = 1, 2, \ldots, k$ and defining

$$\xi_i(x, (\omega_1, \omega_2, \ldots, w_k)) = \xi_{(i)}(x, \omega_i), \quad \omega_i \in \Omega_i, i = 1, 2, \ldots, k,$$

we obtain the validity of the second assertion of the theorem. ∎

We now present two corollaries of the theorem. In view of the proofs given earlier for Theorem 3.4.7 and Corollary 3.4.5, we do not need further proofs for the results here. (Following the proof of Corollary 1 of Theorem 2 of Davies and Shanbhag (1987), one could also arrive at the latter of the two corollaries as a corollary of the former.)

4.2.3   Corollary

If additionally, Condition II of Section 3.4 with μ replaced by $\Sigma_{i=1}^{k} \mu_i$ is met, then Theorem 4.2.2 holds with 'a.s.' in (ii) and (iii) deleted and $\xi_i(., \omega)$ as $\left(\Sigma_{j=1}^{k} \mu_j \right)$-harmonic exponential functions on S for each $\omega \in \Omega$.

4.2.4   Corollary

Let $n \geqslant 1$, $S = \Pi_{i=1}^{n} S_i$ with $S_i = \mathbb{Z}$ or $\mathbb{N}_0$ or $-\mathbb{N}_0$ or $\mathbb{R}$ or $\mathbb{R}_+$ or $-\mathbb{R}_+$ and λ be the restriction to S of a Haar measure on the smallest subgroup of $\mathbb{R}^n$ containing S. Let $k \geqslant 1$, $h_i : S \to \mathbb{R}_+$, $i = 1, 2, \ldots, k$ be Borel measurable functions that are locally integrable with respect to λ and μ_i, $i = 1, 2, \ldots, k$ be σ-finite measures on S satisfying the conditions specified in Theorem 4.2.2. Then

$$h_i(\bullet) = \sum_{j=1}^{k} \int_S h_{i+j-1}(\bullet + y)\mu_j(dy) \quad \text{a.e. } [\lambda], \quad i = 1, 2, \ldots, k,$$

where $h_{k+1}, \ldots, h_{2k-1}$ are defined for convenience and are respectively equal to $h_1, \ldots, h_{k-1}$, implies that

$$h_i(\bullet) = \int_{[-\infty,\infty]^n} \exp\{\langle \bullet, x \rangle\} \alpha_i(dx) \quad \text{a.e. } [\lambda], \quad i = 1, 2, \ldots, k,$$

where α_i, $i = 1, 2, \ldots, k$ are measures on $[-\infty, \infty]^n$ such that

$$\alpha_i\left(\left\{x \in [-\infty, \infty]^n : \int_S e^{\langle x, y \rangle}\left(\sum_{i=1}^{k} \mu_i\right)(dy) \neq 1, \quad \text{or} \quad \langle x, y \rangle \text{ is undefined}\right.\right.$$

$$\left.\left. \text{or infinite for some } y \in S\right\}\right) = 0, \quad i = 1, 2, \ldots, k.$$

4.2.5   Remarks

(i) Versions of Corollaries 4.2.3 and 4.2.4 with Theorem 4.2.1 in place of Theorem 4.2.2 remain valid.

(ii) Theorem 4.2.2 holds even when we take instead of supp $\left[\Sigma_{i=1}^{k}\mu_i\right] \subset S_0^*$, Condition I of Section 3.4 with supp$[\mu]$, $S^*(\mu)$ and S replaced respectively by S_0^* and $S^*\left(\Sigma_{i=1}^{k}\mu_i\right)$. Also, both Theorems 4.2.1 and 4.2.2 remain valid when S is taken merely as a second countable Abelian topological semigroup with zero element or its modified version without the requirement of second countability but instead satisfying $\mu_i((\text{supp}[\mu_i])^c) = 0$, $i = 1, 2, \ldots, k$.

(iii) It is now natural to ask whether Theorem 4.2.2 remains valid when the assumption that supp $\left[\Sigma_{i=1}^{k}\mu_i\right] \subset S_0^*$ is dropped. That the answer to this question is in the negative is shown by the following example.

Example   Let $S = \mathbb{N}_0$, H_1 and H_2 be functions on $\mathbb{N}_0$ such that, for some $c, d > 0$ with $c \neq d$,

$$H_1(x) = \begin{cases} c & \text{if } x \text{ is even} \\ d & \text{if } x \text{ is odd} \end{cases}$$

and

$$H_2(x) = \begin{cases} d & \text{if } x \text{ is even} \\ c & \text{if } x \text{ is odd,} \end{cases}$$

and μ_1 and μ_2 be measures on $\mathbb{N}_0$ such that $\mu_1(\{2\}) = \alpha$, $\mu_2(\{1\}) = 1 - \alpha$, and $\mu_2(\{1\}^c) = \mu_1(\{2\}^c) = 0$, where $0 \leqslant \alpha < 1$. Note that we have here (4.1.1) with $k = 2$ satisfied and the latter assumption in Theorem 4.2.2 is met. However, the conclusions of Theorem 4.2.2 do not hold in the present case.

(iv) From Theorem 4.2.2, it is evident that in the case of $S = \mathbb{R}$ or $\mathbb{R}_+$ or $\mathbb{Z}$ or $\mathbb{N}_0$ (or $-\mathbb{R}_+$ or $-\mathbb{N}_0$), except in the case of supp $\left[\Sigma_{i=2}^{k}\mu_i\right] = \phi$, the functions H_i appearing in the theorem are all equal. In general, however, this situation does not remain valid (although, as implied by the proof of the theorem, it is always true that $H_i(\bullet + y) = H_{i+j-1}(\bullet + y)$ for each $y \in \text{supp}[\mu_j]$, $j = 1, 2, \ldots, k$ and $i = 1, 2, \ldots, k$ with $H_{k+1}, \ldots, H_{2k-1}$ defined as before). This is illustrated by the following example.

Example   Let $S = \mathbb{N}_0^2$ and μ_1 and μ_2 be measures on S such that μ_1 has full support with its restriction to $\{(x, y) : x \in \mathbb{N}_0, y = 0\}$ as a probability measure and μ_2 has $\{(0, 1)\}$ as its support. Define

$$H_1(x, y) = \begin{cases} c & \text{if } x = 0, 1, \ldots, \\ 0 & \text{otherwise} \end{cases} \quad y = 0$$

where c is a fixed positive constant, and take $H_2 \equiv 0$. It trivially follows that, in this case, (4.1.1) with $k = 2$ is satisfied and also all the requirements of Theorem 4.2.2 are met. However, we do not now have $H_1 = H_2$. The same point could obviously be illustrated by several other examples.

(v) Extensions of Theorem 3.3.3 and Corollaries 3.3.4–3.3.6 to the case where k functions are involved are now easy to obtain. In particular, if S is an Abelian topological semigroup, k is an integer $\geqslant 2$, h_i, $i = 1, \ldots, 2k - 1$ are bounded nonnegative Borel measurable functions satisfying $h_{k+1} \equiv h_1, \ldots, h_{2k-1} \equiv h_{k-1}$, and μ_j are measures on S such that $\Sigma_1^k \mu_j$ is a subprobability measure, then it now follows that

$$h_i(x) = \sum_{j=1}^{k} \int_S h_{i+j-1}(x + y)\mu_j(\mathrm{d}y), \quad x \in S, i = 1, 2, \ldots, k \tag{4.2.4}$$

if and only if either $h_i \equiv 0$ for $i = 1, 2, \ldots, k$ or $h_i(x) \neq 0$ for some i and x, $\Sigma_1^k \mu_j(S) = 1$, and

$$h_{i+j-1}(x + y) = h_i(x) \quad \text{for a.a. } [\mu_j]y \in S$$

$$\text{for each } x \in S \quad \text{and} \quad i, j = 1, \ldots, k.$$

This result is immediate from Theorem 3.3.3 on writing (4.2.4) in terms of the function h on $\mathbb{N}_0 \times S$ such that $h(n, x) = h_i(x)$ for each $x \in S$, n of the form $rk + i - 1$ with r as a nonnegative integer, and $i = 1, \ldots, k$.

4.3   STABILITY THEOREMS

We devote this section to discussing stability results concerning integral equations (3.1.1) and (4.1.1). Stability theorems are of importance in characterization theory of probability distributions since they are useful in assessing whether or not the distribution can be taken to be close to a certain distribution when it satisfies a characterization property of that distribution approximately. As the integral equation (3.1.1) or its modified version (4.1.1) is involved in characterizations of several probability distributions, it should therefore be worthwhile to discuss the associated stability results.

4.3.1 Theorem

Let S be a topological semigroup, $H: S \to \mathbb{R}_+$ a Borel measurable function, $c \in [0, 1)$ and μ a σ-finite measure on S such that

$$H(x) = \int_S H(x + y)\mu(dy) + \alpha(x), \quad x \in S \tag{4.3.1}$$

where α is such that $|\alpha(x)| \leqslant \alpha^*(x)$ for all $x \in S$ with α^* as a real-valued Borel measurable function satisfying

$$\int_S \alpha^*(x + y)\mu(dy) \leqslant c\alpha^*(x), \quad x \in S. \tag{4.3.2}$$

Then the H can be expressed as

$$H(x) = H_1(x) + H_2(x), \quad x \in S, \tag{4.3.3}$$

where H_1 is a nonnegative Borel measurable function on S satisfying

$$H_1(x) = \int_S H_1(x + y)\mu(dy), \quad x \in S, \tag{4.3.4}$$

and H_2 is a Borel measurable function on S given by

$$H_2(x) = \alpha(x) + \sum_{m=1}^{\infty} \int_S \alpha(x + y)\mu^{*m}(dy), \quad x \in S,$$

and it is such that $|H_2(x)| \leqslant \alpha^*(x)(1 - c)^{-1}$ for each $x \in S$, where μ^{*m} is the m-fold convolution of μ (i.e. a measure on S such that

$$\mu^{*m}(B) = \mu^m(\{x_1, \ldots, x_m) \in S^m : x_1 + \cdots + x_m \in B\},$$

for every Borel set B of S, μ^m being the product measure $\Pi_{i=1}^m \mu_i$ with all $\mu_i = \mu$).

Proof Using Fubini's theorem, we get successively for $n = 1, 2, \ldots$

$$H(x) = \int_S H(x + y)\mu^{*n}(dy) + \sum_{m=0}^{n-1} \int_S \alpha(x + y)\mu^{*m}(dy), \quad x \in S. \tag{4.3.5}$$

The requirement of the integrability of $\alpha(x + \bullet)$ with respect to the measure μ^{*m} for each $m = 1, 2, \ldots$, and $x \in S$ is met in view of (4.3.2); we use here the notation $\int_S f(x + y)\mu^{*0}(dy)$ for convenience for $f(x)$. Also (4.3.2) implies that

$$\sum_{m=0}^{\infty} \int_S |\alpha(x + y)|\mu^{*m}(dy) \leqslant \alpha^*(x)(1 - c)^{-1}, \quad x \in S. \tag{4.3.6}$$

In view of (4.3.6), it follows that

$$\sum_{m=0}^{n-1} \int_S \alpha^+(x + y)\mu^{*m}(dy)$$

and

$$\sum_{m=0}^{n-1} \int_S \alpha^-(x+y)\mu^{*m}(dy)$$

converge to finite limits as $n \to \infty$ for each $x \in S$ with obviously the limiting functions as Borel measurable. This observation, in turn, implies that

$$\sum_{m=0}^{n-1} \int_S \alpha(x+y)\mu^{*m}(dy)$$

converges as $n \to \infty$ to a Borel measurable function. Denote this function by H_2. Because of (4.3.6), it follows that $|H_2(x)| \leqslant \alpha^*(x)(1-c)^{-1}$ for each $x \in S$. In view of (4.3.5), it follows that $\int_S H(x+y)\mu^{*n}(dy)$ tends as $n \to \infty$ to $H(x) - H_2(x)$, a nonnegative Borel measurable function, for each $x \in S$. Denote this new function by H_1. From (4.3.5) and (4.3.6), it follows that

$$\int_S H(x+y)\mu^{*n}(dy) \leqslant H(x) + \alpha^*(x)(1-c)^{-1}, \quad x \in S, n \geqslant 1.$$

Since

$$\int_S H(x+y)\mu^{*n}(dy) = \int_S \left\{ \int_S H(x+y+z)\mu^{*(n-1)}(dz) \right\} \mu(dy),$$

and $H(x + \bullet)$ and $\alpha^*(x + \bullet)$ are μ-integrable for each $x \in S$, the Lebesgue dominated convergence theorem implies that the H_1 satisfies (4.3.4). (The fact that the H_1 satisfies (4.3.4) could also be seen by noting that, in view of (4.3.1) and (4.3.6), we have for each $x \in S$, $H_1(x + \bullet)$ to be μ-integrable satisfying

$$\int_S H_1(x+y)\mu(dy) = \int_S H(x+y)\mu(dy) - \int_S H_2(x+y)\mu(dy)$$

$$= \int_S H(x+y)\mu(dy) - \sum_{m=1}^{\infty} \int_S \alpha(x+y)\mu^{*m}(dy)$$

$$= H(x) - \alpha(x) - \sum_{m=1}^{\infty} \int_S \alpha(x+y)\mu^{*m}(dy)$$

$$= H_1(x).) \quad \blacksquare$$

4.3.2   Remark

With minor modifications in the proof, it can be seen that if S is a measurable subsemigroup of a locally compact second countable Abelian metric group and λ is the restriction to S of a Haar measure on the group, then the theorem above remains valid if '$x \in S$' in (4.3.1)–(4.3.4) is replaced by 'for a.a. $[\lambda]x \in S$'. This follows on noting that now there exists a sequence $\{B_n : n = 1, 2, \ldots\}$ of Borel sets such that $\bigcup_{n=1}^{\infty} B_n = S$, $\lambda(B_n) < \infty$, $n \geqslant 1$, and, on each B_n, both H and α^* are bounded, and that we have then for each Borel set B

$$\int_{B \cap B_n} H(x)\lambda(\mathrm{d}x) = \int_{B \cap B_n} H_1(x)\lambda(\mathrm{d}x) + \int_{B \cap B_n} H_2(x)\lambda(\mathrm{d}x), \quad x \in S, n \geqslant 1$$

and

$$\int_{B \cap B_n} H_1(x)\lambda(\mathrm{d}x) = \int_{B \cap B_n} \left(\int_S H_1(x + y)\mu(\mathrm{d}y) \right) \lambda(\mathrm{d}x), \quad n \geqslant 1$$

with H_1 as a nonnegative Borel measurable function and H_2 as a Borel measurable function given for a.a. $[\lambda]x \in S$, by

$$H_2(x) = \sum_{m=0}^{\infty} \int_S \alpha(x + y)\mu^{*m}(\mathrm{d}y)$$

and satisfying for all $x \in S$, $|H_2(x)| \leqslant \alpha^*(x)(1 - c)^{-1}$.

4.3.3 Corollary

Let $n \geqslant 1$, $S = \Pi_{i=1}^{n} S_i$ with $S_i = \mathbb{Z}$ or $\mathbb{N}_0$ or $-\mathbb{N}_0$ or $\mathbb{R}$ or $\mathbb{R}_+$ or $-\mathbb{R}_+$, and λ be the restriction to S of a Haar measure on the smallest subgroup of $\mathbb{R}^n$ containing S. Let $h : S \to \mathbb{R}_+$ and $\alpha : S \to \mathbb{R}$ be Borel measurable functions that are locally integrable with respect to λ, and μ be a σ-finite measure on S such that Condition I of Section 3.4 is met and

$$h(\bullet) = \int_S h(\bullet + y)\mu(\mathrm{d}y) + \alpha(\bullet) \quad \text{a.e. } [\lambda]$$

with $|\alpha(x)| \leqslant \alpha^*(x)$ for every $x \in S$ for some λ-locally integrable function α^* satisfying

$$\int_S \alpha^*(x + y)\mu(\mathrm{d}y) \leqslant c\alpha^*(x) \quad \text{for a.e. } [\lambda], x \in S.$$

Then the h can be represented as

$$h(\bullet) = h_1(\bullet) + h_2(\bullet) \quad \text{a.e. } [\lambda]$$

with

$$h_1(\bullet) = \int_{[-\infty,\infty]^n} e^{\langle \bullet, x \rangle} \nu(\mathrm{d}x),$$

ν being a measure on $[-\infty, \infty]^n$ such that

$$\nu\left(\left\{ x \in [-\infty, \infty]^n : \int_S e^{\langle x, y \rangle}\mu(\mathrm{d}y) \neq 1, \quad \text{or} \right. \right.$$

$$\left. \left. \langle x, y \rangle \text{ is infinite or undefined for some } x \in S \right\} \right) = 0,$$

and h_2 as a λ-locally integrable Borel measurable function on S satisfying $|h_2(x)| \leqslant \alpha^*(x)(1-c)^{-1}$ for every $x \in S$ and such that

$$h_2(x) = \sum_{m=0}^{\infty} \int_S \alpha(x+y)\mu^{*m}(dy) \quad \text{for a.a. } [\lambda]x \in S.$$

Proof   In view of Remark 4.3.2, the result is an immediate consequence of Corollary 3.4.5. (Alternatively, one could use essentially the argument of the proof of Corollary 3.4.5 to arrive at the result directly from Theorem 4.3.1. In this case, the H_k defined in the proof of Corollary 3.4.5 satisfies (4.3.1) with α replaced by α_k and α^* replaced by α_k^*, where α_k is such that

$$\alpha_k(\bullet) = \left\{ \int_{Q_k} \alpha(\bullet + y)\lambda(dy) \right\} \Big/ \lambda(Q_k), \quad k = 1, 2, \ldots,$$

and α_k^* is analogously defined. Consequently, we get in obvious notation

$$H_k(x) = H_{1k}(x) + H_{2k}(x), \quad x \in S, \ k = 1, 2, 3, \ldots$$

with H_{1k} and H_{2k} satisfying the conditions corresponding to H_1 and H_2 of the theorem respectively. The continuity theorem used in the proof of Corollary 3.4.5 implies, in view of the local integrability of h and α^*, that the present corollary is then valid.) ∎

4.3.4   Corollary

If the assumptions in the theorem are valid with $\alpha^*(x) = H(x)G(x)$, $x \in S$ for some nonnegative Borel measurable function G satisfying $G(x) < 1 - c$ for each $x \in S$, then the H of the theorem can be expressed as

$$H(x) = H_1(x)H_2(x), \quad x \in S \tag{4.3.7}$$

with H_1 as in the theorem and H_2 as a nonnegative Borel measurable function on S such that $|H_2(x) - 1|$ is bounded by $G(x)(1 - c - G(x))^{-1}$ for each $x \in S$.

Proof   By Theorem 4.3.1, we have

$$H(x) = H_1(x) + H(x)G_2(x), \quad x \in S$$

with H_1 as mentioned in the theorem and G_2 such that

$$|G_2(x)| \leqslant (1-c)^{-1}G(x), \quad x \in S.$$

Consequently, we have (noting in particular that $G(x) < 1 - c$, $x \in S$)

$$H(x) = H_1(x)H_2(x), \quad x \in S$$

with $H_2(x) = (1 - G_2(x))^{-1}$, $x \in S$ meeting the boundedness condition mentioned in the statement of the corollary. ∎

4.3.5 Corollary

Let S be as in the theorem, $c_1 \in [0, 1)$ a constant, $G : S \to [0, c_1]$ and $H : S \to \mathbb{R}_+$ Borel measurable functions, and μ a σ-finite measure on S such that (4.3.1) with α satisfying $|\alpha(x)| \leqslant G(x)H(x)$ for all x is valid and $G(x + y) \leqslant G(x)\xi(y)$ for all $x, y \in S$ for some nonnegative real-valued Borel measurable function ξ. Assume further that there exist constants $c_2, c_3 \in [0, 1)$ and a Borel set B of S such that $G(x + y) \leqslant c_2 G(x)$ for all $x \in S$ and $y \in B^c$ and

$$\int_B (\xi(y) - c_2)H(x + y)\mu(dy) \leqslant c_3 H(x) \quad \text{for all } x \in S.$$

Define $c = c_3 + c_2(1 + c_1)$. Then, provided $c < 1$ and $G(x) < 1 - c$ for all $x \in S$, the representation (4.3.7) of Corollary 4.3.4 with H_1 and H_2 as mentioned holds.

Proof We have

$$\int_S G(x + y)H(x + y)\mu(dy)$$

$$= \int_B G(x + y)H(x + y)\mu(dy) + \int_{B^c} G(x + y)H(x + y)\mu(dy)$$

$$\leqslant G(x) \int_B \xi(y)H(x + y)\mu(dy) + c_2 G(x) \int_{B^c} H(x + y)\mu(dy)$$

$$= G(x) \int_B (\xi(y) - c_2)H(x + y)\mu(dy) + c_2 G(x) \int_S H(x + y)\mu(dy)$$

$$\leqslant c_3 G(x)H(x) + c_2(1 + c_1)G(x)H(x)$$

$$= cG(x)H(x), \quad x \in S.$$

In view of Corollary 4.3.4, we then have the stated result. ∎

4.3.6 Remark

If S and λ are as in Remark 4.3.2 and the assumptions in Theorem 4.3.1 are met with '$x \in S$' in (4.3.1) and (4.3.2) replaced by 'for a.a. $[\lambda]x \in S$' and α^* meeting the further constraint appearing in Corollary 4.3.4, then it follows that (4.3.7) of Corollary 4.3.4 with '$x \in S$' replaced by 'for a.a. $[\lambda]x \in S$' and H_1 satisfying in place of (4.3.4) of Theorem 4.3.1 its modified version with '$x \in S$' replaced by 'for a.a. $[\lambda]x \in S$' holds. The corresponding version of Corollary 4.3.5 is also valid.

We are now led to the following specialized result, which is of particular importance in characterization theory of probability distributions.

4.3.7 Corollary

Let $n \geqslant 1$, $S = \Pi_{i=1}^n S_i$ with $S_i = \mathbb{N}_0$ or $\mathbb{R}_+$ and λ be the restriction to S of a Haar measure on the smallest subgroup of $\mathbb{R}^n$ containing S. Let $h : S \to \mathbb{R}_+$

be a componentwise decreasing Borel measurable function, $\beta : S \to \mathbb{R}$ a Borel measurable function for which

$$|\beta(x)| < K E\{\exp\{-\langle x, Z \rangle\}\} \quad \text{for every } x \in S$$

with K as a nonnegative real number and Z as a random vector such that for some $z^* \in (0, \infty)^n$ we have $Z \geqslant z^*$ almost surely, and μ is a σ-finite measure on S such that Condition I of Section 3.4 is met and

$$h(\bullet)(1 - \beta(\bullet)) = \int_S h(\bullet + y)\mu(dy) \quad \text{a.e. } [\lambda]. \tag{4.3.8}$$

Then, given any $i = 1, 2, \ldots, n$, there exist constants γ and $c^* \geqslant 0$ such that for almost all $[\lambda]x$ with the ith coordinate greater than or equal c^*,

$$h(x) = h_1(x)h_2(x)$$

with h_1 as in Corollary 4.3.3 and h_2 as a nonnegative Borel measurable function on S such that $|h_2(x) - 1| \leqslant K\gamma E\{\exp\{-\langle x, Z \rangle\}\}$ for each x.

Proof   If we take $B = [0, x_0] \cap S$ for some $x_0 > 0$ such that $\mu(B) < 1$ (which clearly exists since $\mu(\{0\}) < 1$), then given an integer $i \geqslant 1$, and $\leqslant n$, there exists a $c^* \in S_i$ such that the criterion of the modified version of Corollary 4.3.5 mentioned in Remark 4.3.6 is met with $\xi \equiv 1$ provided we replace H by h_{c^*} given below:

$$h_{c^*}(x_1, \ldots, x_n) = h(x_1, \ldots, x_{i-1}, \; x_i + c^*, \; x_{i+1}, \ldots, x_n), \quad (x_1, \ldots, x_n) \in S.$$

In view of the corollary, we then arrive at the result. ■

If we resort to operators instead of convolutions, then we get somewhat different versions of Theorem 4.3.1 and Corollary 4.3.4. These results are not only of importance in their own right, but also provide us with useful information about the argument that could be used to obtain other results including certain variants of Corollary 4.3.3. We now give them.

4.3.8   Theorem

Let the assumptions in Theorem 4.3.1 be met but for the modification that μ is not necessarily σ-finite. Then the representation (4.3.3) holds with H_1 as in Theorem 4.3.1 and H_2 as the Borel measurable function on S given by $\mathbb{M}\alpha$, where

$$\mathbb{M} = \sum_{n=0}^{\infty} \mathcal{M}^n$$

with $\mathcal{M}$ as the operator given by

$$(\mathcal{M}f)(x) = \int_S f(x + y)\mu(dy), \quad x \in S,$$

assuming that the operators are defined on the class of Borel measurable functions f on S such that for each f we have $|f|$ dominated by a Borel measurable function β_f^* of the type of α^* (i.e. such that (4.3.2) with β_f^* in place of α^* holds). The function H_2 of the representation satisfies

$$|H_2(x)| \leqslant \alpha^*(x)(1-c)^{-1} \quad \text{for all } x \in S.$$

Proof   Using essentially the argument given in the proof of Theorem 4.3.1 but without involving product measures and hence Fubini's theorem, the result follows. This is especially so because if f is as in the statement of the theorem, then even when μ is not assumed to be σ-finite, we get $\mathcal{M}f$ to be Borel measurable. ■

4.3.9   Theorem

Let the assumptions in Theorem 4.3.7 be met with α^* as in Corollary 4.3.4. Then the function H equals $\nabla_{\hat{G}} H_1$ with H_1 as in Theorem 4.3.7 and

$$\nabla_{\hat{G}} = \sum_{n=0}^{\infty} \Delta_{\hat{G}}^n$$

where $\hat{G}$ is a function such that $\alpha(x) = H(x)\hat{G}(x)$, $x \in S$ and $|\hat{G}(x)| \leqslant G(x)$, $x \in S$, and $\Delta_{\hat{G}}$ is the operator such that $\Delta_{\hat{G}} f = \mathbb{M}(f \bullet \hat{G})$, assuming that the operators are defined on the class of Borel measurable functions f (on S) such that $f \bullet \hat{G}$ lies in the class met in the previous theorem, i.e. the class on which the operators $\mathcal{M}$ and $\mathbb{M}$ were defined. (Here we use the notation $f \bullet \hat{G}$ for the function whose values are $f(x)\hat{G}(x)$.)

Proof   Clearly $\hat{G}$ as required exists; take, for example, $\hat{G}(x) = \alpha(x)/H(x)$ if $H(x) \neq 0$, and $= 0$ otherwise. From Theorem 4.3.7, we have

$$H(x) = H_1(x) + (\mathbb{M}(\hat{G} \bullet H))(x), \quad x \in S$$

which implies that

$$H(x) = H_1(x) + (\Delta_{\hat{G}} H)(x), \quad x \in S.$$

Hence, it follows inductively that

$$H(x) = \left(\left(\sum_{n=0}^{m} \Delta_{\hat{G}}^n \right) H_1 \right)(x) + (\Delta_{\hat{G}}^{m+1} H)(x), \quad m = 0, 1, 2, \ldots, x \in S. \quad (4.3.9)$$

Taking the limit in (4.3.9) as $m \to \infty$, we get the result because the second term of the identity tends to zero as $m \to \infty$ and the first term of the identity tends to $(\Delta_{\hat{G}} H_1)(x)$ as $m \to \infty$. (It is easily seen that the operator $\nabla_{\hat{G}}$ is well defined and H_1 is a member of the class on which the operators $\Delta_{\hat{G}}$ and $\nabla_{\hat{G}}$ are defined.) ■

The result of Corollary 4.3.7 can be improved considerably by relaxing the assumption that h be componentwise decreasing, as we now show; the proof of the theorem takes some hints from Theorem 4.3.9 or its proof.

4.3.10 Theorem

Let the assumptions in Corollary 4.3.7 be met with a modification that we now have the function h such that $h(\bullet) \exp\{\langle \theta, \bullet \rangle\}$ is λ-integrable on S for some $\theta \in \mathbb{R}^n$, instead of it being componentwise decreasing. Then, given any $i = 1, 2, \ldots, n$ and a constant $\gamma > 1$, there exists a constant $c^* \geqslant 0$ such that for almost all $[\lambda]x$ with the ith coordinate greater than or equal to c^*,

$$h(x) = \int_{[-\infty, -\theta]} \exp\{\langle x, y \rangle\}(1 + \xi_x(y))\nu(dy), \qquad (4.3.10)$$

where ν is as defined in Corollary 4.3.3, ∞ is the n-component vector with each of its components to be equal to ∞, and $\xi_x : [-\infty, \infty]^n \to \mathbb{R}$ is a Borel measurable function such that

$$|\xi_x(y)| \leqslant K\gamma E \left\{ \frac{e^{-\langle x, Z \rangle}}{1 - \mu^*(y - Z)} \right\} \qquad \text{for almost all } [\nu]y \leqslant -\theta \qquad (4.3.11)$$

with

$$\mu^*(s) = \int_S e^{\langle s, u \rangle} \mu(du).$$

(In view of the constraints on Z and ν, it follows that the expectation on the right-hand side of (4.3.11) is bounded relative to y on $[-\infty, -\theta]$ but for the points lying in a ν-null set; also, the proof of the theorem implies that the result with a sharper version of (4.3.11) in place of (4.3.11) or in particular, with γ in (4.3.11) replaced by

$$\left(1 - KE \left\{ \frac{e^{-\langle x, z \rangle}}{1 - \mu^*(y - Z)} \right\} \right)^{-1}$$

holds.)

Proof   There is no loss of generality in assuming $\theta = 0$. Define then

$$H(x) = \int_{[x, \infty) \cap S} h(y)\lambda(dy), \qquad x \in S.$$

Observe that H is a componentwise decreasing nonnegative continuous function on S satisfying (4.3.8) with β replaced by another function that has the properties mentioned in Corollary 4.3.7 to be possessed by β. Corollary 4.3.3 (taking into account the information supplied in the proofs of Corollaries 4.3.5 and 4.3.7) implies that given any $i = 1, 2, \ldots, n$, there exists a c^* such that

$$h(x) = h_1(x) + \sum_{m=0}^{\infty} \int_S h(x + y)\beta(x + y)\mu^{*m}(dy) \quad \text{for a.a } [\lambda]x \in A_{ic^*} \quad (4.3.12)$$

together with

$$\sum_{m=0}^{\infty} \int_S \left(\int_{[x,\infty)} h(z + y)\lambda(dz) \right) E\{e^{-\langle x+y, Z\rangle}\}\mu^{*m}(dy)$$

$$\leqslant C \left(\int_{[x,\infty)} h(z)\lambda(dz) \right) E\{e^{-\langle x, Z\rangle}\} \quad \text{for all } x \in A_{ic^*} \quad (4.3.13)$$

and
$KCE\{e^{-\langle x, Z\rangle}\} < 1$ for all $x \in A_{ic^*}$ and some $C > 0$, where h_1 is of the form
specified in Corollary 4.3.3 and A_{ic^*} is the set of x's with the ith coordinate
greater than or equal to c^*. Restricting now to $x \in A_{ic^*}$ (and taking a hint from
(4.3.9)), define for each $r \geqslant 1$, a vector $(h_r(x), H_r(x))$ such that

$$h(x) = h_r(x) + H_r(x) \quad (4.3.14a)$$

and

$$H_{r+1}(x) = \sum_{m=0}^{\infty} \int_S H_r(x + y)\beta(x + y)\mu^{*m}(dy) \quad (4.3.14b)$$

with $h_1(x)$ as in (4.3.12). In view of (4.3.13), the restriction $KCE\{e^{-\langle x, Z\rangle}\} < 1$
following it, and (4.3.12), we get that the vector in (4.3.14) is well defined for a.a.
$[\lambda]x$ and

$$\int_{[x,\infty)} |H_r(y)|\lambda(dy) \to 0 \quad \text{as } r \to \infty. \quad (4.3.15)$$

(Note inductively that we get here

$$\int_{[x,\infty)} |H_r(y)|\lambda(dy) \leqslant (KCE\{e^{-\langle x, Z\rangle}\})^r \int_{[x,\infty)} h(y)\lambda(dy).)$$

Also, we now get that for a.a. $[\lambda]x$ (satisfying obviously the restriction on the ith
coordinate)

$$h_{r+1}(x) = h_1(x) + \sum_{m=0}^{\infty} \int_S h_r(x + z)\beta(x + z)\mu^{*m}(dz), \quad r \geqslant 1. \quad (4.3.16)$$

One can choose the versions of h_r satisfying (4.3.16) inductively; these are of the
form

$$h_r(x) = \int_{[-\infty, -\theta]} \exp\{\langle x, y\rangle\}(1 + \xi_x^{(r)}(y))\nu(dy), \quad (4.3.17)$$

where for a.a $[v]y \leqslant -\theta$ and $r \geqslant 1$

$$\xi_x^{(r+1)}(y) = \sum_{m=0}^{\infty} \int_S e^{\langle z, y \rangle} \beta(x+z)(1 + \xi_{x+z}^{(r)}(y)) \mu^{*m}(dz), \qquad (4.3.18)$$

with $\xi_x^{(1)}(y) = 0$. (This can be seen via a minor manipulation involving Fubini's theorem.) Using in particular (4.3.18), we can conclude via induction that

$$|\xi_x^{(r+1)}(y) - \xi_x^{(r)}(y)| \leqslant K^r \left(E \left\{ \frac{e^{-\langle x, Z \rangle}}{1 - \mu^*(y - Z)} \right\} \right)^r, \quad r \geqslant 1 \text{ for a.a. } [v]y \leqslant -\theta.$$
$$(4.3.19)$$

The theorem is trivially valid if $v([-\infty, -\theta]) = 0$. Assume now that $v([-\infty, -\theta]) > 0$. We can then find a point $y_0 \in [-\infty, -\theta]$

$$E \left\{ \frac{e^{-\langle x, Z \rangle}}{1 - \mu^*(y - Z)} \right\} \leqslant \frac{e^{-\langle x, z^* \rangle}}{1 - \mu^*(y_0 - z^*)} \quad \text{for a.a. } [v]y \leqslant -\theta,$$

where z^* is as defined in the statement of the theorem. Clearly we can choose c^* to be so that we have additionally

$$\frac{K \exp\{-\langle x, z^* \rangle\}}{1 - \mu^*(y_0 - z^*)} \leqslant \frac{\gamma - 1}{\gamma},$$

where γ is as in the statement of the theorem. In that case (4.3.19) implies that $\{\xi_x^{(r)}(y) : r = 1, 2, \ldots\}$ is a convergent sequence for a.a. $[v]y \leqslant -\theta$ with

$$|\xi_x^{(r)}(y)| \leqslant \left[1 - KE \left\{ \frac{\exp\{-\langle x, Z \rangle\}}{1 - \mu^*(y - Z)} \right\} \right]^{-1} KE \left\{ \frac{\exp\{-\langle x, Z \rangle\}}{1 - \mu^*(y - Z)} \right\}$$

$$\leqslant \gamma KE \left\{ \frac{\exp\{-\langle x, y \rangle\}}{1 - \mu^*(y - Z)} \right\} \leqslant \gamma - 1, \quad r \geqslant 1 \quad \text{for a.a. } [v]y \leqslant -\theta.$$
$$(4.3.20)$$

Let ξ_x be a real-valued Borel measurable function on $[-\infty, \infty]^n$ such that for a.a. $[v]y \in [-\infty, -\theta]$, $\xi_x(y)$ agrees with the limit of the sequence $\{\xi_x^{(r)}(y) : r = 1, 2, \ldots\}$. The Lebesgue dominated convergence theorem implies then, in view of (4.3.17), that

$$h_r(x) \to \int_{[-\infty, -\theta]} \exp\{\langle x, y \rangle\}(1 + \xi_x(y)) v(dy). \qquad (4.3.21)$$

Denote the limit in (4.3.21) by $h^*(x)$. From (4.3.14a) and (4.3.15) we hence get on using Fatou's lemma that for a.a $[\lambda]x$ (with obviously the specified constraint on the ith component)

$$h(x) = h^*(x). \qquad (4.3.22)$$

The theorem is then obvious in view of (4.3.20). (The claim as to the validity of the result with a sharper version of (4.3.11) is also clearly seen to be justified.) ∎

In the light of the proof of Theorem 4.3.10, one can make a further observation on the function ξ in (4.3.10). It is now clear that there exists a version of the function such that for each $y \in \text{supp}[\nu]$, the function $\xi_\bullet(y)$ on the set of x's as in the statement of the theorem, is given by

$$\xi_\bullet(y) = \exp\{-\langle \bullet, y \rangle\}\nabla_\beta \exp\{\langle \bullet, y \rangle\}, \tag{4.3.23}$$

where

$$\nabla_\beta = \sum_{m=1}^{\infty} \nabla_\beta^m$$

with ∇_β and Δ_β as the operators on an appropriate class of Borel measurable functions f, such that

$$\Delta_\beta f(\bullet) = \sum_{m=0}^{\infty} \int_S f(\bullet + z)\beta(\bullet + z)\mu^{*m}(dz).$$

(Obviously, we assume f here to be such that the operations are well defined.) We shall have occasions to refer to this crucial information in subsequent chapters.

We now have some additional revelations to make on Theorem 4.3.10 and some of the earlier results through the following remarks.

4.3.11 Remarks

(i) If $n = 1$, the result of Theorem 4.3.8 holds for some θ if the restriction that $h(\bullet)\exp\{\langle \theta, \bullet \rangle\}$ is λ-integrable on S for some $\theta \in \mathbb{R}^n$ is replaced by that h is λ-locally integrable on S; this follows because an argument used in the proof of Theorem 2.2.2 essentially implies that in the presence of other restrictions involved in the statement of Theorem 4.3.10 the two restrictions considered here are equivalent. A special case of this latter result with Z as degenerate has appeared in Gu and Lau (1984). (It may be worth pointing out here that some of the statements appearing in the Gu–Lau paper do not seem to be accurate as stated.)

(ii) One could now ask a question as to how crucial the assumption of $\mu(\{0\}) < 1$ is in Theorem 4.3.10. That without this the situation turns out to be very awkward is shown by the following example.

Example Let $S = \mathbb{N}_0^2$ and μ be that measure on S for which

$$\mu(\{x\}) = \begin{cases} 1 & \text{if } x = (0,0) \text{ or } (1,1) \text{ or } (2,1) \\ 0 & \text{otherwise.} \end{cases}$$

Define h on S such that for each $x \in S$

$$h(x) = e^{-||x||^2}.$$

(One could obviously consider a more general h, but for illustrative purposes our choice is more than adequate.)We have then

$$\int_S h(x + y)\mu(dy) = h(x)(1 - \beta(x)), \quad x \in S$$

with β such that for each x

$$|\beta(x)| < 2\,e^{-\langle(2,2),x\rangle}$$

and μ satisfying Condition I of Section 3.4 except for $\mu(\{0\}) < 1$. Observe that the function h considered here is componentwise decreasing without being of the form arrived at in Corollary 4.3.7. (In the present case, since $1 - \mu^*(y - Z)$ appearing in (4.3.11) equals zero almost surely for a.a. $[v]y$, (4.3.11) and hence the conclusions of Theorem 4.3.8 do not obviously hold.)

(iii) The next example illustrates that Corollary 4.3.7 does not remain valid if the assumption that $Z \geqslant z^*$ almost surely is replaced by the $Z \geqslant 0$ almost surely with $P\{Z = 0\} < 1$.

Example   Let $S = \mathbb{N}_0^2$ and μ be the measure on S such that

$$\mu(\{(0, 1)\}) = \mu(\{(1, 0)\}) = 1 \quad \text{and} \quad \mu((\{(0, 1)\} \cup \{(1, 0)\})^c) = 0.$$

Define h on S such that

$$h(x) = e^{-x_1^2}, \quad x \in S,$$

where x_1 is the first coordinate of x. Note that the h satisfies

$$\int_S h(x + y)\mu(dy) = h(x)(1 - \beta(x)), \quad x \in S$$

with

$$|\beta(x)| < e^{-2x_1} \quad \text{for each } x.$$

Clearly in this example all the assumptions of Corollary 4.3.7 subject to the alteration concerning Z hold, but h is not of the form arrived at in the conclusion of the corollary.

(iv) One may raise a question as to whether Corollary 4.3.7 and Theorem 4.3.10 hold if one or more S_i are taken as $\mathbb{Z}$ or $\mathbb{R}$ and simultaneously (in obvious notation) $\exp\{-\langle x, Z\rangle\}$ is replaced by $\exp\{-\Sigma_{i=1}^n |x_i|Z_i\}$. The following example answers the question in the negative.

Example   Take $S = \mathbb{Z}$ or $\mathbb{R}$ and P to be a probability measure on S with support $[-1, 1] \cap S$ such that $\int_S xP(dx) = 0$. Define the measure μ on S that is such that $\mu(B) = \int_B e^x P(dx)$ for each Borel subset B of S, and the real function h on S for which

$$h(x) = \begin{cases} x\,e^{-x} & \text{if } x \geqslant 1 \\ e^{-x} & \text{otherwise.} \end{cases}$$

Note that

$$\int_S h(x + y)\mu(dy) = h(x)(1 - \beta(x)), \quad x \in S$$

with β such that

$$|\beta(x)| \leqslant K\, e^{-|x|z^*}, \quad x \in S$$

for some $K, z^* > 0$. (Indeed, we have in the present case β to be a bounded function with $\beta(x) = 0$ for x with $|x| \geqslant 2$.) Here the measure ν appearing in the corollary and the theorem is concentrated on $\{-1\}$ and we do not have h of the form specified in the two conclusions in question (though it is decreasing).

(v) A further example illustrates that for $n > 1$ the result of Theorem 4.3.10 does not hold for any $\theta \in \mathbb{R}^n$ if the restriction that $h(\bullet) \exp\{\langle\theta, \bullet\rangle\}$ is λ-integrable is replaced by that h is λ-locally integrable on S.

Example   Define for each $y \in \mathbb{R}$, $\zeta(y)$ such that $\zeta(0) = 0$ and if $y \neq 0$, $\zeta(y) \neq 0$ and

$$\left(\frac{e^y - 1}{y}\right)\left(\frac{e^{\zeta(y)} - 1}{\zeta(y)}\right) = 1,$$

it follows that ζ is continuous with $0 < \zeta(y) < |y|$ if $y < 0$ and $\zeta(y) < 0$ if $y > 0$. If we now take $h : \mathbb{R}_+^2 \to \mathbb{R}_+$ such that

$$h(x_1, x_2) = E\{\exp\langle Y x_1, \zeta(Y)x_2\rangle\}, \quad (x_1, x_2) \in \mathbb{R}_+^2,$$

where Y is a normally distributed random variable, then it follows that h is continuous and hence locally integrable on $\mathbb{R}_+^2$. We have also that

$$h(x_1, x_2) = \int_0^1 \int_0^1 h(x_1 + y_1, x_2 + y_2)\, dy_1\, dy_2, \quad (x_1, x_2) \in \mathbb{R}_+^2.$$

Since, in the present case, we cannot have $\theta \in \mathbb{R}^2$ such that $h(\bullet) \exp\{\langle\theta, \bullet\rangle\}$ is λ-integrable on $\mathbb{R}_+^2$, it is impossible that we have the conclusion of the theorem considered to be valid for some θ.

It may, however, be noted here that if $S = \mathbb{N}_0^n$, we have Corollary 4.3.7 to be valid without the assumption that h be componentwise decreasing (as we have in that case the given proof to be valid on taking $B = \{0\}$) and hence it easily follows that Theorem 4.3.10 holds with the integrability condition on $h(\bullet) \exp\{\langle\theta, \bullet\rangle\}$ deleted, provided in (4.3.10) (and in the sentence in brackets at the end of the statement of the theorem), $[-\infty, -\theta]$ is replaced by $[-\infty, \infty]$ and in (4.3.11), '$\leqslant -\theta$' is deleted. For a version of the ξ of the assertion, the operator representation (4.3.23) also holds.

(vi) In view of what is revealed in Section 4.2, analogues of Corollary 4.3.3 corresponding to multiple integrals that are stability results for Theorems 4.2.1 and 4.2.2 are easy to obtain.

(vii) Suppose n, S and λ are as in Corollary 4.3.7, $h_i : S \to \mathbb{R}_+$, $i = 1, 2, \ldots, k$ are componentwise decreasing Borel measurable functions, and β_i, $i = 1, 2, \ldots, k$ are Borel measurable functions meeting the requirements of β of Corollary 4.3.7 with obviously K, Z and z^* replaced by K_i, Z_i and z_i^* respectively. Suppose further that

$$h_i(\bullet)(1 - \beta_i(\bullet)) = \sum_{j=1}^{k} \int_S h_{i+j-1}(\bullet + y)\mu_j(dy) \quad \text{a.e. } [\lambda], \quad i = 1, 2, \ldots, k$$

with $h_{k+1}(\bullet), \ldots, h_{2k-1}(\bullet)$ as defined in Corollary 4.2.4 and μ_i, $i = 1, 2, \ldots, k$ as σ-finite measures. Then, if either the assumption in Theorem 4.2.1 with $\mu_1(\{0\}) < 1$ replaced by $\Sigma_1^k \mu_i(\{0\}) < 1$ or the assumption in Theorem 4.2.2 is met, using a slightly modified version of the argument leading to Corollary 4.3.7 involving appropriate matrix equations, we can see that for some matrix function Γ

$$\begin{pmatrix} h_1(x) \\ \vdots \\ h_k(x) \end{pmatrix} = (I + \Gamma(x)) \begin{pmatrix} \zeta_1 \\ \vdots \\ \zeta_k \end{pmatrix} \quad \text{for a.a. } [\lambda]x \in S, \tag{4.3.24}$$

where ζ_i are the integrals involved in the resulting representation for h_i in Corollary 4.2.4, the (i, j)th elements $\gamma_{ij}(x)$ of $\Gamma(x)$ are such that if x has at least one of its coordinates to be sufficiently large and I is the identity matrix, we have

$$|\gamma_{ij}(x)| \leqslant \gamma_i E[\exp\{-\langle x, Z_j \rangle\}]$$

with γ_i to be constants (i.e. not depending on x). Clearly if the hypothesis of Theorem 4.2.1 is met with the amendment stated above, then we get the representation (4.3.24) with a further property that $\zeta_1 = \cdots = \zeta_k$.

(viii) Let the assumptions in (vii) above be met with a modification that the functions h_i are now taken to be such that $h_i(\bullet) \exp\{\langle \theta, \bullet \rangle\}$ are λ-integrable on S for some $\theta \in \mathbb{R}^n$ instead of these being componentwise decreasing. Then, essentially via the argument used in the proof of Theorem 4.3.10, it follows that given a constant $\gamma > 1$ there exists a constant $c^* \geqslant 0$ such that for almost all $[\lambda]x$ with at least one coordinate greater than or equal to c^*,

$$h_i(x) = \sum_{j=1}^{k} \int_{[-\infty, -\theta]} \exp\{\langle x, y \rangle\}(\delta_{ij} + \xi_{ij}(x, y))v_j(dy),$$

where v_j are measures on S of the form of v of Corollary 4.3.3, δ_{ij} is the Kronecker delta, ∞ as defined in Theorem 4.3.10, and $\xi_{ij}(x, \bullet) : [-\infty, \infty]^n \to \mathbb{R}$ are Borel measurable functions satisfying

$$|\xi_{ij}(x, y)| \leqslant K_j \gamma E\{q_{ij}(y - Z_j)e^{-\langle x, Z_j \rangle}\} \quad \text{for a.a. } [v_j]y \leqslant -\theta$$

with $q_{ij}(s)$ as the (i, j)th element of

$$
\begin{pmatrix}
1 - \mu_1^*(s) - \mu_2^*(s) \cdots - \mu_k^*(s) \\
-\mu_k^*(s) 1 - \mu_1^*(s) \cdots - \mu_{k-1}^*(s) \\
\vdots \\
-\mu_2^*(s) - \mu_3^*(s) \cdots 1 - \mu_1^*(s)
\end{pmatrix}^{-1},
$$

where

$$
\mu_i^*(s) = \int_S e^{\langle s, u \rangle} \mu_i(du).
$$

(With appropriate modifications, one can also obtain representations of the type of (4.3.23) for a version of (ξ_{ij}); in this case, we get, in place of (4.3.23), a modified version of it which is such that for each $y \in \sup[\nu_j]$ and $j = 1, 2, \ldots, k$

$$
(\xi_{1j}(\bullet, y), \ldots, \xi_{kj}(\bullet y))' = \exp\{-\langle \bullet, y \rangle\} (\nabla_\beta((\exp\{\langle \bullet, y \rangle\}) \bullet I_j))
$$

with ∇_β as a certain matrix operator, I_j as the jth column of the $k \times k$ identity matrix, and the set of x's as that meeting the condition stated above with an appropriate c^*.) Under the assumptions of Theorem 4.2.1 (with obvious amendment of $\Sigma_1^k \mu_i(\{0\}) < 1$ in place of $\mu_1(\{0\}) < 1$), we have the stated result to be valid with ν_j as independent of j. Also, if $n = 1$, essentially from the argument mentioned in (i) above, we get that the result is valid if the restriction that $h_i(\bullet) \exp\{\langle \theta, \bullet \rangle\}$ are λ-integrable on S for some $\theta \in \mathbb{R}^n$ is replaced by that h_i are λ-locally integrable on S.

(ix) In view of the example in (ii) above, it follows that the concerned results in (vii) and (viii) do not remain valid if the modification in the assumptions of Theorem 4.2.1 of having $\Sigma_1^k \mu_i(\{0\}) < 1$ in place of $\mu_1(\{0\}) < 1$ is not carried out; this can be seen by taking $h_1 = h_2 = h$, $\mu_1 = \mu_2 = \frac{1}{2}\mu$ and $\beta_1 = \beta_2 = \beta$ and noting that the form of h_1 given in the example is not the same as that in the conclusions of the two results.

(x) In Corollary 4.3.7 (assuming that S, μ and Z are given) or Theorem 4.3.10 (assuming that S, μ, Z and θ are given), if we let K be sufficiently small, then we can choose $c^* = 0$. Similarly, in (viii) above (assuming that K, S, μ_j's, Z_j's and θ are given), if we let K_j's be sufficiently small, then we can choose $c^* = 0$.

4.4 POTENTIAL THEORETIC RESULTS: SOME OBSERVATIONS

In what follows we look briefly at the potential theoretic arguments based on the Choquet theory for arriving at the main results of Chapters 3 and 4. In particular, we give here for Theorem 3.3.1, which forms the basis of the findings in the two chapters, an alternative proof based on such arguments. This would be of interest to readers specializing in potential theory.

Let us begin our discussion by giving the following theorem, which has effectively been proved on pages 168 and 169 in Williams (1979). The theorem is somewhat more general than the one required to meet our immediate needs. However, since the proof of the result given here essentially involves the arguments to be used even for proving the special case and also the more general result is of interest in its own right, we have opted for this result.

4.4.1   Theorem

Let I be a countable set and Π be a nonnegative (in the sense of Seneta (1973)) $I \times I$ matrix. Assume that there exists a reference point b in I such that $\sup_n \Pi^n(b, j) > 0$ for each $j \in I$, where $\Pi^n(b, j)$ denotes the (b, j)th element of Π^n. If we denote by $\mathcal{F}$ the class of functions $f : I \to \mathbb{R}_+$ such that $f(b) = 1$ and $\Pi f \leqslant f$, and by $\mathcal{F}_e$ the set of extreme points of $\mathcal{F}$, then the set $\mathcal{F}$ is a compact convex metrizable subset of the locally convex linear topological space $\mathbb{R}^I$ and $\mathcal{F}_e$ is a G_δ subset of $\mathcal{F}$. Moreover, given any $f \in \mathcal{F}$, there exists a probability measure ν on $\mathcal{F}_e$ (i.e. on the Borel σ-field of $\mathcal{F}_e$) such that for each $i \in I$

$$f(i) = \int_{\mathcal{F}_e} \xi(i)\nu(\mathrm{d}\xi). \qquad (4.4.1)$$

(The map $\xi \to \xi(i)$ is continuous on $\mathcal{F}$.)

Proof   Denote $\sup_n \Pi^n(b, j)$ by $\theta(j)$; by the assumption in the theorem, we have $\theta(j) > 0$ for all j. Supposing $f \in \mathcal{F}$, we have

$$f \geqslant \Pi f \geqslant \Pi^2 f \geqslant \cdots,$$

and hence

$$f(b) \geqslant \theta(j)f(j) \quad \text{for all } j. \qquad (4.4.2)$$

In view of Fatou's lemma, on appealing to (4.4.2), it then follows that $\mathcal{F}$ satisfies the hypothesis of Theorem 1.3.8. Consequently, we have the present theorem. ∎

To see the relevance of Theorem 4.4.1 to the results arrived at in Chapter 3 as well as in the present chapter, note that if (3.1.1) is met with S as countable and the smallest subsemigroup of S with zero element containing $\{x \in S : \mu(\{x\}) > 0\}$ equals S, then unless $H \equiv 0$, we have that $H(\bullet)/H(0)$ is a member of a class $\mathcal{F}$ of the type in the theorem, where $I = S$, 0 is the reference point and

$$\prod(i, j) = \mu(\{x \in S : i + x = j\})$$

for all i, j. In that case, $H(\bullet)/H(0)$ has a representation of the type (4.4.1) based on extreme points ξ such that a.a. $[\nu]\xi$ satisfy (3.1.1) with $\xi(0) = 1$ (i.e. satisfy $\xi(i) = \Sigma_{j \in I}\xi(i + j)\mu(\{j\})$, $i \in I$, $\xi(0) = 1$); the result follows on noting amongst other things that $\xi(j) = 0 \Rightarrow \xi(i + j) = 0$ for all $i \in I$, and $\xi(j) \neq 0$, $\mu(\{j\}) > 0$ implies as

$$\xi(i) = \frac{\xi(i + j)}{\xi(j)}\xi(j)\mu(\{j\}) + \sum_{k(\neq j) \in I} \xi(i + k)\mu(\{k\}) \quad i \in I$$

that $\xi(i + j) = \xi(i)\xi(j)$, $i \in I$. Consequently, we have as a corollary of Theorem 4.4.1, the special case of Theorem 3.4.7 when S is countable and $S^*(\mu) = S$ (with the topology as discrete).

The special case of Theorem 3.4.7 arrived at above is indeed a result of Ressel (1975) and gives, amongst other things, Theorem 2.2.1 as a corollary. As mentioned in Remark 3.3.7(i), Shanbhag (1991) has given a short proof, based on Theorem 1.2.7, for Theorem 3.3.1 in the case when S is a Polish Abelian semigroup with zero element. Kendall (1967) and others have observed that Theorem 1.2.7 itself is a by-product of the Choquet theory. It is now natural to enquire whether it is possible to obtain the general version of Theorem 3.3.1 via Theorem 1.2.7 so that there is a further link between potential theoretic results and the results on integral equations given in the present monograph. The following proof of Theorem 3.3.1 based on exchangeability provides us with some useful information in this connection.

4.4.2   A proof based on exchangeability

As in the existing proof in Section 3.3 of the theorem, $S^*(\mu) \subset \hat{S}$ and (3.3.1) follows. If $\{\xi_n(x, \bullet)\}$ is a regular martingale, then due to the exchangeability property of $\{X_n : n = 1, 2, \ldots\}$, we have for each $y_1, y_2 \in S^*(\mu)$

$$H\left(x + \sum_{1}^{2} y_r\right) = H(0) \lim_{m\to\infty} \lim_{n\to\infty} E\left\{\left(\prod_{r=1}^{2} \zeta_n^{(m)}(y_r, \bullet)\right)\xi(x, \bullet)\right\}, \qquad (4.4.3)$$

where

$$\zeta_n^{(m)}(y_r, \bullet) = \frac{1}{n}\sum_{i=1}^{n} I_{\{X_i \in A_m(y_r)\}}, \qquad m, n = 1, 2, \ldots$$

with $\{A_m(y_r) : m = 1, 2, \ldots\}$ as a sequence of open spheres with centre y_r and radii converging to zero. Hence, on using Corollary 1.2.6 and Lemma 3.2.3 (together with the trivial fact that $0 \in \hat{S}$ or that $S^*(\mu) \subset \hat{S}$), we see that

$$0 = \lim_{m\to\infty} \lim_{n\to\infty} E(\xi(x + 2y) - 2\zeta_n^{(m)}(y, \bullet)\xi(x + y, \bullet) + (\zeta_n^{(m)}(y, \bullet))^2\xi(x, \bullet))$$

$$\geqslant \lim_{m\to\infty} \lim_{n\to\infty} E\left(\left(\frac{\xi(x + y, \bullet)}{(\xi(x, \bullet))^{1/2}} - \zeta_n^{(m)}(y, \bullet)(\xi(x, \bullet))^{1/2}\right)^2\right)$$

$$\geqslant 0, \quad x \in \hat{S}, y \in S^*(\mu). \tag{4.4.4}$$

where $\xi(x+y, \bullet)/(\xi(x, \bullet))^{1/2}$ is defined to be equal to zero on $\{\xi(x, \bullet) = 0\}$. (Note that for arriving at (4.4.4) we have noted, amongst other things, that $\xi(x+y, \bullet) = 0$ a.s. $[P_H]$ on $\{\xi(x, \bullet) = 0\}$.) As $\xi(0, \bullet) = 1$ a.s. $[P_H]$, (4.4.4) implies that given any $x \in \hat{S}$ and $y \in S^*(\mu)$, there exists a sequence $\{n_m : m = 1, 2, \ldots\}$ of positive integers such that $\{\zeta_{n_m}^{(m)}(y, \bullet) : m = 1, 2, \ldots\}$ converges in L_2 to $\xi(y, \bullet)$ and

$\{\zeta_{n_m}^{(m)}(y, \bullet)(\xi(x, \bullet))^{1/2} : m = 1, 2, \ldots\}$ converges in L_2 to $\xi(x + y, \bullet)/(\xi(x, \bullet))^{1/2}$. This, in turn, implies that

$$\xi(x + y, \bullet)/(\xi(x, \bullet))^{1/2} = \xi(y, \bullet)(\xi(x, \bullet))^{1/2}, \quad x \in \hat{S}, y \in S^*(\mu).$$

Since $\xi(x + y, \bullet) = 0$ a.s. $[P_H]$ on $\{\xi(x, \bullet) = 0\}$ for $y \in S^*(\mu)$, we have then (3.3.2) to be valid. The validity of (3.3.4) and (3.3.3) follows on using exactly the same arguments as in the previous proof. ∎

It may be of relevance to note now that Ressel's result met above (i.e. the special case of Theorem 3.4.7 with S countable and $S^*(\mu) = S$) also follows directly from de Finetti's theorem (or Theorem 1.2.7). The theorem in question implies immediately that

$$E\{(\zeta(x, \bullet)\zeta(y.\bullet) - \zeta(x + y, \bullet))^2\} = 0, \quad x, y \in X,$$

where $\zeta(x, \bullet) = P\{X = x|\mathcal{I}\}/\mu_2(\{x\})$, $x \in S$ and $\mathcal{I}$ as the invariant σ-field relative to $\{X_n\}$. A properly chosen version of $\zeta(x, \bullet)$, $x \in S$ then meets the requirements of the product measurable function of Theorem 3.4.7, justifying the claim made.

4.4.3 Remarks

(i) If S is a second countable Abelian topological semigroup with zero element (or more generally an Abelian topological semigroup with zero element satisfying $\mu((\text{supp}[\mu])^c) = 0$, provided in this latter case μ_2 is defined as in Remark 3.3.7(iv)), then, given any $x \in S$ for which $\{\xi_n(x, \bullet)\}$ is a regular martingale and $y_1, y_2 \in S^*(\mu)$, we can choose neighborhoods $O_m(y_r)$ of y_r such that (4.4.3) is valid with the x as given as well as with this replaced by 0 and $A_m(y_r)$ replaced by $O_m(y_r)$. Consequently it follows that with appropriate alterations, Proof 4.4.2 also applies to the general S considered here.

(ii) In view of what is shown in Section 4.2, implications of the proof appearing under 4.4.2 to the problem of solving (4.1.1) are obvious. The proof obviously provides us with alternative approaches for arriving at various results which assume Theorem 3.3.1.

(iii) Lau and Zeng (1990) have studied the problem of solving (3.1.1) via a potential theoretic approach under a different and somewhat involved setup. However, in crucial cases such as those with S as in Corollaries 3.4.4 or 3.4.5, their results follow easily from those appearing in this monograph.

(iv) Extending the extreme point argument given above for arriving at Ressel's (1975) result, one can produce an alternative proof for Corollary 3.4.4. First apply Ressel's result to the restriction of H to $S^*(\mu)$ and using it in conjunction with Corollary 1.2.6 observe that, in view of Condition I, we have, for each $x \in S$, the existence of a σ-finite measure ν_x concentrated on $S^*(\mu)$ such that $x + \text{supp}[\nu_x] \subset$

$S^*(\mu)$ and

$$H(x + y) = \int_{\mathrm{supp}[\nu_x]} H(x + y + z)\nu_x(\mathrm{d}z), \quad y \in S.$$

Then, defining

$$\xi^*(x) = \begin{cases} \xi(x) & \text{if } x \in S^*(\mu) \\ \int_{\mathrm{supp}[\nu_x]} \xi(x + z)\nu_x(\mathrm{d}z) & \text{if } x \in (S^*(\mu))^c \end{cases}$$

corresponding to each extreme point ξ in the representation for the restriction of $H(\bullet)/H(0)$ to $S^*(\mu)$, it can easily be seen that ξ^* are μ-harmonic exponential functions for a.a. $[\nu]\xi$ and there is an integral representation for $H(\bullet)$ in terms of ξ^* as required.

(v) The simpler proof based on exchangeability for Theorem 3.3.1 is not only of interest in its own right but also has some further implications. In view of what is revealed in Remarks 3.2.4(iv) and (v), if we consider the hypothesis of Theorem 2.2.2 or Corollary 2.3.2, then it follows trivially that the conclusions of Theorem 3.3.1 with $\hat{H}$ in place of H, where $\hat{H}(x) = \int_x^\infty \mathrm{e}^{-\delta y} H(y)\mathrm{d}y$, $x \in S$ so constructed that it is real, and $\hat{S} = S$ hold. Then we can find versions of $\xi(x, \bullet)$ such that for each ω, $\xi(\bullet, \omega)$ is decreasing and right continuous satisfying

$$\hat{H}(x) = \hat{H}(0)E(\xi(x, \bullet)), \quad x \in S$$

and

$$\xi(x, \bullet)\xi(y, \bullet) = \xi(x + y, \bullet), \quad x \in S, y \in S^*(\mu),$$

implying, in view of Marsaglia and Tubilla (1985), that the conclusions of Theorem 2.2.2 and Corollary 2.3.2 hold when the respective $\hat{H}$ is taken in place of H. From what we have observed, it can easily be seen that Theorem 2.2.2 and Corollary 2.3.2 are valid, and hence we have alternative arguments to arrive at these results. In view of the role of Theorem 3.3.1 in these arguments, it is clear that the simpler proof of Theorem 3.3.1 is of vital importance.

(vi) The potential theoretic approach described in this section to solve (3.1.1) is of relevance to the problem of solving the multiple integral equations (4.1.1) since we have reduced (4.1.1) to (3.1.1) (with notational alterations).

CHAPTER 5

Mean Residual Life Function and Hazard Measure

5.1 INTRODUCTION

Amongst numerous characterizations of the distribution of a random variable or a random vector, those based on the mean residual life function or the hazard measure (or their different versions in the multivariate case) have received prominence in recent years. Relevant contributions in this respect, in the univariate case, have come from Cox (1961, 1972), Barlow and Prochan (1965, 1975), Meilijson (1972), Swartz (1973), Laurent (1974), Jacod (1975), and Kotz and Shanbhag (1980) amongst others and, in the multivariate case, from Basu (1971), Puri and Rubin (1974), Marshall (1975), Johnson and Kotz (1975), Zahedi (1985), Shanbhag and Kotz (1987) and others. See also Galambos and Kotz (1978) for an account of the literature existing prior to its publication. Some of these results have important applications in reliability and allied topics.

Attempts have also been made at characterizing distributions via conditional expectations $E\{h(X)|X \geqslant x\}$ or $E\{h(X - x)|X \geqslant x\}$ with X as a random variable or a random vector and h meeting certain conditions. Hamdan (1972), Kotlarski (1972), Sahobov and Geshev (1974), Shanbhag and Bhaskara Rao (1975), Gupta (1975), Shimizu (1979), Kotz and Shanbhag (1980), Klebanov (1980) and some others have produced various results belonging to this category. These latter results have obvious links with, or in many cases are generalizations of, the earlier results on the mean residual life function.

There are situations in which both the mean residual life function and hazard measure have stability versions for the corresponding characterizations. These were studied in Kotz and Shanbhag (1980) and Shanbhag and Kotz (1987). Stability theorems corresponding to characterizations of the exponential distributions based on properties of the type $E\{h(X - x)|X \geqslant x\} = c$ or related characterizations of the Pareto distributions have been established by Shimizu (1980), Gu and Lau (1984) and Yanushkyavichyus (1988) amongst others.

The present chapter is devoted to the discussion of a major portion of the aforementioned work on characterization and stability problems. Although an attempt is made in places to extend the existing results, in most of the presentation, we restrict our attention to unifying the literature following the treatment of Kotz and Shanbhag (1980) and Shanbhag and Kotz (1987).

5.2   THE MEAN RESIDUAL LIFE FUNCTION AND ITS MODIFIED VERSIONS: UNIVARIATE CASE

We begin by first defining the mean residual life function formally as follows.

5.2.1   Definition

Let X be a real-valued random variable with $E(X^+) < \infty$. Define a real-valued Borel measurable function s on $\mathbb{R}$ satisfying $s(x) = E\{X - x | X \geq x\}$ for all x such that $P\{X \geq x\} > 0$. This function is called the mean remaining life function (m.r.l. function for short).

The restriction of the m.r.l. to $(-\infty, b)$ where b is the right extremity of the distribution of X is clearly left continuous and hence is determined by its knowledge on a dense subset of $(-\infty, b)$.

There are many interesting properties of the m.r.l. function and two of its modified versions which are given, for x such that $P\{X \geq x\} > 0$, respectively by $E\{h(X) - h(x-)|X \geq x\}$ and $E\{h(X - x)|X \geq x\}$ with h satisfying certain constraints. The following theorems describe some of these.

5.2.2   Theorem

Let $b(\leqslant \infty)$ denote the right extremity of the distribution function (d.f.) F of a random variable X with $E(X^+) < \infty$ and let s be its m.r.l. function. Further, let $A = \{y : \lim_{x \uparrow y} s(x) \text{ exists and equals } 0\}$. Then $b = \infty$ if A is empty and $b = \inf\{y : y \in A\}$ if A is non-empty. Moreover, for every $-\infty < y < x < b$

$$\frac{1 - F(x-)}{1 - F(y-)} = \frac{s(y)}{s(x)} \exp\left\{-\int_y^x \frac{dz}{s(z)}\right\}, \tag{5.2.1}$$

and for every $-\infty < x < b$, $1 - F(x-)$ is given by the limit of the right-hand side of (5.2.1) as $y \to -\infty$.

Proof   The first part follows trivially from the left-continuity of the restriction of s to $(-\infty, b)$ and the fact that $0 < s(x) \leqslant b - x$ for every $x \in (-\infty, b)$. The second part can be arrived at by applying Lemma 1.3.2 as follows. Observe that $s(x) = \{\int_x^\infty (1 - F(z)) \, dz\}\{1 - F(x-)\}^{-1}$ if $x \in (-\infty, b)$. Consequently,

$$\int_y^z \frac{dz}{s(z)} = \int_y^x f(G^*(z))G(dz) \quad \text{for every } -\infty < y < x < b$$

with

$$G(z) = G^*(z) = -\int_z^\infty (1 - F(u)) \, du,$$

$z \in \mathbb{R}$, and $f(z) = -1/z$ if $z < 0$ and $= 0$ otherwise. Hence from the lemma

$$\int_y^z \frac{dz}{s(z)} = \log \left\{ \int_y^\infty (1 - F(z)) \, dz \Big/ \int_x^\infty (1 - F(z)) \, dz \right\}$$

$$= \log \left\{ \frac{s(y)}{s(x)} \bullet \frac{1 - F(y-)}{1 - F(x-)} \right\}, \qquad -\infty < y < x < b,$$

which yields the assertion immediately. The remainder of the theorem is an obvious corollary of this assertion. ∎

5.2.3   Corollary

Let X be a nonnegative random variable with d.f. F and $E(X) < \infty$ and let b be the right extremity of F. Then, for every $x \in [0, b)$,

$$1 - F(x-) = \frac{s(0)}{s(x)} \exp \left\{ - \int_0^x \frac{dz}{s(z)} \right\},$$

where s is as defined in Theorem 5.2.2.

5.2.4   Theorem

Let g be a left-continuous and nondecreasing function from $[c, \infty)$ into $(0, \infty)$, where c is a real number. Define $G : \mathbb{R} \to [0, 1]$ by the relation

$$G(x) = \begin{cases} 0 & \text{if } x < c \\ 1 - \dfrac{g(c)}{g(x+)} \exp \left\{ - \displaystyle\int_c^x (g(y))^{-1} \, dy \right\} & \text{if } x \geq c. \end{cases} \qquad (5.2.2)$$

Then G is a d.f. (of a random variable) with right extremity equal to ∞ and finite mean. Moreover, the m.r.l. function s which corresponds to G is given by

$$s(x) = \begin{cases} g(x) \left[1 - \exp \left\{ - \displaystyle\int_x^\infty (g(y))^{-1} \, dy \right\} \right] & \text{if } x \in [c, \infty) \\ s(c) + c - x & \text{otherwise.} \end{cases} \qquad (5.2.3)$$

Proof   It is easy to check that G is a d.f. with right extremity equal to ∞. Denote by Y a random variable with this d.f. Then it easily follows that for $x \in [c, \infty)$

$$E\{Y - x | Y \geq x\} = \frac{\displaystyle\int_x^\infty (1 - G(y)) \, dy}{1 - G(x-)}$$

$$= g(x) \int_x^\infty f(G_x(y)) G_x(dy),$$

where $f(y) = e^{-y}$, $y \in \mathbb{R}$ and $G_x(y) = \int_x^y (g(u))^{-1} \, du$ if $y \geq x$ and $= 0$ otherwise. Lemma 1.3.2 then yields that $E\{Y - x | Y \geq x\}$ for $x \in [c, \infty)$ is equal to the right-hand side of the first equation of (5.2.3). This implies that Y is integrable and also that the corresponding m.r.l. function is given by (5.2.3) completely. Hence we have the theorem. ∎

5.2.5   Theorem

Let F be a d.f. on $\mathbb{R}$ and let (α, β) be a nonempty interval and $b \in (\alpha, \beta]$ such that $F(\beta-) - F(x) > 0$ for all $x \in (\alpha, b)$. (Here we allow $\alpha = -\infty$ and/or $\beta = \infty$.) Let h be a real-valued Borel measurable function on $\mathbb{R}$ such that it is increasing on (α, β) and $h(x) < h(b-)$ for all $x \in (\alpha, b)$. Denote by X a random variable with d.f. F. Then either $E\{h^+(X)|X \in (\alpha, \beta)\} = \infty$ or this expectation is finite and

$$\frac{F(\beta-) - F(x-)}{F(\beta-) - F(y-)} = \frac{g(y)}{g(x)} \left\{ \prod_{x_r \in D_{yx}} g^*(x_r) \right\} \exp\left\{ -\int_y^x \frac{h_c(dz)}{g(z)} \right\}$$

$$\text{if } \alpha < y < x < b, \tag{5.2.4}$$

and $(F(\beta-) - F(x-))/(F(\beta-) - F(\alpha))$ for every $x \in (\alpha, b)$ equals the limit of the right-hand side of (5.2.4) as $y \to \alpha$ from above, where D_{yx} is the set of discontinuity points of h in the interval $[y, x)$, $g(z) = E\{h(X)|X \in [z, \beta)\} - h(z-)$, $z \in (\alpha, b)$, h_c denotes the continuous part of the restriction of h to (α, β) (i.e. of its right-continuous version), and

$$g^*(x_r) = \left\{ \frac{g(x_r) - (h(x_r) - h(x_r-))}{g(x_r+) + (h(x_r+) - h(x_r))} \right\} \frac{g(x_r+)}{g(x_r)}.$$

Proof   Clearly only the case of $E\{h^+(X)|X \in (\alpha, \beta)\} < \infty$ requires verification. In this case, the definition of g implies that

$$g(z) = (F(\beta-) - F(z-))^{-1} \int_{[z,\beta)} \left\{ \int_{[z,x)} h_+(dy) + (h(x) - h(x-)) \right\} F(dx),$$

$$z \in (\alpha, b),$$

where $h_+(y) = h(y+)$. Using Fubini's theorem, we then obtain

$$g(z) = (F(\beta-) - F(z-))^{-1} L(z), \quad z \in (\alpha, b), \tag{5.2.5}$$

where

$$L(z) = \int_{[z,\beta)} (F(\beta-) - F(y))h_+(dy) + \sum_{x_r \in D_z} (h(x_r) - h(x_r-)) P(\{x_r\})$$

with $P(\{x_r\}) = F(x_r) - F(x_r-)$ and D_z as the set of discontinuity points of h in $[z, \beta)$. Consequently we have that if $\alpha < y < x < b$,

$$\int_y^x \frac{h_c(dz)}{g(z)} = \int_y^x f(G_y(z-))G_y^c(dz), \tag{5.2.6}$$

where $f(z) = -1/z$ if $z < 0$ and $= 0$ otherwise,

$$G_y(z) = \begin{cases} -L(z+) & \text{if } y < z < b \\ 0 & \text{if } z \geqslant b \\ -L(y+) & \text{if } z \leqslant y, \end{cases}$$

and G_y^c is the continuous part of G_y. Applying Lemma 1.3.2 to (5.2.6) and carrying out simple manipulations, we arrive at

$$\frac{L(x)}{L(y)} = \left\{ \prod_{x_r \in D_{yx}} \frac{L(x_r+)}{L(x_r)} \right\} \exp\left(-\int_y^x \frac{h_c(dz)}{g(z)} \right), \quad \alpha < y < x < b. \quad (5.2.7)$$

Since (5.2.5) implies $L(z) = (F(\beta-) - F(z-))g(z)$, $z \in (\alpha, b)$, and

$$\frac{L(x_r+)}{L(x_r)} = \frac{F(\beta-) - F(x_r)}{F(\beta-) - F(x_r-)} \frac{g(x_r+)}{g(x_r)} = g^*(x_r), \quad x \in D_{yx},$$

(5.2.4) is obvious from (5.2.7). The last assertion of the theorem is immediate from (5.2.4). ∎

Before we consider the theorem on characterization of probability distributions based on $E\{h(X - x)|X \geqslant x\}$, we observe a curious phenomenon of the function through the following two examples.

Example 1   Let h be a real-valued function on $\mathbb{R}$ such that

$$h(x) = x \quad \text{for all } x \in \mathbb{R}$$

or

$$h(x) = e^{cx} \quad \text{for all } x \in \mathbb{R}$$

with c real. Also, let X denote a real-valued random variable on $\mathbb{R}$ with $E(h^+(X)) < \infty$. Then from Theorem 5.2.2 it easily follows that $E\{h(X - x)|X \geqslant x\}$, $x \in \mathbb{R}$ (where the expectation is defined arbitrarily on the set of points x at which $P\{X \geqslant x\} = 0$) determines the distribution of X.

Example 2   Let h be an increasing right continuous real-valued function on $\mathbb{R}$ such that $h(0) = h(1)$ and X be a nonnegative random variable with distribution having support $[0,3/2]$. Then, irrespective of the behavior of the h and the d.f. of X on $(1, \infty)$, we find that $E\{h(X - x)|X \geqslant x\}$, $x \in \mathbb{R}$ does not determine the distribution of X. (Note that the truncated distribution $P\{X \leqslant x|\frac{1}{2} < X < 1\}$, $x \in (\frac{1}{2}, 1)$ is not determined by the set of conditional expectations.) However, interestingly enough, if $h((3/2)-) > h(x)$ for all $x < 3/2$ and $P\{X = 3/2\} = 0$, then by Theorem 5.2.4, $E\{h(X)|X \geqslant x\}$, $x \in \mathbb{R}$ (or its modified version $E\{h(X)|X \geqslant x\} - h(x-)$, $x \in \mathbb{R}$) determines the distribution of X. (It is worth pointing out at this stage that if we take, for instance, $h(1) = h(3/2)$ in place of $h(0) = h(1)$, then the latter set of conditional expectations do not determine the truncated distribution $P\{X \leqslant x|X \geqslant 1\}$, $x \in [1, \infty)$, illustrating that Theorem 5.2.4 does not remain valid if the assumption that $h(b-) > h(x)$ for all $x \in (\alpha, b)$ in it is dropped.)

It is evident from the results of Theorem 5.2.4 that, under a fairly general hypothesis, the conditional expectations $E\{h(X)|X \geqslant x\}$ determine the distribution of X. However, the situation is not the same for the conditional expectations $E\{h(X - x)|X \geqslant x\}$. In this latter case, we can have only a partial characterization result, though, in general, under some weak conditions, there exists an integral representation, in terms of extreme points, for a distribution with the given conditional expectations. (See 5.4.8(v), for some details of the representation.) The characterization result is given by the following.

5.2.6   Theorem

Let X be a nonnegative random variable with $P\{X = 0\} < 1$ and h be a monotonic right continuous function on $\mathbb{R}_+$ such that $E(|h(X)|) < \infty$ and $E(h(X)) \neq h(0)$. Then

$$E\{h(X - x)|X \geqslant x\} = E(h(X)), \quad x \in \mathbb{R}_+ \quad \text{with} \quad P\{X \geqslant x\} > 0 \quad (5.2.8)$$

if and only if either h^* is nonarithmetic and X is exponential, or for some $\lambda > 0$, h^* is arithmetic with span λ and $P\{X \geqslant n\lambda + x\} = P\{X \geqslant x\}(P\{X \geqslant \lambda\})^n, n \in \mathbb{N}_0, x \in \mathbb{R}_+$, where

$$h^*(x) = \begin{cases} (h(x) - h(0))/(E(h(X)) - h(0)), & x \geqslant 0 \\ 0, & x < 0 \end{cases}$$

(We define h^* to be arithmetic or nonarithmetic according to whether the measure determined by it on $\mathbb{R}$ is arithmetic or nonarithmetic.)

Proof   Note that (5.2.8) is equivalent to, writing $\overline{F}(x) = P\{X \geqslant x\}, x \in \mathbb{R}_+$,

$$\int_{\mathbb{R}_+} \overline{F}(x + y)\mu_{h^*}(\mathrm{d}y) = \overline{F}(x), \quad x \in \mathbb{R}_+,$$

where μ_{h^*} is the measure determined by h^*. Theorem 2.2.2 then establishes the 'only if' part of the assertion. As the 'if' part of the assertion is trivial, we then have the theorem. ∎

5.2.7   Remarks

(i) Let s^* be a real-valued Borel measurable function on $\mathbb{R}$ defined for a random variable X with $E(X^+) < \infty$ such that $s^*(x) = E\{X - x|X > x\}$ whenever $P\{X > x\} > 0$. Clearly $s^*(x) = s(x+)$ on $(-\infty, b)$, where b is the right extremity of the d.f. of X, and hence Theorem 5.2.2 with $F(x-)$ and $F(y-)$ replaced respectively by $F(x)$ and $F(y)$ is valid when s is replaced by s^*. Consequently s^* determines the corresponding distribution uniquely. However, it may be noted that if $F(0-) = 0$, then F is uniquely determined by the restriction of s to $[0, \infty)$ while this is not so for s^*. In the latter case, the unique determination is possible only under some additional restriction such as $s^*(0) = E(X)$ as in Shanbhag's (1970) characterization of the exponential distributions.

(ii) Theorem 2.2.3 yields easily that if s^{**} is a positive left-continuous measurable function on $\mathbb{R}$ and F is a d.f. such that

$$\frac{1 - F(x-)}{1 - F(y-)} = \frac{s^{**}(y)}{s^{**}(x)} \exp\left\{ -\int_y^x (s^{**}(z))^{-1}\,dz \right\}$$

for all x, y such that $-\infty < y < x < b$, where b is the right extremity of F, then it is not necessary that s^{**} be an m.r.l. function.

(iii) The analogue of Theorem 5.2.5 with $E\{h(X)|X \in (z, \beta)\}$ in place of $E\{h(X)|X \in [z, \beta)\}$ or when h is taken as decreasing instead of increasing is obvious.

(iv) Glänzel *et al* (1984) have characterized Pearson-type distributions via versions of Theorem 5.2.5.

(v) If X is a nonnegative random variable with $P\{X = 0\} = 0$ and h is an increasing or decreasing real function on $[1, \infty)$ for which, for some points $x_1, x_2 > 1$ such that $\log x_1 / \log x_2$ is irrational, we have $h(x + x_i) \neq h(x_i-)$, $i = 1, 2$ for each $x > 0$, then, according to Theorem 5.2.6, the condition that for almost all $a > 0$

$$E\{h(X/a)|X \geqslant a\} = \text{constant}$$

implies that X is Pareto. This latter result is an improved version of Theorem 6.2 in Rao (1983) and was reported in Rao and Shanbhag (1986).

(vi) A somewhat different version of Theorem 5.2.6 without assuming h to be monotonic was established by Klebanov (1980). This result involves a certain assumption based on a Fourier transform and, is for a monotonic h, substantially weaker than Theorem 5.2.5; see, also, Shimizu (1979) for a restrictive result.

(vii) If Y is a positive random variable, then defining

$$h(x) = P\{Y \leqslant x\}, \quad x \in \mathbb{R}_+,$$

we can obtain from Theorem 5.2.6 Shimizu's (1978) and Ramachandran's (1979) characterizations of the exponential and geometric distributions, via the strong memoryless property. A version of the characterization of the geometric distributions has also appeared in Zijlstra (1983).

Fosam and Shanbhag (1994) have recently shown that the results in 5.2.7(vii), in turn, give the following two interesting results as immediate corollaries.

5.2.8   Corollary

Let $n \geqslant 2$, $1 \leqslant k \leqslant n - 1$ and $Y_1, \ldots, Y_n$ be independent positive random variables such that $P\{Y_i \leqslant y\} > 0$ for each $y > 0$ and $i = 2, \ldots, n$. Then

$$P\{Y_i - Y_{i+1} > y|Y_1 > Y_2 > \cdots > Y_n\}$$
$$= P\{Y_i > y|Y_1 > Y_2 > \cdots > Y_i\}, y > 0; i = 1, 2, \ldots, k \quad (5.2.9)$$

(where the right-hand side of the identity is to be read as $P\{Y_1 > y\}$ for $i = 1$) if and only if Y_i, $i = 1, 2, \ldots k$ are exponential random variables. (The result also holds if '>' in (5.2.9) is replaced by '$\geqslant$'.)

Proof   Defining for each $i = 1, 2, \ldots, k$, $X_1^{(i)}$ and $X_2^{(i)}$ to be independent positive random variables with distribution functions

$$P\{Y_i \leqslant x | Y_1 > Y_2 > \cdots > Y_i\}, \quad x \in \mathbb{R}_+$$

and

$$P\{Y_{i+1} \leqslant x | Y_{i+1} > Y_{i+2} > \cdots > Y_n\}, \quad x \in \mathbb{R}_+,$$

we can, on using the result of 5.2.7(vii) in the case when the distribution of Y is nonarithmetic, see that (5.2.9) is valid if and only if the distribution functions $P\{Y_i \leqslant x | Y_1 > Y_2 > \cdots > Y_i\}$, $x \in \mathbb{R}_+$ are those corresponding to exponential distributions for $i = 1, 2, \ldots, k$. It is an easy exercise to see inductively that we have the distribution functions $P\{Y_i \leqslant x | Y_1 > Y_2 > \cdots > Y_i\}$, $x \in \mathbb{R}_+$, for $i = 1, 2, \ldots, k$ as those corresponding to exponential distributions if and only if the random variables $Y_1, \ldots, Y_k$ are exponential. Hence we have the corollary. ∎

5.2.9   Corollary

Let $n \geqslant 2$, $1 \leqslant k \leqslant n - 1$ and $Y_1, \ldots, Y_n$ be independent nonnegative integer-valued random variables such that $P\{Y_1 \geqslant 1\} > 0$, $P\{Y_i = 1\} > 0$ for $2 \leqslant i \leqslant k + j$, and $P\{Y_i = 0\} > 0$ for $k + j < i \leqslant n$ for some $j \geqslant 1$. Then

$$P\{Y_i - Y_{i+1} \geqslant y | Y_1 \geqslant Y_2 \geqslant \cdots \geqslant Y_n\}$$
$$= P\{Y_i \geqslant y | Y_1 \geqslant Y_2 \geqslant \cdots \geqslant Y_i\} y = 0, 1, \ldots; i = 1, 2, \ldots, k \quad (5.2.10)$$

(where the right-hand side of the identity is to be read as $P\{Y_1 \geqslant y\}$ for $i = 1$) if and only if Y_i, $i = 1, 2, \ldots, k$ are geometric random variables.

Proof   The result follows on using essentially the argument in the proof Corollary 5.2.8, but obviously with the result in 5.2.7(vii) when the distribution of Y is arithmetic in place of that when the distribution is nonarithmetic. ∎

Corollary 5.2.8 provides, as revealed by Fosam and Shanbhag (1994), a result stronger than that appearing in Liang and Balakrishan (1992); Corollary 5.2.8 is obviously a discrete version of this corollary. Fosam and Shanbhag (1994) have also made some further relevant observations on these corollaries.

We conclude the present section by giving a stability theorem relative to the m.r.l. function. The general stability theorem to appear in Section 5.4 subsumes this result (as well as a stability result corresponding to the characterizations based on $E\{h(X) | X \geqslant x\}$). However, in view of the importance of the m.r.l. function in practice, we consider this special case separately here. This result may be useful in problems related to approximating the m.r.l. function by that associated with some standard distribution where the m.r.l. function is particularly appealing and of easily manageable form, or to fitting a distribution to data on the basis of a rough estimate of the corresponding m.r.l. function.

5.2.10   Theorem

Let $\{F_n : n = 1, 2, \ldots\}$ be a sequence of d.f.'s on $\mathbb{R}$ such that

$$\lim_{x \to -\infty} \limsup_{n \to \infty} F_n(x) = 0, \ 1 - F_n(x) \leqslant G(x) \quad \text{for all } n \geqslant 1 \quad \text{and} \quad x > 0$$

$$(5.2.11)$$

where G is a Borel-measurable function integrable with respect to Lebesgue measure on $(0, \infty)$. Let $\{s_n : n = 1, 2, \ldots\}$ denote the corresponding sequence of m.r.l. functions (which clearly is well-defined in view of the latter of the conditions in (5.2.11)) and let F and s be a d.f. on $\mathbb{R}$ such that $\int_0^\infty y F(\mathrm{d}y) < \infty$ and the corresponding m.r.l. function respectively. Denote by b the right extremity of F. Then $s_n(x) \to s(x)$ as $n \to \infty$ for every continuity point $x \in (-\infty, b)$ of s (and hence that of F) if and only if $\{F_n\}$ converges weakly to F.

Proof   Assume that $\{F_n\}$ converges weakly to F. If $x \in (-\infty, b)$ is a continuity point of F, then the identity $(1 - F_n(x-))s_n(x) = \int_x^\infty (1 - F_n(y))\,\mathrm{d}y$, $n \geqslant 1$ implies in view of the dominated convergence theorem (which we can obviously apply to the right-hand side of the identity because F_n's are d.f.'s satisfying the boundedness assumption in (5.2.11)) that $\lim_{n \to \infty} s_n(x)$ exists and equals $s(x)$. Hence we have the 'if' part of the proposition. To establish the 'only if' part, we assume that $\{F_n\}$ does not converge weakly to F, while the corresponding sequence $\{s_n\}$ does converge weakly to s on $(-\infty, b)$. Then, in view of Helly's first theorem (cf. Lukacs 1970, pp. 44–45) and the first condition in (5.2.11), it follows easily that there exists a subsequence $\{F_{n_i}\}$ of $\{F_n\}$ converging weakly to an increasing, nonnegative, bounded and right-continuous function F^* different from F satisfying $\lim_{x \to -\infty} F^*(x) = 0$. Applying the dominated convergence theorem as in the above case, we see that $\lim_{i \to \infty} s_{n_i}(x)$ exists and equals $\left\{ \int_x^\infty (1 - F^*(y))\,\mathrm{d}y \right\} \{1 - F^*(x-)\}^{-1}$ at each continuity point x of F^* such that $F^*(x) < 1$. Since $s_n(x) \to s(x)$ at each continuity point $x < b$ of F, it then easily follows that

$$s(x) = \left\{ \int_x^\infty (1 - F^*(y))\,\mathrm{d}y \right\} \{1 - F^*(x-)\}^{-1} \quad \text{for all } x \in (-\infty, b^*) \cdot (5.2.12)$$

where $b^* = \min(b, \sup\{x : F^*(x) < 1\})$. Hence F^* should be a proper d.f. with an m.r.l. function. In view of Theorem 5.2.2, (5.2.12) implies moreover that $b = b^*$ and consequently $F(x) \equiv F^*(x)$. The contradiction to the assumption that F is different from F^* establishes the validity of the 'only if' part and hence we have the theorem. ∎

5.2.11   Corollary

Theorem 5.2.10 remains valid with s, $\{s_n : n \geqslant 1\}$ replaced by s^*, $\{s_n^* : n \geqslant 1\}$, where s^* and $\{s_n^*\}$ are as defined in 5.2.7(i).

Proof   The corollary follows trivially from the above proposition and the fact that for every $x \in \mathbb{R}$, we have $(1 - F_n(x))s_n^*(x) = (1 - F_n(x-))s_n(x)$ and $(1 - F(x))s^*(x) = (1 - F(x-))s(x)$. ∎

5.2.12   Remarks

(i) The following two examples illustrate that the stability theorem does not remain valid if the tightness assumption as given by (5.2.11) is omitted.

Example 1   Let $\{F_n\}$ be such that $F_n((-n)-) = 0$, $F_n(-n) = 1 - (d/n)$ and

$$F_n(x) = 1 - \frac{d}{n} + \frac{d}{n}(1 - e^{-\lambda x})$$

if $x \geqslant 0$, where $0 < d < 1$ and $\lambda > 0$ and let F be the exponential d.f. with mean λ^{-1}. Note that in this example (5.2.11) is not met. It is easy to check that here $s_n \to s$ pointwise on $\mathbb{R}$, while $\{F_n\}$ does not converge to F at any point of $(0, \infty)$.

Example 2   Consider $\{F_n\}$ such that for each $n \geqslant 1$,

$$F_n(0) - F_n(0-) = 1 - (d/n) \quad \text{and} \quad F_n(n) - F_n(n-) = (d/n)$$

with $0 < d < 1$, and let F be the d.f. of the distribution that is degenerate at 0. Here $b = 0$, and on $(-\infty, 0)$, $s(x) = -x$ and $s_n(x) = d - x$ for each $n \geqslant 1$. Consequently in this example at no point of $(-\infty, 0)$ do the sequence $\{s_n\}$ converges to s. However, here $\{F_n\}$ converges weakly to F on $\mathbb{R}$. Clearly, in this example also, (5.2.11) is not met.

(ii) In view of Theorem 4.3.10, one can arrive at a stability theorem of a somewhat different nature relative to the characterization obtained in Theorem 5.2.6. Suppose we replace the right-hand side of (5.2.8) by $E(h(X))(1 - \beta(x))$, where $|\beta(x)| \leqslant K \exp\{-\eta x\}$ with $K|E(h(X)) \bullet (E(h(X)) - h(0))^{-1}| < 1$ and $\eta > 0$. Then, this equation implies that if h^* is nonarithmetic, we have for some $\alpha > 0$

$$P\{X \geqslant x\} \propto (1 - \beta^*(x))\exp\{-\alpha x\}, \quad x \in \mathbb{R}_+,$$

and if h^* is arithmetic with some span $\lambda > 0$, we have for some $\theta > 0$ (with [.] as the integral part)

$$P\{[X/\lambda] \geqslant n\} = P\{X \geqslant n\lambda\} \propto (1 - \beta^*(n\lambda))\theta^n, \quad n = 0, 1, 2, \ldots,$$

where β^* is such that $|\beta^*(x)| \leqslant K\gamma \exp\{-\eta x\}$ for some $\gamma > 1$[†]. (Using (4.3.23) or otherwise, one could give an explicit expression in terms of an operator for $\beta^*(x)$, or give a sharper bound for $|\beta^*(x)|$, when x is sufficiently large.) A stability version of the characterization of the Pareto distribution of Remark 5.2.7(v) is also now easy to see. Shimizu (1980) and Gu and Lau (1984) have studied stability results of the type considered here.

[†] We have given a somewhat restrictive version of the stability result, for simplicity.

5.3   HAZARD MEASURE: UNIVARIATE CASE

The hazard measure of a probability distribution on $\mathbb{R}$ (i.e. in the univariate case) is defined as follows.

5.3.1   Definition

Let F be a probability distribution function (d.f.) on $\mathbb{R}$ and ν_F be a measure on $\mathbb{R}$ such that for each Borel set B

$$\nu_F(B) = \int_B \frac{F(\mathrm{d}x)}{1 - F(x-)}.$$

Then ν_F is called the hazard measure relative to F.

From Lemma 1.3.1, it is evident that $\nu_F(\mathbb{R}) < \infty$ if and only if the right extremity of F is its discontinuity point. Observe that if F is absolutely continuous with respect to the Lebesgue measure so also is ν_F. In that case, the Radon–Nikodym derivative of ν_F with respect to the Lebesgue measure is equal to the standard hazard function almost everywhere with respect to the Lebesgue measure on $\{x : F(x) < 1\}$. Further, if F is purely discrete, then the Radon–Nikodym derivative of ν_F with respect to the counting measure on $\mathbb{R}$ equals the hazard function in the discrete case. In these two special cases, especially for a distribution concentrated on $\mathbb{R}_+$, the concept of hazard measure is well known (at least through its Radon–Nikodym derivative) and is widely used in practice in areas such as reliability, insurance statistics and biometrics to mention just a few of the relevant branches of applied probability and statistics. The general definition of a hazard measure given here, however, is of recent origin and was first introduced in this form by Kotz and Shanbhag (1980).

In what follows, we give two theorems. The first of these gives a representation for a d.f. in terms of the hazard measure and the second gives a partial stability result concerning the measure. The representation in the first theorem under some restrictions including, especially, the one that the distribution does not possess a singular continuous component was given earlier by Cox (1972) (see also Barlow and Proschan (1965) and (1975) for a corollary and Jacod (1975) for a representation under slightly different conditions).

5.3.2   Theorem

Let ν_F be as defined above and ν_F^c be the continuous (nonatomic) part of ν_F and let $H_c(x) = \nu_F^c((-\infty, x])$. Denote by b the right extremity of F. Then $b = \sup\{x : \nu_F((x, x + \delta)) > 0 \text{ for some } \delta > 0\}$, and the survivor function $S(x) \equiv 1 - F(x-)$ is given by

$$S(x) = \left[\prod_{x_r \in D_x} (1 - \nu_F(\{x_r\})) \right] \exp\{-H_c(x)\}, \quad x < b, \qquad (5.3.1)$$

where D_x is the set of all points $y \in (-\infty, x)$ such that $\nu_F(\{y\}) > 0$. (If we take $e^{-\infty} = 0$, then (5.3.1) trivially implies the validity of the representation for all $x \in \mathbb{R}$.)

Proof   The first assertion of the theorem is trivial. To prove the second assertion, observe that

$$H_c(x) = \int_{-\infty}^{x} f(G^*(z)) G_c(dz), \quad x < b,$$

with G_c as in Lemma 1.3.2 when $G(z) = F(z)$, $z \in \mathbb{R}$, $G^*(z) = F(z-)$, $z \in \mathbb{R}$, and $f(z) = 1/(1 - z)$ if $z < 1$ and $= 0$ otherwise. Hence, applying Lemma 1.3.2, we get

$$H_c(x) = \log \left\{ \frac{1}{S(x)} \prod_{x_r \in D_x} \left(\frac{1 - F(x_r)}{1 - F(x_r-)} \right) \right\}, \quad x < b.$$

The assertion is now immediate because $\{(1 - F(x_r))/(1 - F(x_r-))\} = 1 - \nu_F(\{x_r\})$, $x_r- \in D_x$. ∎

5.3.3   Corollary

If $-\infty < \alpha \leqslant \infty$ and the restriction of F to $(-\infty, \alpha)$ is continuous (i.e. if ν_F is continuous or nonatomic), then

$$1 - F(x) = \exp\{-H(x)\} \quad \text{for all } x \in (-\infty, \alpha),$$

where $H(x) = \nu_F((-\infty, x])$ and we define $\exp\{-\infty\} = 0$.

5.3.4   Theorem

Let $\{F_n : n \geqslant 1\}$ be a sequence of d.f.'s on $\mathbb{R}$ and let F be a continuous d.f. on $\mathbb{R}$. Also, let b denote the right extremity of F. Then

$$F_n(x) \to F(x) \quad \text{as } n \to \infty \quad \text{for all } x \in \mathbb{R} \tag{5.3.2}$$

if and only if

$$\nu_{F_n}((-\infty, x]) \to \nu_F((-\infty, x]) \quad \text{as } n \to \infty \quad \text{for all } x \in (-\infty, b), \tag{5.3.3}$$

where ν_{F_n} and ν_F are hazard measures relative to F_n and F respectively.

Proof   First assume that (5.3.2) is valid. Let $x \in (-\infty, b)$ and $n_0(x)$ be a positive integer such that $F_n(x) < 1$ for all $n \geqslant n_0(x)$. For the x considered and all $n \geqslant n_0(x)$, we have using the continuity of F

$$|\nu_{F_n}((-\infty, x]) - \nu_F((-\infty, x])| \leqslant \left| \int_{-\infty}^{x} \frac{F_n(dy)}{1 - F_n(y-)} - \int_{-\infty}^{x} \frac{F_n(dy)}{1 - F(y)} \right| + \Delta_n(x)$$

$$\leqslant \int_{-\infty}^{x} \frac{|F_n(y-) - F(y-)|}{(1 - F_n(y-))(1 - F(y))} F_n(dy) + \Delta_n(x)$$

$$\leqslant \frac{\left\{ \sup_{y \leqslant x} |F_n(y) - F(y)| \right\}}{(1 - F_n(x))(1 - F(x))} + \Delta_n(x), \tag{5.3.4}$$

where

$$\Delta_n(x) = \left| \int_{-\infty}^x \frac{F_n(\mathrm{d}y)}{1 - F(y)} - \int_{-\infty}^x \frac{F(\mathrm{d}y)}{1 - F(y)} \right|.$$

In view of Problem 4 on page 160 of Breiman (1968) and Helly's second theorem (see for example, Lukacs, 1970, p. 45), we conclude from (5.3.4) that $v_{F_n}((-\infty, x]) \to v_F((-\infty, x])$ as $n \to \infty$ and hence we have the 'only if' part of the theorem.

To establish the 'if' part of the theorem, we assume that (5.3.3) is valid but not (5.3.2). Using Helly's first theorem (see, for example, Lukacs, 1970, pp. 44–45), we have the existence of a right-continuous nonnegative increasing function G different from F to which some subsequence $\{F_{n_i} : i \geq 1\}$ of $\{F_n\}$ converges weakly. Since

$$v_{F_{n_i}}((-\infty, x]) = F_{n_i}(x) + \int_{-\infty}^x \frac{F_{n_i}(y-)F_{n_i}(\mathrm{d}y)}{1 - F_{n_i}(y-)}, \quad -\infty < x < \infty, \quad i \geq 1.$$

and (5.3.3) is valid, we get consequently the existence of a nonnegative increasing right-continuous function H on $(-\infty, b)$ such that

$$G(x) + H(x) = v_F((-\infty, x]) \quad \text{for all} \ -\infty < x < b, \tag{5.3.5}$$

which implies, in view of the continuity of F, that the restriction of G to $(-\infty, b)$ is continuous. From (5.3.5) it is also evident that $\lim_{x \to -\infty} G(x) = 0$. Using the validity of (5.3.3) and the argument of the 'only if' part (which we can even when G is not proved to be a proper continuous d.f.) we can conclude that

$$v_F((-\infty, x]) = \lim_{i \to \infty} v_{F_{n_i}}((-\infty, x]) = \int_{-\infty}^x \frac{G(\mathrm{d}y)}{1 - G(\mathrm{d}y)}, \quad -\infty < x < \min(b, b^*),$$

$$\tag{5.3.6}$$

where $b^* = \sup\{x : G(x) < 1\}$. Since both F and G are continuous on $(-\infty, b)$, $F(b-) = 1$ and (5.3.6) is valid, Corollary 5.3.3 yields easily that $F(x) \equiv G(x)$. We have thus a contradiction, and hence the validity of the 'if' part of the theorem follows. This completes the proof of the theorem. ∎

5.3.5 Remarks

(i) Theorem 5.3.2 implies that the hazard measure v_F relative to F uniquely determines F.

(ii) Let $c \in \mathbb{R}$. If F is a d.f. on $\mathbb{R}$ such that $F(c-) = 0$, $\int_c^\infty x \, \mathrm{d}F(x) < \infty$, and b is the right extremity of F, then the m.r.l. function s corresponding to F equals the reciprocal of the hazard function of the d.f.

$$G(x) = \int_c^x (1 - F(y)) \, \mathrm{d}y \Big/ \int_c^\infty (1 - F(y)) \, \mathrm{d}y, \quad x \in [c, \infty),$$

almost everywhere with respect to the Lebesgue measure on $[c, b)$. Meilijson (1972) essentially used this fact to arrive at his representation for the distributions concentrated on $\mathbb{R}_+$ with $b = \infty$ and finite mean, in terms of the m.r.l. function from the representation for an absolutely continuous distribution (with respect to the Lebesgue measure) on $\mathbb{R}_+$ in terms of the corresponding hazard function.

(iii) Using the Lebesgue decomposition theorem (see, for example, Burrill (1972), p. 130), one can easily see from (5.3.1) or otherwise that if F and b are as in Theorem 5.3.2, then for every $x \in (-\infty, b)$,

$$(1 - F(x)) = (1 - F_d(x))(1 - F_s(x))(1 - F_{ac}(x))$$

where F_d, F_s, and F_{ac} are respectively discrete, singular continuous and absolutely continuous (possibly improper) d.f.'s Moreover, the above decomposition is unique in the sense that the restrictions of F_d, F_s, and F_{ac} to $(-\infty, b)$ are uniquely determined.

(iv) The following two examples illustrate respectively that the 'if' part and the 'only if' part of Theorem 5.3.4 do not remain valid if the assumption that F is continuous is dropped. It is evident that even when we replace 'for all $x \in$' appearing in (5.3.2) and (5.3.3) by 'for every continuity point x of F in' this remains the case.

Example 1   Choose $0 < b < \infty$ and $\lambda > 0$ and define F such that $F(x) = 0$ if $x < 0$, $= 1$ if $x \geqslant b$, and $F(x) = 1 - \exp(-\lambda x)$ if $0 \leqslant x < b$. Also, for each $n \geqslant 1$, define F_n such that $F_n(x) = 0$ if $x < 0$, $= 1$ if $x \geqslant b\beta$, and $= 1 - \exp(-\lambda x)$ if $0 \leqslant x < b\beta$, where $\beta > 1$. Observe that here b is the right extremity of F and $v_{F_n}((-\infty, x]) \to v_F((-\infty, x])$ for every $x \in (-\infty, b)$. However, in this example, at no point of $(b, b\beta)$ does the sequence $\{F_n\}$ converge to F.

Example 2   Let F be the d.f. corresponding to a nondegenerate discrete probability distribution whose support is a subset of the set of natural numbers and a be its left extremity. Let $\{F_{nj} : n, j = 1, 2, \ldots\}$ be a sequence of continuous d.f.'s such that each F_{nj} satisfies $F_{nj}(j - (1/n)) = 0$ and $F_{nj}(j) = 1$. Define

$$F_n(x) = \sum_{j=1}^{\infty} \{F(j) - F(j - 1)\} F_{nj}(x), \quad -\infty < x < \infty, \quad n \geqslant 1.$$

It is easy to check that here the sequence $\{F_n\}$ converges pointwise to F but at no point $x \in [a, \infty)$ does the sequence $\{v_{F_n}((-\infty, x])\}$ converge to $v_F((-\infty, x])$.

(v) Clearly Theorem 5.3.4 with (5.3.2) replaced by the condition that $F_n(x) \to F(x)$ as $n \to \infty$ for all $x \in (-\infty, b)$ and/or (5.3.3) replaced by the condition that $v_{F_n}((-\infty, x]) \to v_F((-\infty, x])$ as $n \to \infty$ for all $x \in \mathbb{R}$ remains valid.

(vi) Let F, $\{F_n : n \geqslant 1\}$ be a sequence of d.f.'s and b be the right extremity of F. Example 3 given below reveals that if $v_{F_n}((-\infty, x]) \to v_F((-\infty, x])$ as $n \to \infty$ for every $x \in \mathbb{R}$, then it does not even follow that $\{F_n\}$ converges weakly to F on $(-\infty, b)$, i.e., at each continuity point of F in $(-\infty, b)$.

Example 3   Let $0 < \alpha < 1 - e^{-1}$, $0 < b < \infty$ and F be the d.f. for which $F(0) - F(0-) = \log(1-\alpha)^{-1}$ and $F(b) - F(b-) = 1 + log(1-\alpha)$. Let $\{G_n : n \geqslant 1\}$

be a sequence of continuous d.f.'s such that for each $n \geqslant 1$, $G_n(-1/n) = 0$ and $G_n(0) = 1$ and G be the d.f. degenerate at b.

Define $F_n(x) = \alpha G_n(x) + (1 - \alpha)G(x)$, $n \geqslant 1$, $x \in \mathbb{R}$. Clearly b is the right extremity of F and $\nu_{F_n}((-\infty, x]) \to \nu_F((-\infty, x])$ as $n \to \infty$ for every $x \in \mathbb{R}$ but here $\{F_n\}$ does not converge weakly to F on $(0, b)$.

(vii) Kupka and Loo (1989) have revealed several interesting properties of the hazard measure ν_F in its present general form.

(viii) Stability results different from the one appearing in Theorem 5.3.4 have been dealt with by Shimizu (1986) and Brown (1983) amongst others. In particular, these authors address the problem of obtaining bounds, in terms of moments, for the Kolmogorov–Smirnov distance between a d.f. F and the exponential d.f. with mean the same as F, assuming that F is an absolutely continuous d.f. (with respect to the Lebesgue measure) concentrated on $\mathbb{R}_+$ and that ν meets certain constraints.

5.4 M.R.L. FUNCTION AND HAZARD MEASURE: MULTIVARIATE CASE

If $X = (X_1, \ldots, X_p)$ is a random vector, then the survivor function $P\{X \geqslant x\}$, $x \in \mathbb{R}^p$ satisfies

$$P\{X \geqslant x\} = P\{X_p \geqslant x_p\} \prod_{i=1}^{p-1} P\{X_i \geqslant x_i | X_{i+1} \geqslant x_{i+1}, \ldots, X_p \geqslant x_p\},$$

$$x = (x_1, \ldots, x_p) \in \mathbb{R}^p. \qquad (5.4.1)$$

In view of this, it follows that, provided $E(X_i^+) < \infty$, $i = 1, 2, \ldots, p$,

$$(E\{X_1 - x_1 | X \geqslant x\}, E\{X_2 - x_2 | X_2 \geqslant x_2, \ldots, X_p \geqslant x_p\}, \ldots, E\{X_p - x_p | X_p \geqslant x_p\}),$$

$$x \in \mathbb{R}^p \qquad (5.4.2)$$

determines the distribution of X. This in turn implies that under the stated condition $E\{X - x | X \geqslant x\}$, $x \in \mathbb{R}^p$ determines the distribution of X; the function $E\{X - x | X \geqslant x\}$, $x \in \mathbb{R}^p$ or any Borel measurable function on $\mathbb{R}^p$ that agrees with this function on $\{x \in \mathbb{R}^p : P\{X \geqslant x\} > 0\}$ may be referred to as the m.r.l. function corresponding to the random vector X. It is now an easy exercise to extend the representation of Theorem 5.2.2 to arrive at a representation for the survivor function of X in terms of the corresponding m.r.l. function. (Note that the representation can be given in terms of (5.4.2) avoiding some of the redundancies existing in the m.r.l. function.) Several authors including Jupp and Mardia (1982) have characterized specific distributions via the corresponding m.r.l. functions; these results are clearly obvious corollaries of the general representation or uniqueness theorem. (If we define the m.r.l. function with '>' in place of '$\geqslant$' as some authors do, then the uniqueness and representation theorems still hold in view of the analogue of (5.4.1) for that case.)

In view of (5.4.1), one can extend the characterizations of the univariate case based on $E\{h(X)|X \geqslant x\}$ and the hazard measure as well. Indeed, taking a clue from this, Shanbhag and Kotz (1987) suggested the collection of hazard measures corresponding to all the survivor functions appearing on the right-hand side of (5.4.1) to be a possible generalization of the definition of a univariate hazard measure to the multivariate case. This definition is more general than those appearing in Johnson and Kotz (1975) and Marshall (1975) and it removes some of the redundancies existing in the earlier definitions.

To illustrate how in the aforementioned cases (5.4.1) leads us immediately to results for distributions on $\mathbb{R}^p$ from those for distributions on $\mathbb{R}$, we give below two theorems (in which the undefined notations $x_{(p+1)}$, $X_{(p+1)}$ are used only for convenience and should be ignored, i.e. $s_p(\bullet, x_{(p+1)})$ should be read as $s_p(\bullet)$ and so on).

5.4.1   Theorem

Let $X = (X_1, \ldots, X_p)$ be a p-component random vector such that $E(X_i^+) < \infty$, $i = 1, 2, \ldots, p$ with m.r.l. function s and survivor function $\overline{F}$. Then, we have

$$
\overline{F}(x) = \begin{cases}
0 \quad \text{if } x_j > b_j^* \quad \text{for some } j \geqslant 1 \quad \text{and} \quad \leqslant p \\[2mm]
\displaystyle\prod_{i=1}^{p} \lim_{y_i \to -\infty} \frac{s_i(y_i, x_{(i+1)})}{s_i(x_{(i)})} \exp\left\{-\int_{y_i}^{x_i} \frac{dz}{s_i(z, x_{(i+1)})}\right\} \quad \text{if } x_j < b_j^* \\[3mm]
\hspace{5cm} \text{for all } j \geqslant 1 \quad \text{and} \quad \leqslant p,
\end{cases}
$$

where

$$
x_{(i)} = (x_i, \ldots, x_p),
$$
$$
s_i(x_{(i)}) = E\{X_i - x_i \mid X_{(i)} \geqslant x_{(i)}\}
$$

and

$$
b_i^* = \begin{cases}
\infty & \text{if } A_i \text{ is empty} \\
\inf\{y : y \in A_i\} & \text{otherwise}
\end{cases}
$$

with $A_i = \{y : \lim_{x_i \uparrow y} s_i(x_{(i)}) \text{ exists and equals } 0\}$ and $X_{(i)} = (X_i, \ldots, X_p)$.

5.4.2   Theorem

Let $v_F^{(i)}(\bullet \mid x_{(i+1)})$ denote the hazard measure of the conditional distribution of X_i given $X_{(i+1)} \geqslant x_{(i+1)}$ for each $i = 1, 2, \ldots, p$, where $X_{(i+1)}$ and $x_{(i+1)}$ are as defined in Theorem 5.4.1. Then the survivor function corresponding to F is represented by

$$
\overline{F}(x) = \prod_{i=1}^{p} \left\{ \left[\prod_{y_i \in D_i(x_{(i)})} \left\{ 1 - v_F^{(i)}(\{y_i\} \mid x_{(i+1)}) \right\} \right] \right.
$$
$$
\left. \times \exp[-v_F^{(c,i)}((-\infty, x_i] \mid x_{(i+1)})] \right\}, \quad x \in \mathbb{R}^p,
$$

and for a continuous F the representation is

$$\overline{F}(x) = \exp\left\{-\sum_{i=1}^{p} v_F^{(i)}((-\infty, x_i]|x_{(i+1)})\right\}, \quad x \in \mathbb{R}^p,$$

where $e^{-\infty}$ is defined to be zero, $D_i(x_{(i)})$ is the set of real points $y_i < x_i$ at which $v_F^{(i)}(\{y_i\}|x_{(i+1)})$ is positive, and $v_F^{(c,i)}(\bullet|x_{(i+1)})$ is the continuous (non-atomic) part of $v_F^{(i)}(\bullet|x_{(i+1)})$. Furthermore, if F is continuous and $\{F_n : n = 1, 2, \ldots,\}$ is a sequence of d.f.'s on $\mathbb{R}^p$, then using the same notation

$$v_{F_n}^{(i)}((-\infty, x_i]|x_{(i+1)}) \rightarrow v_F^{(i)}((-\infty, x_i]|x_{(i+1)})$$

for each x such that $\overline{F}(x) > 0$ and $i = 1, 2, \ldots, p$ if and only if $\{F_n\}$ converges to F.

In view of (5.4.1), Theorem 5.4.1 follows trivially from Theorem 5.2.2 and Theorem 5.4.2 follows easily from Theorems 5.3.2 and 5.3.4. (To obtain the last part of Theorem 5.4.2, note, in particular, that for every x such that $\overline{F}(x) > 0$, the conditional distributions of X_i given $X_{i+1} \geqslant x_{i+1}, \ldots, X_p \geqslant x_p$ are all continuous.)

Theorem 5.2.6 also has a natural extension to the multivariate case with $p \geqslant 2$ (where p is the number of variables as before) though here without some further condition a characterization is not assured. This is given by the following theorem.

5.4.3   Theorem

Let $S = \Pi_{i=1}^{p} S_i$ with $S_i = \mathbb{R}_+$ or $\mathbb{N}_0$ and X be a p-component random vector with distribution concentrated on S. Also, let h be a function on S satisfying

$$h(x) = \mu((0, x]) \quad x \in S,$$

where μ is a Lebesgue–Stieltjes measure on $\mathbb{R}^p$ that is concentrated on $(0, \infty)^p \cap S$. Then, provided $E(h(X)) \neq 0$ and the smallest closed subgroup of $\mathbb{R}^p$ containing supp $[\mu]$ equals that containing S, we have

$$E\{h(X - x)|X \geqslant x\} = E(h(X)), \quad x \in S \quad \text{with} \quad P\{X \geqslant x\} > 0 \quad (5.4.3)$$

if and only if

$$P\{X \geqslant x\} = E(\exp\{-\langle x, Y\rangle\}), \quad x \in S, \quad (5.4.4)$$

where Y is a random vector such that, for some $c > 0$,

$$\int_S \exp\{-\langle x, Y\rangle\}\mu(dx) = c \quad \text{a.s.}$$

Proof   Note that (5.4.3) is equivalent to

$$\int_{(0,\infty)^p \cap S} \overline{F}(x + y)\mu^*(dy) = \overline{F}(x), \quad x \in S, \quad (5.4.5)$$

where

$$\overline{F}(x) = P\{X \geqslant x\}$$

and

$$\mu^*(\bullet) = (E(h(X)))^{-1}\mu(\bullet).$$

Corollary 3.4.5 and the left continuity of $\overline{F}$ on $(0, \infty)^n$ imply that if (5.4.5) holds, then the equation in (5.4.4) holds at least on $(0, \infty)^n \cap S$. Appealing to (5.4.5), we can then see that (5.4.4) is valid and hence we have the 'only if' part of the theorem. The 'if' part of the theorem is trivial. ∎

5.4.4   Remarks

(i) Suppose Z is a positive p-component random vector with distribution concentrated on S. Then, considering

$$h(x) = P\{Z \leqslant x\}, \quad x \in S,$$

where S is as in the statement of Theorem 5.4.3, we can see that for any random vector X as in Theorem 5.4.3, provided $P\{X \geqslant Z\} > 0$ and the distribution of Z meets the condition in the statement of the theorem concerning the support of μ,

$$P\{X \geqslant Z + x | X \geqslant Z\} = P\{X \geqslant x\}, \quad x \in S \qquad (5.4.6)$$

if and only if (5.4.4) holds with $E(\exp\{-\langle Z, Y \rangle\})$ in place of

$$\int_S \exp\{-\langle x, Y \rangle\}\mu(dx).$$

(ii) Theorem 5.4.3 remains valid if $\mu((0, x])$ is changed to $\mu((0, x))$ provided, in (5.4.3) and (5.4.4), $X \geqslant x$ is replaced by $X > x$. In view of this, we have that in (i) above if we replace $P\{X \geqslant Z\} > 0$ by $P\{X > Z\} > 0$, then (5.4.6) with '>' in place of '$\geqslant$' holds if and only if (5.4.4) with $P\{X > x\}$ in place of $P\{X \geqslant x\}$, and Y as in 5.4.4, (i) above, is valid. This latter result gives as a corollary the extension of Ramachandran's (1979) characterization of the exponential distributions given earlier by Davies and Shanbhag (1987) and Shanbhag (1991).

(iii) A stability version of Theorem 5.4.3 and of its variant in (ii) above hold. If we replace the right-hand side of (5.4.6) by $(1 - \beta(x))P\{X \geqslant x\}$, where β is such that $|\beta(x)| \leqslant KE(\exp\{-\langle x, Z^* \rangle\})$ for all $x \in S$ with $K < 1$ and Z^* meeting the requirements of Z of Corollary 4.3.7, then we have from the corollary referred to the validity of the identity only if

$$P\{X \geqslant x\} \propto (1 - \beta^*(x))E(\exp\{-\langle x, Y \rangle\}), \quad x \in S \qquad (5.4.7)$$

with β^* such that $|\beta^*(x)| \leqslant \gamma KE(\exp\{-\langle x, Z^* \rangle\})$ for all $x \in S$ and some $\gamma > 1$. (The stability version of the variant of the theorem, in (ii), is now obvious.) If we only ask for the behavior of the survivor function $P\{X \geqslant x\}$ of X, for sufficiently large $\|x\|$, then, in view of Theorem 4.3.10, a stronger result incorporating the information given by (4.3.23) or otherwise, can be obtained. Because of what is observed in Remark 4.3.11(x), it also follows that, under appropriate conditions, a stronger result than that given above holds if K is sufficiently small.

(iv) We are not dealing here with a stability version of Theorem 5.4.1 separately as it follows as a corollary to a general stability theorem to be given later.

(v) Several authors including Mukherjee and Roy (1986), Roy (1990) and Osaki and Li (1988) have characterized specific univariate or multivariate distributions via properties involving both the m.r.l. function and hazard measure of a distribution. (Here by the hazard measure in the multivariate case we mean the measure considered by Johnson and Kotz (1975) and Marshall (1975)).

For extending a hazard measure to the multivariate case, attempts have also been made through a different approach. In what follows we shall study the approach and point out some known results in this connection. Let, as in the previous discussion, F be a d.f. on $\mathbb{R}^p$, X be a p-component random vector with this distribution and $\overline{F}$ be the corresponding survivor function. Denote by P_F the measure determined by F on $\mathbb{R}^p$. Since, in the case $p > 1$, we can have an F such that $P_F(\{x : \overline{F}(x) = 0\}) > 0$ (e.g., if we take F to be continuous such that

$$P_F(\{x(=(x_1, \ldots, x_p)) : x_1 = -x_2\}) = 1,$$

we obtain $P_F(\{x : \overline{F}(x) = 0\}) = 1)$, the definition of a hazard measure of Section 5.3 is not extendable as it stands. However, if we restrict ourselves only to the set $\mathcal{C}$ (say) of distributions F for which $\overline{F}(\bullet) > 0$ almost surely $[P_F]$, the definition in question admits an obvious extension: for each $F \in \mathcal{C}$, we can define the measure ν_F on $\mathbb{R}^p$ as that given by

$$\nu_F(B) = \int_B \frac{1}{\overline{F}(x)} P_F(\mathrm{d}x) \quad \text{for all } B \in \mathcal{B}_p, \tag{5.4.8}$$

where $\mathcal{B}_p$ is the Borel σ-field of $\mathbb{R}^p$. Let $F \in \mathcal{C}$. If we denote by $\mathcal{D}_F$ the set of all d.f.'s G on $\mathbb{R}^p$ that are members of $\mathcal{C}$ for which $\nu_G(\bullet) = \nu_F(\bullet)$, then it follows easily that $\mathcal{D}_F$ is convex although not necessarily closed relative to weak convergence. Consider now the set of all d.f.'s on the compactified Euclidean space $\overline{\mathbb{R}}^p$, where $\overline{\mathbb{R}} = [-\infty, \infty]$. There exists a metrizable locally convex topological vector space of which this is a compact subset with the corresponding relative metric as a metric of weak convergence. Then, as a subset of this compact set, the closure $\overline{\mathcal{D}}_F$ of the set $\mathcal{D}_F$ is also compact. (For simplicity we abuse the notation slightly here and elsewhere in what follows in this chapter by using the same notation to denote a set of d.f.'s on $\overline{\mathbb{R}}^p$ that are concentrated on $\mathbb{R}^p$, and the set of the restrictions to $\mathbb{R}^p$, of its members.) Since $\overline{\mathcal{D}}_F$ is seen also to be convex, Theorem 1.3.9 implies that each $F^* \in \overline{\mathcal{D}}_F$ can be represented as the barycentre of a probability measure on $\overline{\mathcal{D}}_F$ that is concentrated on the set of extreme points of $\overline{\mathcal{D}}_F$. It may also be noted here that if we take, in place of $\mathcal{D}_F$, the set, $\mathcal{D}'_F$, of all d.f.'s G on $\mathbb{R}^p$ satisfying

$$P_G(B) = \int_B \overline{G}(x) v_F(\mathrm{d}x), \quad B \in \mathcal{B}_p, \tag{5.4.9}$$

then the above conclusion (i.e. the assertion with obviously the closure of $\overline{\mathcal{D}}'_F$ in place of $\overline{\mathcal{D}}_F$) still holds and we have $\mathcal{D}_F \subset \mathcal{D}'_F$.

Before discussing other properties of $\mathcal{D}_F$ and $\mathcal{D}'_F$, the following specific examples illustrating some important points are worth mentioning.

Example 1 Let

$$F(x) = \prod_{i=1}^{p} F_i(x_i), \quad x = (x_1, \ldots, x_p) \in \mathbb{R}^p,$$

where F_i are continuous d.f.'s on $\mathbb{R}$. Then, from Corollary 3.4.5 or its specialized version considered by Puri and Rubin (1974), we see that each member of $\mathcal{D}_F$ (or $\mathcal{D}'_F$) is given by

$$F^*(x) = \int_{\mathbb{R}^p} \prod_{i=1}^{p} \{1 - (1 - F_i(x_i))^{\lambda_i}\} v(\mathrm{d}\lambda), \quad x \in \mathbb{R}^p \tag{5.4.10}$$

where v is a probability measure concentrated on

$$\{\lambda(=(\lambda_1, \ldots, \lambda_p)) : \lambda_i > 0, \quad i = 1, 2, \ldots, p, \prod_{1}^{p} \lambda_i = 1\}.$$

(A stability version of this result follows easily from Theorem 4.3.10 which the reader may easily verify.) In view of this, it is clear that the extreme points of $\mathcal{D}_F$ are given precisely by $\prod_{i=1}^{p} \{1 - (1 - F_i(x_i))^{\lambda_i}\}$, $(x_1, \ldots, x_p) \in \mathbb{R}^p$ and we can consider (5.4.10) to be the integral representation for F^* in terms of the extreme points of $\mathcal{D}_F$. Looking at an arbitrary member F^* of $\mathcal{D}_F$ given by (5.4.10) for $p > 1$, one can also observe a curious property that if it has any $p - 1$ of its univariate marginals agreeing with those of F, then it equals F.

Example 2 Let $p \geqslant 2$, k be a real number and S be a countable subset of $\mathbb{R}^{p-1}$. Also let $\mathcal{G}$ denote the set of d.f.'s on $\mathbb{R}^{p-1}$ that are concentrated on S giving a positive probability mass to each point of S. For each $G \in \mathcal{G}$, let F_G denote the d.f. on $\mathbb{R}^p$ which is concentrated on $\{x : x \in \mathbb{R}^p, \Sigma_1^p x_i = k\}$ with

$$F_G(x_1, \ldots, x_{p-1}, \infty) = G(x_1, \ldots, x_{p-1}), (x_1, \ldots, x_{p-1}) \in \mathbb{R}^{p-1}$$

(in the usual notation). It is easily seen that here v_{F_G} are all (well defined and) identical. If we now consider $p \geqslant 4$ and any of the F_G's to be F, then it is clearly seen that the condition $F^* \in \mathcal{D}_F$ does not imply $F^* = F$ even if it is given that

F^* has all of its univariate marginals or bivariate marginals the same as those of F. However, for the F in question, the condition $F^* \in \mathcal{D}_F$ together with

$$F^*(x_1, \ldots, x_{p-1}, \infty) = F(x_1, \ldots, x_{p-1}, \infty), \quad (x_1, \ldots, x_{p-1}) \in \mathbb{R}^{p-1}$$

implies that $F^* = F$. Note also that here we have the set of extreme points of $\mathcal{D}_F$ to be empty and the set of extreme points of $\overline{\mathcal{D}}_F$ to be the closure (relative to weak convergence) of the set of the degenerate d.f.'s on $\overline{\mathbb{R}}^p$ that are concentrated on $\{x : x \in \mathbb{R}^p, \Sigma_1^p x_i = k\}$; clearly here the stipulation of the last example that each $F^* \in \mathcal{D}_F$ has an integral representation in terms of the extreme points of $\mathcal{D}_F$ is not valid.

In spite of certain isolated cases, such as that of Fréchet's distribution or of a d.f. F that satisfies for some $b \in \mathbb{R}^p$ the conditions $F(b) = 1$ and $\overline{F}(b) = P_F(\{b\}) > 0$, in which case F is characterized by v_F, it now follows that, in general, unless at least one of the $(p-1)$-variate marginals of the distribution (or something equivalent to it) is given, v_F does not characterize F. One might then be interested to know whether F is characterized by v_F given any one of the $(p-1)$-variate marginals. The question is as yet unanswered.

We now present the following results revealing some interesting properties of $\mathcal{D}_F$ and $\mathcal{D}'_F$.

5.4.5   Theorem

If $F^* \in \mathcal{D}'_F$ and, for each $i = 1, 2, \ldots, p$, we have (in the standard notation)

$$F^*(x_1, \ldots, x_{i-1}, \infty, x_{i+1}, \ldots, x_p) = F(x_1, \ldots, x_{i-1}, \infty, x_{i+1}, \ldots, x_p)$$

$$\text{for all } x_i \in \mathbb{R}, \quad j = 1, 2, \ldots, i-1, i+1, \ldots, p \tag{5.4.11}$$

then $F^* = F$. Furthermore, given an $F^* \in \overline{\mathcal{D}}'_F$, there exists a probability measure μ^* on the set of all d.f.'s, G, on $\overline{\mathbb{R}}^p$, such that

$$F^*(x) = \int_{\mathcal{K}} G(x)\mu^*(dG), \quad x \in \mathbb{R}^p \tag{5.4.12}$$

where $\mu^*(\mathcal{K}) = 1$ and $\mathcal{K}$ is the closure (relative to weak convergence) of the set of the d.f.'s $K_t(\bullet)$ for $t \in \text{supp}[F]$ such that $\overline{F}(t) > 0$ ($\overline{F}$ being the survivor function of F as stated before), where each of the $K_t(\bullet)$ is defined to be a d.f. on $\mathbb{R}^p$ such that it is the degenerate d.f. at t if $P_F(\{t\}) = \overline{F}(t)$ and the d.f. satisfying the following otherwise

$$K_t(x) = \frac{k(x, t)}{k(t, t)}, \quad x \in (-\infty, t] \tag{5.4.13}$$

where

$$k(y, t) = \Delta_t(y) + \sum_{n=1}^{\infty} \int_{(-\infty, y]} \int_{[y_1, t]} \cdots \int_{[y_{n-1}, t]} v_F(dy_n) \cdots v_F(dy_1) \tag{5.4.14}$$

with $\Delta_t(\bullet)$ as the d.f. degenerate at t. (The proof of the theorem asserts that $K_t(\bullet)$ is well defined.)

Proof   Let $\{X_{(n)} : n = 0, 1, \ldots\}$ be a temporally homogeneous Markov chain with state space $\mathbb{R}^p$ and transition function $P(\bullet, \bullet)$ such that for each Borel set B of $\mathbb{R}^p$

$$
P(x, B) = \begin{cases} P_F(B \cap [x, \infty))/\overline{F}(x) & \text{if } \overline{F}(x) > 0 \\ 1 & \text{if } \overline{F}(x) = 0 \quad \text{and} \quad x \in B \\ 0 & \text{if } \overline{F}(x) = 0 \quad \text{and} \quad x \notin B. \end{cases}
$$

Consider first the case with F as the distribution function of $X_{(0)}$. For each t with $\overline{F}(t) > P_F(\{t\})$, and $y \leqslant t$ and $n \geqslant 1$, define $k_n(y, t)$ to be the multiple integral as that under the summation sign on the right-hand side of (5.4.14). We have then for each t and n as above

$$
k_n(t, t) = E\{(\overline{F}(X_{(n-1)}))^{-1} I_{\{X_{(n-1)} \leqslant t\}}\}
$$

$$
\leqslant P\{X_{(n-1)} \leqslant t\}(\overline{F}(t))^{-1}
$$

$$
(= (\overline{F}(t))^{-1} E\{(\overline{F}(X_{(n-2)}))^{-1} P_F([X_{(n-2)}, t]) I_{\{X_{(n-2)} \leqslant t\}}\}
$$

$$
\leqslant \alpha_t P\{X_{(n-2)} \leqslant t\}(\overline{F}(t))^{-1} \quad \text{if } n \geqslant 2)
$$

$$
\vdots
$$

$$
\leqslant \alpha_t^{n-1} P\{X_{(0)} \leqslant t\}(\overline{F}(t))^{-1}
$$

$$
= \alpha_t^{n-1} F(t)(\overline{F}(t))^{-1}, \tag{5.4.15}
$$

where $\alpha_t = F(t)/\{F(t) + \overline{F}(t) - P_F(\{t\})\} < 1$. (To arrive at (5.4.15), we have noted amongst other things that $\alpha_t \geqslant P_F([y, t])/\overline{F}(y)$ for $y \leqslant t$.) (5.4.15) establishes that $K_t(\bullet)$ in the statement of the theorem is well defined. Now, for each d.f. F^* on $\mathbb{R}^p$ such that $F^* \in \mathcal{D}_F'$ and t as in (5.4.13), we have, in view of the relation $P_{F^*}(B) = \int_B F^*(x) \nu_F(\mathrm{d}x)$ with B as an arbitrary Borel set and F^* as the survivor function corresponding to F^*,

$$
\overline{F}^*(t) = \xi(t) + \sum_{n=0}^{\infty} (-1)^{(n+1)p} E\left\{ \frac{\xi(X_{(0)})}{\overline{F}(X_{(n)})} I_{\{X_{(0)} < X_{(1)} < \cdots < X_{(n)} < t\}} \right\}, \tag{5.4.16}
$$

where $\xi(y) = \overline{F}^*(y) - (-1)^p F^*(y-)$, $y \leqslant t$ with $F^*(y-) = P_{F^*}((-\infty, y))$. To see the validity of (5.4.16), note in particular that the expectation under the summation sign on its right-hand side with $\xi(X_0)$ deleted is bounded by $k_{n+1}(t, t)$ and

$$
E\left\{ \frac{F^*(X_{(0)}-)}{\overline{F}(X_{(n)})} I_{\{X_{(0)} < X_{(1)} < \cdots < X_{(n)} < t\}} \right\}
$$

$$
= E\left\{ \frac{\overline{F}^*(X_{(0)})}{\overline{F}(X_{(n+1)})} I_{\{X_{(0)} < X_{(1)} < \cdots < X_{(n+1)} < t\}} \right\}, \quad n = 0, 1, \ldots,
$$

and that $\overline{F}^*(X_{(0)})$, $F^*(X_{(0)}-)$ and $\xi(X_{(0)})$ are bounded random variables. Clearly the ξ in (5.4.16) is determined by the $(p-1)$-dimensional marginals of F^*. This, in turn, implies that if (5.4.11) is valid, then

$$\overline{F}^*(t) = \overline{F}(t) \quad \text{for each } t \text{ such that } \overline{F}(t) > P_F(\{t\}). \tag{5.4.17}$$

In view of the left continuity of $\overline{F}$ and $\overline{F}^*$ and the fact that

$$\{x : x \in \mathbb{R}^p, \overline{F}(x) = 0\} = \{x : x \in \mathbb{R}^p, \overline{F}^*(x) = 0\},$$

we can conclude that if (5.4.16) is valid, then we have $\overline{F}^* = \overline{F}$ or equivalently $F^* = F$. This establishes the first part of the theorem.

To establish the second part of the theorem, define

$$B^* = \{t : t \in \operatorname{supp}[F], F(t), \overline{F}(t) > 0\},$$

$$B_0 = \{t : t \in \mathbb{R}^p, \overline{F}(t) = P_F(\{t\}) > 0\},$$

and

$$B_m = \left\{ t : t \in B^*, \overline{F}(t) \geqslant P_F(\{t\}) + \frac{1}{m} \right\}, \quad m = 1, 2, \ldots .$$

Let $F^* \in \mathcal{D}'_F$. We shall get the required result by first showing that for the F^* in question (5.4.12) holds.

Consider now that $X_{(0)}$ has d.f. F_0 which is not necessarily the same as F, such that it is concentrated on $\{x : \overline{F}(x) > 0\}$. Define for $m = 1, 2, \ldots$

$$\eta_m = \inf\{n : X_{(n)} \in B^* \setminus B_m\}.$$

Essentially due to (5.4.15), it follows that $\eta_m < \infty$ for $m = 1, 2, \ldots$ almost surely. Taking F_0 to be such that $E(\overline{F}^*(X_{(0)})/\overline{F}(X_{(0)})) < \infty$, we see that

$$\left\{ \frac{\overline{F}^*(X_{(n)})}{\overline{F}(X_{(n)})}, \quad \sigma(X_{(m)}, m \leqslant n) \right\}$$

is a martingale, i.e., bounded on each B_m for $m \geqslant 1$, and hence

$$E\left(\frac{\overline{F}^*(X_{(0)})}{\overline{F}(X_{(0)})} \right) = \lim_{m \to \infty} E\left(\frac{\overline{F}^*(X_{(\eta_m)})}{\overline{F}(X_{(\eta_m)})} \right)$$

$$= E\left(\frac{\overline{F}^*(X_{(1)})}{\overline{F}(X_{(1)})} I_{\{X_{(1)} \in B_0\}} \right) + \lim_{m \to \infty} E\left(\frac{\overline{F}^*(X_{(\eta_m)})}{\overline{F}(X_{(\eta_m)})} I_{\{\eta_m \geqslant 2\}} \right)$$

$$= \lim_{m \to \infty} \int_{B^*} E(\overline{K}_t(X_{(0)})/\overline{F}(X_{(0)})) \mu_m(dt), \tag{5.4.18}$$

where $\{\mu_m : m = 1, 2, \ldots\}$ is a sequence of measures on $\mathbb{R}^p$ such that for each m, the measure μ_m is concentrated on $B_0 \cup B_m$ with

$$\mu_m(\{t\}) = \overline{F}^*(t), \quad t \in B_0$$

and for each Borel set B of $\mathbb{R}^p$,

$$\mu_m(B \cap B_m) = \int_{B \cap B_m} \left(\int_{B^* \setminus B_m} \frac{\overline{F}^*(y)}{\overline{F}(y)} P(t, \mathrm{d}y) \right) k(t, t) F(\mathrm{d}t).$$

Note that for each m, $\mu_m(B_m \cup B_0) = P_{F*}(B_m \cup B_0)$, implying that μ_m is a subprobability measure. It is also now clear that $\mu_m(\mathbb{R}^p) \to 1$ as $m \to \infty$. Suppose x is a point such that $\overline{F}(x) > 0$. If we now take F_0 such that it is degenerate at x, then (5.4.18) leads us to

$$\overline{F}^*(x) = \lim_{m \to \infty} \int_{B^*} \overline{K}_t(x) \mu_m(\mathrm{d}t). \tag{5.4.19}$$

Also, we have trivially the equation (5.4.19) to be valid for each x such that $\overline{F}(x) = 0$. Hence, (5.4.19) is valid for each $x \in \mathbb{R}^p$. Since $\mathcal{K}$ is compact, using Parthasarathy's (1967) Theorem 6.4 on page 45, amongst other things, it can then be easily seen that there exists a probability measure μ^* on $\mathcal{K}$ such that

$$\overline{F}^*(x) = \int_{\mathcal{K}} \overline{G}(x) \mu^*(\mathrm{d}G), \quad x \in \mathbb{R}^p.$$

(One can obtain the result, for example, by first seeing via Billingsley's (1968) Theorem 1.2 that there exists a measure μ^* such that the identity with $\overline{F}^*(x)$ and $\overline{G}(x)$ replaced by $P_{F*}(A)$ and $P_G(A)$ respectively, in obvious notation, holds for each closed set A.) Since $\overline{\mathcal{D}}'_F$ is the closure of $\mathcal{D}'_F$, a further application of the same argument then yields the validity of the second part of the theorem. ∎

5.4.6   Corollary

If F is continuous, then we have d.f. F^* on $\mathbb{R}^p$ to be a member of $\mathcal{D}'_F$ if and only if it has a representation

$$F^*(x) = \int_{\mathcal{K} \cap \mathcal{D}'_F} G(x) \mu(\mathrm{d}G), \quad x \in \mathbb{R}^p, \tag{5.4.20}$$

for some probability measure μ on $\mathcal{K} \cap \mathcal{D}'_F$. (In the present case, we also have $\mathcal{K} \cap \mathcal{D}'_F$ to be a G_δ set of the space of all d.f.'s on $\mathbb{R}^p$.)

Proof   In this case, if $G \in \overline{D}'_F$, then we have (5.4.9) with '=' replaced by '$\geqslant$' to be valid. Consequently, it follows easily by Fubini's theorem that $F^* \in \mathcal{D}'_F$ if and only if (5.4.20) holds. ∎

5.4.7   Corollary

The hazard measure ν_F jointly with the analogous hazard measures relative to all the univariate and multivariate marginals of F determines F uniquely. (This corollary can be verified by induction.)

5.4.8 Remarks

(i) There is some analogy between the proof of the second part of Theorem 5.4.7 and the proofs of Williams (1979) and Seneta (1981) of the Poisson–Martin integral representation theorem for a super regular vector corresponding to a nonnegative matrix.

(ii) The right-hand side of (5.4.16) could also be expressed in terms of an operator, giving a representation for $F^*(t)$ appearing somewhat similar to the Riemann representation for the solution of a hyperbolic partial differential equation.

(iii) The set $\mathcal{K}$ of Theorem 5.4.5 may also be defined alternatively taking $F(t)$, $\overline{F}(t) > 0$ in place of $\overline{F}(t) > 0$; the result obtained in that case is a slight improvement over Theorem 3 of Shanbhag and Kotz (1987). (Incidentally, the proof given by these latter authors for the second part of their theorem assumes implicitly that ν_F is concentrated on $\{x : F(x) > 0\}$; however, using a minor argument one could easily see that there is no loss of generality in making such an assumption.) It may also be pointed out here that in Corollary 1 and Remark 11 of Shanbhag and Kotz (1987), $\mathcal{D}_F$ should read $\mathcal{D}'_F$.

(iv) If we redefine for each $t \in B_0$, where B_0 is as in the proof of Theorem 5.4.5, K_t as, instead of the d.f. that is degenerate at t, the d.f. that satisfies

$$K_t(x) = \frac{\hat{k}(x, t)}{\hat{k}(t, t)}, \qquad x \in (-\infty, t],$$

where $\hat{k}(y, t)$ for each $y \leqslant t$ is given by the expression on the right-hand side of (5.4.14), but with $(-\infty, y]$, $[y_1, t]$, ..., $[y_{n-1}, t]$ replaced respectively by $(-\infty, y] \setminus \{t\}$, $[y_1, t] \setminus \{t\}$, ..., $[y_{n-1}, t] \setminus \{t\}$, the second assertion of Theorem 5.4.5 still holds. Now, if F is concentrated on a countable set and $\mathcal{K}$ is as in the result just stated, then one can easily show that there is a discrete analogue to Corollary 5.4.6; an alternative representation for members of $\mathcal{D}'_F$ could also be given, using essentially the argument of William's (1979, pp. 171–172).

(v) We had hinted immediately before Theorem 5.2.5 regarding the existence of an integral representation under some constraints for a distribution with specified $E\{h(X - x)|X \geqslant x\}$. As far as an application of Theorem 1.3.9 is concerned, the situation here is exactly the same as that in the case of members of $\mathcal{D}'_F$ and hence the validity of a representation in terms of extreme points is obvious.

A general stability theorem subsuming crucial parts of the stability theorems given earlier for the m.r.l. function and hazard measure in Sections 5.2 and 5.3 exists. This also gives the corresponding results for the multivariate generalizations of the m.r.l. function and hazard measure contained in Theorem 5.4.1 and Corollary 5.4.7 respectively, and of characterizations based on $E\{h(X)|X \geqslant x\}$ in both univariate and multivariate cases. To state the theorem, we require some notation.

Let S be a metric space, T an index set, $\mathcal{L}$ the Borel σ-field on S, $\mathcal{P}, \mathcal{P}_1, \mathcal{P}_2, \mathcal{P}_3$ families of probability measures on $(S, \mathcal{L})$, $\{\mathcal{Q}_t : t \in T\}$ a family of collections of sets with $Q_t \subset \mathcal{L}$ for every $t \in T$, and $\{h(\bullet|t, A_t, P) : A_t \in Q_t, P \in \mathcal{P}, t \in T\}$ a family of real-valued Borel measurable functions on $(S, \mathcal{L})$ satisfying the following conditions in which the notation $D(t, A_t, P)$ stands for the set of discontinuity points of $h(\bullet|t, A_t, P)$:

(i) $\mathcal{P}_1, \mathcal{P}_2, \mathcal{P}_3 \subset \mathcal{P}$, also $\mathcal{P}_2$ is closed (under weak convergence).

(ii) $P_1^{(1)}, P_2^{(1)}, P_3^{(1)}, \ldots \in \mathcal{P}_1$ and $\{P_n^{(1)} : n \geqslant 1\}$ converges weakly to

$$P^* \in \mathcal{P} \Rightarrow h(\bullet|t, A_t, P_n^{(1)}) \to h(\bullet|t, A_t, P^*)$$

as $n \to \infty$ uniformly almost surely $[P^*]$ on $A_t \cap D^c(t, A_t, P^*)$ and

$$\sup_{n \geqslant 1} E_{P_n^{(1)}} \{|h(\bullet|t, A_t, P_n^{(1)})| I_{A_t \cap \{|h(\bullet|t, A_t, P_n^{(1)})| \geqslant \alpha\}}\} \to 0$$

as $\alpha \to \infty$ for each $t \in T$ and P^*-continuity set A_t in Q_t with $P^*(A_t)$ positive.

(iii) $P_1^{(2)}, P_2^{(2)} \in \mathcal{P}_2$ and are distinct $\Rightarrow$ there exist $t \in T$ and $A_t \in Q_t$ such that $P_1^{(2)}(A_t)$, $P_2^{(2)}(A_t)$ are both positive, A_t is both $P_1^{(2)}$-continuity set and $P_2^{(2)}$-continuity set and

$$E_{P_1^{(2)}} \{h(\bullet|t, A_t, P_1^{(2)})|A_t\} \neq E_{P_2^{(2)}} \{h(\bullet|t, A_t, P_2^{(2)})|A_t\}.$$

(iv) $P^{(3)} \in \mathcal{P}_3 \Rightarrow D(t, A_t, P^{(3)})$ has zero $P^{(3)}$-measure for every t in T and $P^{(3)}$-continuity set A_t in Q_t.

Further, let $P \in \mathcal{P}$ and $\{P_n : n \geqslant 1\}$ be a sequence of members of $\mathcal{P}_1$ such that $\{P_n : n = 1, 2, \ldots\}$ is relatively compact. Then we have the following stability theorem.

5.4.9   Theorem

The condition that

$$P \in \mathcal{P}_3, \quad \{P_n : n \geqslant 1\} \text{ converges weakly to } P \tag{5.4.21}$$

implies that

$$E_{P_n} \{h(\bullet|T, A_t, P_n)|A_t\} \to E_P \{h(\bullet|t, A_t, P)|A_t\} \tag{5.4.22}$$

as $n \to \infty$ for every $t \in T$ and P-continuity set $A_t \in Q_t$ with $P(A_t) > 0$.

Moreover, if additionally $P, P_1, P_2, \ldots \in \mathcal{P}_2$ and the set of cluster points of $\{P_n : n = 1, 2, \ldots\}$ (relative to weak convergence) is a subset of $\mathcal{P}_3$, then the converse assertion is valid.

Proof   Assume first that (5.4.21) is valid. Since $P \in \mathcal{P}_3$, it is obvious that the set of discontinuity points of $h(\bullet|T, A_t, P) I_{A_t}$ has zero P-measure for every $t \in T$ and P-continuity set $A_t \in Q_t$. Now, let $t \in T$ and P-continuity set $A_t \in Q_t$ be arbitrarily fixed. Since $P_n \in \mathcal{P}_1$, $n \geqslant 1$, the requirements of Billingsley's (1968) Theorem 5.5 are clearly met with $h(\bullet|t, A_t, P)I_{A_t}$ as h and $h(\bullet|t, A_t, P_n)I_{A_t}$ as h_n. This theorem implies that $\{P_n h_n^{-1}, n = 1, 2, \ldots\}$ converges weakly to $P h^{-1}$. If we now consider X_n, $n \geqslant 1$, and X to be some random variables having distributions $P_n h_n^{-1}$, $n \geqslant 1$, and $P h^{-1}$, respectively, we have $\{X_n : n = 1, 2, \ldots\}$ converging to X in distribution. Also, the fact that $P_n \in \mathcal{P}_1$, $n \geqslant 1$, implies that $\{X_n : n = 1, 2, \ldots\}$ considered here is uniformly integrable. Since Billingsley's (1968) Theorem 5.4 yields that $E\{X_n\} \to E\{X\}$ as $n \to \infty$ in such a situation, we can conclude that

$$E_{P_n}\{h(\bullet|t, A_t, P_n)I_{A_t}\} \to E_P\{h(\bullet|t, A_t, P)I_{A_t}\} \quad \text{as } n \to \infty. \qquad (5.4.23)$$

In view of the assumptions that $\{P_n\}$ converges weakly to P and A_t is a P-continuity set, it follows that $P_n(A_t) \to P(A_t)$ as $n \to \infty$. If $P(A_t) > 0$, we then have (5.4.22) as an obvious consequence of (5.4.23). Hence we have the first part of the stability theorem to be valid.

To establish that the second part of the theorem holds, assume that $P, P_1, P_2, \ldots \in \mathcal{P}_2$ and the set of cluster points of $\{P_n : n = 1, 2, \ldots\}$ is a subset of $\mathcal{P}_3$ and also that (5.4.22) is valid. Since each cluster point of $\{P_n : n = 1, 2, \ldots\}$ is an element of $\mathcal{P}_3$ and $\{P_n : n = 1, 2, \ldots\}$ is relatively compact, we should have a subsequence $\{P_{n_r} : r = 1, 2, \ldots\}$ of $\{P_n : n = 1, 2, \ldots\}$ converging weakly to $Q \in \mathcal{P}_3$ with $Q \neq P$ unless (5.4.21) is valid. If Q^* denotes the (weak) limit of a subsequence of $\{P_n\}$, then clearly we have $Q^* \in \mathcal{P}_3$ and hence the first part of the theorem and the validity of (5.4.22) lead us to

$$E_P\{h(\bullet|t, A_t, P)|A_t\} = E_{Q^*}\{h(\bullet|t, A_t, Q^*)|A_t\} \qquad (5.4.24)$$

for every $t \in T$ and $A_t \in \mathcal{Q}_t$ such that A_t is a P-continuity set with $P(A_t) > 0$ as well as a Q^*-continuity set with $Q^*(A_t) > 0$. We have assumed that $P \in \mathcal{P}_2$ and for each $n \geqslant 1$, $P_n \in \mathcal{P}_2$, and also we have $\mathcal{P}_2$ closed. In that case, we have $P, Q^* \in \mathcal{P}_2$ and hence, in view of (5.4.24), $Q^* = P$. It is therefore impossible that (5.4.21) be not valid. Hence we have the second part of the theorem. ∎

5.4.10   Remarks

(i) In the case of $h(\bullet|t, A_t, P)$ being independent of P, obviously the part of condition (ii) that $h(\bullet|t, A_t, P_n^{(1)}) \to h(\bullet|t, A_t, P^*)$ uniformly almost surely $[P^*]$ on $A_t \cap D^c(t, A_t, P^*)$ for every $t \in T$ and P^*-continuity set A_t with $P^*(A_t) > 0$ is trivially met. Also, if $h(\bullet|t, A_t, P)$ are all continuous, then the condition (iv) above is obviously satisfied with $\mathcal{P}_3 = \mathcal{P}$. If S is a Polish space or in particular, if it is a Euclidean space, we have a sequence $\{P_n : n = 1, 2, \ldots\}$ of members of $\mathcal{P}$ to be relatively compact if and only if it is tight in the sense of Billingsley (1968, p. 37) (cf. Theorems 6.1 and 6.2 in the cited monograph). Thus, it is evident that in various specialized situations, the theorem given above has simplified and perhaps more appealing versions.

(ii) To illustrate that the stability theorem given above does not remain valid if the assumptions $P \in \mathcal{P}_3$ and the set of cluster points of $\{P_n : n = 1, 2, \ldots\}$ is a subset of $\mathcal{P}_3$ respectively appearing in the two parts of the theorem are omitted, it is sufficient to consider the following example.

Example   Let $\{x_n : n = 1, 2, \ldots\}$ be a sequence of strictly increasing, real numbers converging to a real number x'. Let x'' be a real number greater than x'. Define P, P', $\{P_n : n = 1, 2, \ldots\}$ to be a sequence of probability measures on $\mathbb{R}$ such that for some $0 < \alpha < 1$,

$$P_n(\{x\}) = \begin{cases} \alpha & \text{if } x = x_n, \\ 1 - \alpha & \text{if } x = x'', \end{cases}$$

$$P(\{x\}) = \begin{cases} \alpha & \text{if } x = x', \\ 1 - \alpha & \text{if } x = x'', \end{cases}$$

and

$$P'(\{x\}) = \begin{cases} \alpha + \dfrac{\alpha(d - c)}{x'' - x'} & \text{if } x = x' \\ 1 - \alpha - \dfrac{\alpha(d - c)}{x'' - x'} & \text{if } x = x'', \end{cases}$$

where c and d are given real numbers such that $c < d$ and $\{\alpha(d - c)/(x'' - x')\} < 1 - \alpha$. Also, define h on R such that

$$h(x) = \begin{cases} c & \text{if } x < x' \\ d + (x - x') & \text{if } x \geqslant x''. \end{cases}$$

If we take $T =$ the singleton $\{1\}$, $\mathcal{Q}_1 = \{(-\infty, x) : -\infty < x < x''\}$, $\mathcal{P} = \{P, P', P_1, P_2, \ldots\}$ and $h(\bullet | 1, A, P^*) = h(\bullet)$ for every member A of $\mathcal{Q}_1$ and $P^* \in \mathcal{P}$, then it follows that $\mathcal{P}$ itself satisfies the requirements on $\mathcal{P}_1$ and $\mathcal{P}_2$ mentioned above. However, in this case we cannot have a nonempty subset $\mathcal{P}_3$ of $\mathcal{P}$ satisfying condition (iv) as required. Consequently, it follows that in this example neither the assumption of $P \in \mathcal{P}_3$ nor the assumption of the set of cluster points of $\{P_n : n = 1, 2, \ldots\}$ being a subset of $\mathcal{P}_3$ is met. Observe that here $\{P_n : n = 1, 2, \ldots\}$ converges to P weakly, $P \neq P'$ and (5.4.22) is not valid (since it is not true that $E_{P_n}\{h(\bullet) | A\} \to E_P\{h(\bullet) | A\}$ whenever $A = (-\infty, x)$ with $x < x'$) but (5.4.22) with P replaced by P' is valid. This implies that with the deletions mentioned above neither the first part of the theorem nor its second part remain valid.

As mentioned earlier, Theorem 5.4.9 has several interesting corollaries. The details in this respect are given in Shanbhag and Kotz (1987). It may be noted, in particular, that included in corollaries of the theorem is the following.

5.4.11 Corollary

Let $p \geqslant 1$ and $\{F_n : n = 1, 2, \ldots\}$ be a sequence of d.f.'s on $\mathbb{R}^p$ and F be a continuous d.f. on $\mathbb{R}^p$. Assume that F and for each n, F_n, are members of the set

$\mathcal{C}$ defined earlier. Then

$$F_n(x) \to F(x) \quad \text{for all } x \in \mathbb{R}^p$$

if and only if

$$v^*_{F_n}(x) \to v^*_F(x) \quad \text{for all } x \quad \text{with} \quad \overline{F}(x) > 0,$$

where the notation $v^*_G(x)$ stands for the vector whose elements (given in some specified order) are $v_G(\prod_{i=1}^P(-\infty, x_i])$, x_i being the ith coordinate of x, and its counterparts relative to all the univariate and multivariate marginals of G, with appropriate subvectors in place of x, and $\overline{F}$ stands for the survivor function corresponding to F as before.

CHAPTER 6

Properties Based on Fourier or Mellin Transforms

6.1 INTRODUCTION

Many of the characterization results in distribution theory are expressed or have proofs involving Fourier or Mellin transforms. Some of these are discussed in the present chapter.

Amongst the problems treated herein are a famous one of Dugué of identifying the class of pairs of distributions for which convolutions are the same as mixtures, a problem of characterizing the symmetry of a random variable via that of a linear function of its independent copies, and some identifiability or characterization problems concerning elliptical and stable distributions, including those studied by Letac (1981), Zinger (1956), Alzaid *et al* (1990), and Gine and Hahn (1983). Although in some cases the results themselves are expressed in terms of transforms, in some other cases it is more due to the proofs rather than the statements of the results that these are allocated to this chapter.

The Cauchy functional equation or its integrated versions such as those appearing in Theorem 2.2.2 and Corollary 2.3.2, and the uniqueness theorems concerning the characteristic functions and Mellin transforms are amongst the tools that are used herein. A major portion of the chapter represents the material appearing in Rao and Shanbhag (1986), Alzaid *et al* (1990) and Rao and Shanbhag (1992b); the chapter also contains some new results.

6.2 THE DUGUÉ AND BEHBOODIAN PROBLEMS

We devote this section to studying the problems whose initial versions were posed by Dugué and Behboodian.

6.2(a) *The Dugué problem*

Rossberg (1972) and in an unpublished article Wolinska-Welcz and Szyanal (1984) have considered the problem of identifying characteristic functions ϕ_1 and ϕ_2 (of probability distributions on $\mathbb{R}$) for which the following equation holds:

$$(1 - c)\phi_1(t) + c\phi_2(t) = \phi_1(t)\phi_2(t), \quad t \in \mathbb{R} \tag{6.2.1}$$

with $0 < c < 1$. This is indeed an extended version of the problem posed earlier by Dugué for $c = 1/2$.

Rossberg solved the problem when at least one of the ϕ_i's is nonarithmetic, as did Wolinska-Welcz and Szynal when both ϕ_1 and ϕ_2 are arithmetic. In both these papers, there is an assumption that the left extremity of the distribution corresponding to ϕ_1 is nonnegative and the right extremity of the distribution corresponding to ϕ_2 is nonpositive. Rao and Shanbhag (1986) showed that under the assumptions made by these authors, the problem of identifying the solution to (6.2.1) reduces to a straightforward application of Theorem 2.2.2. If at least one of the ϕ_i's is identically equal to 1, then we have trivially that (6.2.1) is satisfied if and only if ϕ_1 and ϕ_2 are both identically equal to 1. Hence it is sufficient if we restrict ourselves to the case when neither of the ϕ_i's is identically equal to 1 to solve (6.2.1); in this latter case, if it is assumed that the extremity conditions concerning ϕ_1 and ϕ_2 stated above are met, then Theorem 2.2.2 leads one to the following theorem.

6.2.1   Theorem

(6.2.1) holds if and only if $F_1(x) = 1 - \exp\{-bx\}$, $x \in \mathbb{R}_+$ and $F_2(-x) = \exp\{-ax\}$, $x \in \mathbb{R}_+$ with $a > 0$ and b such that $b = ac/(1 - c)$, where F_1 and F_2 are d.f.'s corresponding to distributions with characteristic functions ϕ_1 and ϕ_2 respectively, or

$$\phi_1(t) = c\frac{\alpha - \beta + (1 - \alpha)\exp\{ibt\}}{\alpha - c\beta - (\alpha - c)\exp\{ibt\}}, \quad t \in \mathbb{R} \tag{6.2.2}$$

and

$$\phi_2(t) = 1 - \alpha + \alpha\frac{(1 - \beta)\exp\{-ibt\}}{1 - \beta\exp\{-ibt\}}, \quad t \in \mathbb{R} \tag{6.2.3}$$

for some $b > 0$, $\beta \in [0, 1)$ and $\alpha \in [\max\{\beta, c\}, 1]$.

Proof   As in the statement of the theorem, let F_1 and F_2 denote the d.f.'s corresponding to distributions with characteristic functions ϕ_1 and ϕ_2 respectively. In view of the extremity conditions, we get

$$F_1(0-) = 1 - F_2(0) = 0. \tag{6.2.4}$$

In the remainder of the proof, we assume the validity of (6.2.4) without any further reference to it. Suppose (6.2.1) is met. Then, it is obvious that

$$cF_2(-x) = \int_{\mathbb{R}_+} F_2(-x - y)F_1(dy), \quad x \in \mathbb{R}_+ - \{0\}. \tag{6.2.5}$$

If F_1 is nonarithmetic, (6.2.5) implies, in view of Theorem 2.2.2, that $F_2(-x) \propto \exp\{-ax\}$ for $x > 0$ and some $a > 0$, and consequently (6.2.1) yields $F_2(-x) = \exp\{-ax\}$, $x \in \mathbb{R}_+$ for some $a > 0$. This follows since under the given assumptions, the fact that $F_2(-x) \propto \exp\{-ax\}$, $x \in \mathbb{R}_+ - \{0\}$ when used in (6.2.1) gives the following equation relative to probability values of $\{0\}$ on both sides:

$$(1 - c)F_1(0) + c[F_2(0) - F_2(0-)] = F_1(0)[F_2(0) - F_2(0-)]$$

and hence $F_2(0) = F_2(0-)$. Note that we have now F_2 to be nonarithmetic. Consequently, by symmetry we have that if (6.2.1) is valid and at least one of the F_i is nonarithmetic, then

$$F_1(x) = 1 - \exp\{-bx\}, \quad x \in \mathbb{R}_+, \qquad F_2(-x) = \exp\{-ax\}, \quad x \in \mathbb{R}_+$$

with $a > 0$ and b such that $b = ac/(1 - c)$. The converse of the assertion is obvious. On the other hand, if (6.2.1) is valid with both F_1 and F_2 as arithmetic, then, assuming without loss of generality the span of F_2 to be greater than or equal to that of F_1, in view of (6.2.4) and Theorem 2.2.2, we get immediately that

$$\phi_2(t) = 1 - \alpha + \alpha \frac{(1 - \beta)\exp\{-ibt\}}{1 - \beta\exp\{-ibt\}}, \quad t \in \mathbb{R} \tag{6.2.6}$$

with $\alpha \in (0, 1]$, $\beta \in [0, 1)$ and b as the span of F_1. If (6.2.6) is valid, then clearly we have (6.2.1) to be valid if and only if

$$\phi_1(t)\{(c - \alpha)\exp\{ibt\} - c\beta + \alpha\} = c\{(1 - \alpha)\exp\{ibt\} - \beta + \alpha\}, \quad t \in \mathbb{R}. \tag{6.2.7}$$

(6.2.7) gives that

$$(c - \alpha)(1 - F_1(0)) = (c\beta - \alpha)(1 - F_1(b)) \tag{6.2.8}$$

and

$$(\alpha - c\beta)F_1(0) = c(\alpha - \beta). \tag{6.2.9}$$

Since $F_1(0) \neq 1$, (6.2.8) and (6.2.9) imply that $\alpha \geq c$ and $\alpha \geq \beta$ respectively and hence that $\alpha \geq \max\{\beta, c\}$. Thus, it follows that if (6.2.7) holds, then $\alpha \geq \max\{\beta, c\}$ and (6.2.2) is valid. As ϕ_1 given by (6.2.2) is trivially seen to be a characteristic function satisfying (6.2.7), we have then that if both F_1 and F_2 are arithmetic, (6.2.1) holds if and only if ϕ_1 and ϕ_2 are given by (6.2.2) and (6.2.3) respectively with b, β and α as stated. Hence we have the theorem. ∎

The example to appear below shows that Theorem 6.2.1 does not remain valid if the assumption concerning the extremities of F_1 and F_2 i.e. the one that $F_1(0-) = 1 - F_2(0)$, $= 0$ is dropped.

Example 1   For a real $\theta \neq 0$, 1, let

$$\phi_1(t) = \left[(1 + it)(1 - \theta it)\left(1 - \frac{\theta it}{\theta - 1}\right)\right]^{-1}, \quad t \in \mathbb{R},$$

and

$$\phi_2(t) = \phi_1(-t), \quad t \in \mathbb{R}.$$

Observe that (6.2.1) is satisfied with $c = 1/2$ and ϕ_1 and ϕ_2 are nonarithmetic. However, ϕ_1 and ϕ_2 are not of the form given in the theorem.

In the above counterexample, we have $F_1(0-) = 1 - F_2(0) > 0$. It may be noted that there also exist examples illustrating the point with either $F_1(0-) = 0$ or $1 - F_2(0) = 0$. In particular, if we take $c = \alpha/(1 + \alpha)$,

$$\phi_1(t) = (1 - it)^{-2}, \quad t \in \mathbb{R},$$

and

$$\phi_2(t) = (1 + \beta\sqrt{\alpha}it)^{-1}(1 - \beta^{-1}\sqrt{\alpha}it)^{-1}, \quad t \in \mathbb{R},$$

with $\alpha = (\beta^2 - 1)^2/4\beta^2$ and $\beta > 1$, we have an example with $F_1(0-) = 0$. (The existence of an example with $1 - F_2(0) = 0$ follows by symmetry.)

The next theorem throws further light on the mechanism of the characterization in question.

6.2.2 Theorem

Let F_1 be a d.f. concentrated on $\mathbb{R}_+$ such that $F_1(0) < 1$. Then there exist a constant $c \in (0, 1)$ and d.f.'s F_2 and F_3 concentrated on $(-\infty, 0]$, with $F_3(0-) > c$ and F_2 and F_3 as both nonarithmetic, or arithmetic having the same span, such that

$$(1 - c)F_1(x) + cF_2(x) = (F_1 * F_3)(x), \quad x \in \mathbb{R} \tag{6.2.10}$$

if and only if the distribution corresponding to F_1 is either exponential, or is concentrated on $\{0, b, 2b, \ldots\}$ satisfying

$$\overline{\overline{F}}_1(nb) = \overline{\overline{F}}_1(0)\beta^n, \quad n = 0, 1, \ldots \tag{6.2.11}$$

for some $b > 0$ and some $0 < \beta \leqslant \overline{\overline{F}}_1(0)$, where $\overline{\overline{F}}_1(x) = 1 - F_1(x)$, $x \in \mathbb{R}$. Also, we cannot have the existence of c, F_2 and F_3 unless F_1, F_2 and F_3 are all arithmetic with the same span or all nonarithmetic.

Proof Assume first that c, F_2 and F_3 exist. We have then

$$(1 - c)\overline{\overline{F}}_1(x) = \int_{(-\infty,0]} \overline{\overline{F}}_1(x - y)F_3(dy), \quad x \in \mathbb{R}_+ \tag{6.2.12}$$

and

$$(1 - c) + c\overline{\overline{F}}_2(0-) = \int_{(-\infty,0]} \overline{\overline{F}}_1((-y)-)F_3(dy), \tag{6.2.13}$$

where $\overline{\overline{F}}_2(0-) = 1 - F_2(0-)$. In view of Theorem 2.2.2, (6.2.12) implies that

$$\overline{\overline{F}}_1(x) = \overline{\overline{F}}_1(0)e^{-\lambda x}, \quad x \in \mathbb{R}_+ \tag{6.2.14}$$

with $\lambda > 0$, if F_3 is nonarithmetic, and that $\overline{\overline{F}}_1$ satisfies

$$\overline{\overline{F}}_1(x + nb) = \overline{\overline{F}}_1(x)\beta^n, \quad x \in \mathbb{R}_+, \quad n \in \mathbb{N}_0 \tag{6.2.15}$$

with $0 < \beta < 1$ and b as the span of F_3, if F_3 is arithmetic.

In the first of the two cases, (6.2.12) and (6.2.13) lead us to

$$(1 - c)F_1(0) + c\overline{\overline{F}}_2(0-) = F_1(0)(1 - F_3(0-)),$$

which contradicts the requirement that $F_3(0-) > c$ unless $F_1(0) = 0$ and $\overline{\overline{F}}_2(0-) = 0$; consequently in this case we get the distribution corresponding to F_1 to be exponential. In the second case, (6.2.12) and (6.2.13) imply that

$$(1 - c) + c\overline{\overline{F}}_2(0-) = \overline{\overline{F}}_3(0-) + \beta^{-1}((1 - c)\overline{\overline{F}}_1(b-) - \overline{\overline{F}}_1(b-)\overline{\overline{F}}_3(0-)),$$

where $\overline{\overline{F}}_3(0-) = 1 - F_3(0-)$, and hence, in view of the constraint $F_3(0-) > c$, we get $\overline{\overline{F}}_1(b-) \geqslant \beta$. (6.2.10) implies that

$$F_2(x) = \begin{cases} c^{-1} \displaystyle\int_{(-\infty,0]} F_1(x - y)F_3(dy) & \text{if } x < 0 \\ 1 & \text{if } x \geqslant 0 \end{cases} \qquad (6.2.16)$$

If F_3 is arithmetic with span b, then the validity of (6.2.15) implies, in view of the first equation in (6.2.16), that F_2 cannot be concentrated on $\{0, -b, -2b, \ldots\}$ unless F_1 is concentrated on $\{0, b, 2b, \ldots\}$. Since, in the present case, we require F_2 to be concentrated on $\{0, -b, -2b, \ldots\}$, this completes the proof of the 'only if' part of the first assertion, as well as the proof of the second assertion.

To prove the 'if' part of the first assertion, note that if F_1 is the d.f. of an exponential or a geometric distribution as in the statement of the theorem, we have the existence of c and F_3 satisfying (6.2.12) with $F_3(0-) > c$ and $F_3(0) = 1$, and such that F_1 and F_3 are either both arithmetic with span b or both nonarithmetic. If we now define F_2 by (6.2.16), we see that F_2 is a d.f. concentrated on $(-\infty, 0]$. (To see that F_2 is a d.f., it is sufficient to verify that $F_2(0-) \leqslant 1$; (6.2.12) and the constraints to be met by F_1 give that this condition is valid.) As c, F_3 and F_2 defined can easily be seen to satisfy our requirements, we have then the 'if' part of the first assertion. ∎

6.2(b)   Behboodian's problem

Let for $n \geqslant 2$, $X_1, \ldots, X_n$ be independent identically distributed random variables with a nonvanishing characteristic function. Then from an appropriate specialized version of Corollaries 3.3.4 or 3.4.5, it follows that under some mild conditions, the distribution of $\Sigma_{r=1}^n \alpha_r X_r$ with X_r real, is symmetric about zero only if the distribution of X_r is symmetric about some point. (Note that $\Sigma_{r=1}^n \alpha_r X_r$ has a distribution that is symmetric about zero if and only if

$$\left(\prod_{\{r:\alpha_r>0\}} \phi(|\alpha_r|t) \right) \left(\prod_{\{r:\alpha_r<0\}} \phi(-|\alpha_r|t) \right)$$

$$= \left(\prod_{\{r:\alpha_r>0\}} \phi(-|\alpha_r|t) \right) \left(\prod_{\{r:\alpha_r<0\}} \phi(|\alpha_r|t) \right), \qquad t > 0,$$

where $\phi(t)$ is the characteristic function of X_r.)

Behboodian considered a version of the problem with $n = 3$, $\alpha_1 = 1$, $\alpha_2 = m$ and $\alpha_3 = -(1 + m)$ with $0 < m \leqslant 1$ and conjectured that the characterization result holds here even without the restriction that the characteristic function of X_r is nonvanishing. That the conjecture is not valid is shown by the following example.

Example 2

$$\tau(x) = \begin{cases} 1 - |x| & \text{if } |x| \leqslant 1 \\ 0 & \text{otherwise.} \end{cases}$$

This satisfies the criterion of problem 3 on page 647 in Feller (1971). From the problem, it is clear that we have a characteristic function ϕ such that

$$\phi(t) = \tau(t) + \tfrac{1}{2}e^{-i}\tau(t - 3) + \tfrac{1}{2}e^{i}\tau(t + 3), \quad t \in \mathbb{R}.$$

Note that if we take $0 < m < \infty$, we get here $\phi(t)\phi(mt)\phi(-(1 + m)t)$ to be always real. However, we do not have in the example X_i to be symmetric about a point.

In the case of $m = 1$, Behboodian gives a theorem implying that if the characteristic function of X_r is nonvanishing, then the result holds. However, the proof produced by the author for the theorem is based on an erroneous argument, and the following example of Kochar (1992) reveals that the theorem is in fact false.

Example 3   For every $\gamma \in \mathbb{R}_+$, let ϕ_γ denote an infinitely divisible characteristic function with or without a Gaussian factor, such that the Lévy (canonical) measure is concentrated on

$$\{-2^n : n = 0, \pm1, \pm2, \ldots\} \cup \{2^{n+1/2} : n = 0, \pm1, \pm2, \ldots\}$$

and is such that

$$\mu(\{-2^n\}) = 2^{-n}, \quad n = 0, \pm1, \pm2, \ldots$$

and

$$\mu(\{2^{n+1/2}\}) = \gamma 2^{-n}, \quad n = 0, \pm1, \pm2, \ldots \ .$$

It is easily checked that

$$(\phi_\gamma(t))^2(\phi_\gamma(2t))^{-1} = \exp\{ia_\gamma t\}\eta_\gamma(t), t \in \mathbb{R}, \gamma \in \mathbb{R}_+ \tag{6.2.17}$$

with η_γ as a real even function for each γ and a_γ as a continuous function of γ such that $a_0 < 0$ and $\lim_{\gamma \to \infty} a_\gamma = \infty$. Clearly, there exists a $\gamma_0 \in \mathbb{R}_+$ for which $a_{\gamma_0} = 0$ and hence from (6.2.17) we have that $(\phi_{\gamma_0}(t))^2\phi_{\gamma_0}(-2t)$, $t \in \mathbb{R}$ is a real function. If X_r, $r = 1, 2, 3$ are independent identically distributed random variables with characteristic function ϕ_{γ_0}, then from what is observed it follows that $X_1 + X_2 - 2X_3$ has its distribution to be symmetric about zero and the distribution of X_r is not symmetric about any point.

Does it then mean that the situation concerning Behboodian's theorem is hopeless? The answer to the question is provided by the theorem that follows.

6.2.3 Theorem

Let X_r, $r = 1, 2, 3$ be independent identically distributed integrable random variables with a nonvanishing characteristic function. Then $X_1 + mX_2 - (m + 1)X_3$ has for some $m \in (0, \infty)$, its distribution to be symmetric about some point if and only if the distribution of X_r is symmetric about some point.

Proof Let ϕ denote the characteristic function of X_r and $m \in (0, \infty)$ be fixed. It follows that $\phi(t)\phi(mt)\phi(-(1+m)t)$, $t \in \mathbb{R}$, the characteristic function of $X_1 + mX_2 - (m + 1)X_3$, has the corresponding distribution to be symmetric about some point if and only if

$$(m + 1)\mathrm{Im}\left(\frac{\phi'((1+m)t)}{\phi((1+m)t)}\right) = m\,\mathrm{Im}\left(\frac{\phi'(mt)}{\phi(mt)}\right) + \mathrm{Im}\left(\frac{\phi'(t)}{\phi(t)}\right), \quad t \in \mathbb{R}_+.$$

(6.2.18)

Since the function $\mathrm{Im}(\phi'(t)/\phi(t))$, $t \in \mathbb{R}_+$, is continuous and is bounded on every bounded subinterval of $\mathbb{R}_+$, Theorem 2.2.2 after some minor manipulation (or a specialized version of Corollary 3.3.4) yields that (6.2.18) implies

$$\mathrm{Im}\left(\frac{\phi'(t)}{\phi(t)}\right) = \lim_{n\to\infty} \mathrm{Im}\left(\frac{\phi'\left(\frac{t}{(1+m)^n}\right)}{\phi\left(\frac{t}{(1+m)^n}\right)}\right) = \mathrm{Im}\left(\frac{\phi'(0)}{\phi(0)}\right), \quad t \in (0, \infty),$$

which, in turn, implies that the function $\mathrm{Im}(\phi'(t)/\phi(t))$, $t \in \mathbb{R}_+$ is identically equal to a constant. On the other hand if the function $\mathrm{Im}(\phi'(t)/\phi(t))$, $t \in \mathbb{R}_+$ is identically equal to a constant, (6.2.18) is met. Hence we have the theorem. ∎

The above theorem obviously reveals that Behboodian's theorem is valid if we impose a further restriction that X_r be integrable. (Indeed, in this latter case, i.e. when $m = 1$, the result follows trivially without any reference to Theorem 2.2.2 on Corollary 3.3.4.) Moreover the example of Kochar (1992) met above shows that the theorem does not hold if the assumption of the integrability of X_r is dropped; another example illustrating this point is that of a stable distribution with exponent 1.

While we are on the topic of characterizing distributions via symmetry, we may discuss briefly another curious problem that warrants some attention. If X and Y are independent random variables such that at least one of them is symmetric about 0, then, since

$$XY \stackrel{\mathrm{d}}{=} -(XY),$$

it is obvious that XY is symmetric about 0. Recently, Duncan and Kadane (1980) proved that if X and Y are independent nondegenerate random variables that are symmetric about some nonzero points, with $E(|X|^3) < \infty$ and $E(|Y|^3) < \infty$, then XY is not a symmetric random variable. These authors further conjectured that the result holds if the condition that $E(|X|^3) < \infty$ and $E(|Y|^3) < \infty$ is dropped. As revealed in Shanbhag (1988), some authors have made unsuccessful attempts to settle the conjecture. However, Shanbhag (1988) also hints that a partial result in this connection holds; the result is given by the next theorem.

6.2.4   Theorem

If X and Y are independent nondegenerate random variables that are symmetric about some points, with $E(X^2) < \infty$ and $E(Y^2) < \infty$, then XY is not a symmetric random variable.

Proof   It is easily seen that the assertion is equivalent to that if X and Y are independent nondegenerate square integrable random variables that are symmetric about the origin, then $XY + X + Y$ is not a symmetric random variable. Assume then that X and Y are as in this latter assertion. Since $E(XY + X + Y) = 0$, it is then sufficient if we show that the following does not hold:

$$E(\sin(t(XY + X + Y))) = 0, \quad t \in \mathbb{R}. \qquad (6.2.19)$$

Since X and Y are independent and symmetric about 0, it follows that (6.2.19) is valid if and only if

$$E(\sin(tXY)\sin(tX)\sin(tY)) = 0, \quad t \in \mathbb{R}. \qquad (6.2.20)$$

We have here $0 < E(X^2Y^2) < \infty$, and, hence, in view of the Lebesgue dominated convergence theorem, (6.2.20) gives that

$$0 = \lim_{t \downarrow 0} E\left\{ \frac{1}{t^3}(\sin(tXY)\sin(tX)\sin(tY)) \right\} = E(X^2Y^2) > 0,$$

leading us to a contradiction. Hence it follows that (6.2.20) cannot be valid. This, in turn, implies that (6.2.19) is not valid and hence that the theorem holds. ∎

It may also be worth pointing out here that there exist specialized situations in which the Duncan–Kadane conjecture holds even when the restriction that X and Y be square integrable is dropped. The following is an illustration.

Example   Let X and Y be independent nondegenerate random variables that are symmetric about the origin, satisfying additionally the conditions that $E(|Y|) < \infty$ and the distribution of X has at most three support points. In this case, (6.2.20) is equivalent to

$$E(\sin(ctY)\sin(tY)) = 0, \quad t \in \mathbb{R} \qquad (6.2.21)$$

with $c > 0$ as fixed; clearly (6.2.21) is equivalent to

$$E(\cos(t(c+1)Y)) = E(\cos(t(c-1)Y)), \quad t \in \mathbb{R}. \qquad (6.2.22)$$

Observe that (6.2.22) contradicts the assumption that Y is nondegenerate; consequently, we get that if we have X and Y as in the example (6.2.20) does not hold. From this it is easily seen that the Duncan–Kadane conjecture holds if one of the variables is integrable and the other one has at most three support points in its distribution.

6.3   CERTAIN CHARACTERIZATIONS BASED ON SPHERICITY AND ELLIPTICITY

This section studies the characterizations of probability distributions based on sphericity and ellipticity, given recently by Letac (1981) and Alzaid *et al* (1990).

Letac (1981) established the result that if X_1, X_2 and X_3 are three independent real random variables such that $P\{X_j = 0\} = 0$, $j = 1, 2, 3$ then $(R^{-1}X_1, R^{-1}X_2, R^{-1}X_3)$, where $R = (X_1^2 + X_2^2 + X_3^2)^{1/2}$ is uniformly distributed on the sphere $\{x \in \mathbb{R}^3 : \|x\| = 1\}$ if and only if X_j's are all distributed as $N(0, \sigma^2)$ for some σ^2.

We now refer to the aforementioned result as Letac's theorem. Rao and Shanbhag (1989) have shown that Letac's theorem follows from Corollary 2.3.2; we reproduce below their proof.

6.3.1   The Rao–Shanbhag proof of Letac's theorem

Since the 'if' part is trivial, it is sufficient if we prove just the 'only if' part of the result. Assume then that the vector has the spherical distribution. We have, in view of the structure of the spherical distribution satisfying $|X_j| > 0$ a.s., for every $-1 < \alpha < 1$

$$E \left\{ \frac{X_1|X_1|^\alpha}{(|X_2||X_3|)^{(1+\alpha)/2}} \right\} = 0$$

with the function under expectation well defined a.s., which implies that $E\{X_1|X_1|^\alpha\} = 0$. Consequently, we have X_1 and hence by symmetry all X_i to be symmetric. The assumption of sphericity (on appealing to $|X_j| > 0$ a.s.) yields

$$E \left\{ \frac{|X_1|^\alpha}{|X_3|^\alpha} \right\} = E \left\{ \frac{|X_2|^\alpha}{|X_3|^\alpha} \right\} < \infty, \quad -1 < \alpha < 1, \qquad (6.3.1)$$

(with the functions under the expectations well defined a.s.) and hence that $E\{|X_1|^\alpha\} = E\{|X_2|^\alpha\} < \infty$, $-1 < \alpha < 1$, implying, in view of the uniqueness theorem of Mellin transforms, X_1 and X_2 to be identically distributed. By symmetry then we have X_1 and X_3 and hence X_1, X_2, X_3 to be identically distributed. Observe now that (6.3.1) remains valid with X_2 replaced by $(2X_1 + X_2)/\sqrt{5}$; we can hence claim that

$$X_1 \stackrel{\mathrm{d}}{=} \frac{2X_1 + X_2}{\sqrt{5}},$$

which implies that the characteristic function f of X_1 satisfies the equation

$$f(t) = f\left(\frac{2t}{\sqrt{5}}\right) f\left(\frac{t}{\sqrt{5}}\right), \quad -\infty < t < \infty,$$

which in turn implies in view of the symmetry of X_1 that f is a positive (real) even function. Defining $g(x) = -\log f(e^x)$, $x \in \mathbb{R}$, and noting that it is a nonnegative continuous (real) function, we can then see from Corollary 2.3.2 that $f(t) = e^{-at^2}$,

$t \in \mathbb{R}$ for some $a > 0$. Consequently, we have X_i to be independent identically distributed normal random variables with zero mean. ∎

Letac's theorem given here implies a result due to Zinger (1956). Suppose $X_1, \ldots, X_n$ are independent identically distributed continuous real random variables. Define $\overline{X} = (1/n)\Sigma_{i=1}^n X_i$ and $S^2 = \Sigma_{i=1}^n (X_i - \overline{X})^2$. Then, the result of Zinger is that for $n \geqslant 6$, the vector

$$Y = \left(\frac{X_1 - \overline{X}}{S}, \ldots, \frac{X_n - \overline{X}}{S} \right)$$

is uniformly distributed on the $(n-2)$-dimensional sphere

$$\Phi = \{ x \in \mathbb{R}^n : \sum_{i=1}^n x_i = 0, \|x\| = 1 \}$$

if and only if X_i's are normal. Zinger's result clearly does not hold for $n = 2$ and it has not been settled for $n = 3, 4$ and 5. We now show using essentially the argument of Section 6.3.1, that the result holds for $n = 5$. (The validity of the result for $n = 5$ implies clearly its validity for $n \geqslant 5$; also note that we have assumed X_i's continuous to have Y well defined almost surely.)

We need a proof only for the 'only if' part of the assertion. Suppose Y is uniform on Φ. We have then

$$E\left\{ \frac{\left(\frac{2}{\sqrt{3}} \left(X_1 - \frac{X_2 + X_3}{2} \right)^+ \right)^\theta}{|X_4 - X_5|^\theta} \right\} = E\left\{ \frac{((X_1 - X_2)^+)^\theta}{|X_4 - X_5|^\theta} \right\} < \infty, \quad 0 \leqslant \theta < 1$$

implying that

$$E\left\{ \left(\frac{2}{\sqrt{3}} \left(X_1 - \frac{X_2 + X_3}{2} \right)^+ \right)^\theta \right\} = E\{((X_1 - X_2)^+)^\theta\} < \infty, \quad 0 \leqslant \theta < 1,$$

and hence (on using the uniqueness property of Mellin transforms) that

$$\frac{2}{\sqrt{3}} \left(X_1 - \frac{X_2 + X_3}{2} \right)^+ \overset{\mathrm{d}}{=} (X_1 - X_2)^+.$$

Similarly, we have

$$\frac{2}{\sqrt{3}} \left(X_1 - \frac{X_2 + X_3}{2} \right)^- \overset{\mathrm{d}}{=} (X_1 - X_2)^-.$$

Hence, it follows that

$$\frac{2}{\sqrt{3}} \left(X_1 - \frac{X_2 + X_3}{2} \right) \stackrel{\text{d}}{=} X_1 - X_2. \tag{6.3.2}$$

Letting f denote the characteristic function of X_1, we see from (6.3.2) that f^*, where $f^*(t) = f(t)\overline{f}(t)$, $t \in \mathbb{R}$ ($\overline{f}$ being the complex conjugate of f) satisfies

$$f^* \left(\frac{2}{\sqrt{3}} t \right) \left(f^* \left(\frac{1}{\sqrt{3}} t \right) \right)^2 = (f^*(t))^2, \quad t \in \mathbb{R},$$

which implies that f^* is a positive (real) even function. Defining $g(x) = -\log f^*(e^x)$, $x \in \mathbb{R}$, we can then see that f^* is normal and hence $X_1 - X_2$ is normal. Using Cramér's theorem or, since (6.3.2) implies that $(2/\sqrt{3})(X_1 - (X_2 + X_3)/2)$ is normal and hence that X_1 has the moments of a normal random variable, a moment argument implies that X_1 (or each X_r) is normal.

We now turn our attention to the characterization results arrived at by Alzaid *et al* (1990) on elliptical distributions. In this discussion, a distribution on $\mathbb{R}^n$ is referred to as an elliptically symmetric distribution (an e.s.d., for short) if its characteristic function $\phi(t)$, $t \in \mathbb{R}^n$, is a real function of $t \Sigma t'$ for some nonnegative definite real symmetric matrix Σ. (Note that we denote here an e.s.d. random vector by a row.) A special case of the distribution when $\Sigma = I$ is referred to as a spherically symmetric distribution (an s.s.d., for short). From the work of Schoenberg (1938) and subsequent authors it is known that an n-component random vector X has an e.s.d. if and only if it has the representation

$$X \stackrel{\text{d}}{=} TUC, \tag{6.3.3}$$

where C is an $n \times n$ real matrix such that $C'C = \Sigma$, U is an n-component random vector that is uniformly distributed on $\{x \in \mathbb{R}^n : \|x\| = 1\}$ and T is a nonnegative real random variable independent of U. It is implicit here that the distribution of X in question depends on C only through Σ. In what follows, we use the notation $\mathcal{E}_n(\Sigma)$ for the class of e.s.d.s defined by (6.3.3). The class of spherical distributions is denoted by $\mathcal{E}_n(I)$.

To present the results of Alzaid *et al* we need some notation.

Define now the following polynomial functions

$$Q_1(X) = \Sigma_1 a_{i_1,\dots,i_n} \prod_{r=1}^{n} x_r^{i_r}, \quad X = (x_1, \dots, x_n) \in \mathbb{R}^n \tag{6.3.4}$$

$$Q_2(X) = \Sigma_2 b_{i_1,\dots,i_n} \prod_{r=1}^{n} x_r^{i_r}, \quad X = (x_1, \dots, x_n) \in \mathbb{R}^n, \tag{6.3.5}$$

where Σ_1 and Σ_2 denote summations over

$$\{(i_1, \dots, i_n) : i_1 + \dots + i_n = k\}$$

and

$$\{(i_1, \ldots, i_n) : i_1 + \cdots + i_n = m\},$$

respectively, with $k, m \geqslant 1$ and fixed. We shall assume Q_1 and Q_2 to be such that

$$P\{Q_1(UC) = 0\} \neq 1 \quad \text{and} \quad P\{Q_2(UC) = 0\} \neq 1, \tag{6.3.6}$$

where U is an n-component random vector with uniform distribution on the unit sphere, for some (and hence all) $n \times n$ real C satisfying the condition $C'C = \Sigma$. Note that the assumption (6.3.6) is equivalent to

$$P\{Q_1(UC) = 0\} = 0 \quad \text{and} \quad P\{Q_2(UC) = 0\} = 0$$

for some (and hence all) C satisfying the condition $C'C = \Sigma$. (We consider the problem of verifying the claim as an exercise for the reader; a solution to the exercise is given by Note A_2 in the appendix of Alzaid *et al* (1990).)

We can now give the results referred to above:

6.3.2   Theorem

Let $b > 0$, Q_1 and Q_2 be as defined in (6.3.4) and (6.3.5) and F be a specified nondegenerate member of $\mathcal{E}_n(\Sigma)$ for which the property

$$E\{|Q_1(X)|^b \| Q_2(X)|\} = c \text{ a.s.} \tag{6.3.7}$$

holds, where $X \sim F$. Then the property (6.3.7) characterizes F except for a change of scale of the random variable involved, in the class of nondegenerate distributions in $\mathcal{E}_n(\Sigma)$. (That is, if there is a nondegenerate random vector Y whose distribution belongs to $\mathcal{E}_n(\Sigma)$ and satisfies (6.3.7) then there exists a constant $\lambda > 0$ such that $\lambda Y \sim F$, i.e., distributed with d.f. F).

Proof   Let $X \sim F \in \mathcal{E}_n(\Sigma)$ and $Y \sim G \in \mathcal{E}_n(\Sigma)$ both have the property (6.3.7). By the representation (6.3.3), $X \stackrel{\mathrm{d}}{=} T_1 UC$ and $Y \stackrel{\mathrm{d}}{=} TUC$, where T_1 and T are nonnegative r.v.'s. We show that $T \stackrel{\mathrm{d}}{=} \lambda T_1$, which establishes the theorem.

The property (6.3.7) for Y, with $c = 1$, without loss of generality, implies

$$E\{|Q_1(UC)|^b|Q_2(UC)|^\theta T^{bk+m\theta}\} = E\{|Q_2(UC)|^\theta T^{m\theta}\} \quad \text{for } \theta \geqslant 0 \tag{6.3.8}$$

with $T > 0$ a.s. (Note that $P\{T = 0\} > 0$ is impossible, since it contradicts (6.3.7) in view of the fact that $P\{Q_i(UC) = 0\} = 0$, $i = 1, 2$. Also, in (6.3.8) we take $\theta \geqslant 0$ only because it is sufficient for our purpose; indeed equation (6.3.8) is valid for all real θ with possibly infinite value for both sides for some θ values.) In view of the independence of T and U, (6.3.8) implies

$$E\{|Q_1(UC)|^b|Q_2(UC)|^\theta\}E\{T^{bk+m\theta}\} = E\{|Q_2(UC)|^\theta\}E\{T^{m\theta}\} \quad \text{for } \theta \geqslant 0. \tag{6.3.9}$$

On taking $\theta = 0, m^{-1}bk, 2m^{-1}bk, \ldots$, successively, (6.3.9) yields inductively

$$E\{T^{abk}\} < \infty \quad \text{for all integers } a \geqslant 0$$

(since $E(T^0) = 1$, and the identity (6.3.9) gives $E\{T^{(a+1)bk}\} < \infty$ whenever $E\{T^{abk}\} < \infty$) and hence that $E\{T^\theta\} < \infty$ for all nonnegative real θ. Consequently, we have from (6.3.9) the following identity, with both sides well defined and finite,

$$E\{T^{bk+m\theta}\} = \frac{E\{|Q_2|^\theta\}}{E\{|Q_1|^b|Q_2|^\theta\}} E(T^{m\theta}), \quad \theta \geqslant 0, \tag{6.3.10}$$

where $Q_i = Q_i(UC)$, $i = 1, 2$. We now use the following lemma.

6.3.3   Lemma

If Z is a positive random variable and ψ is a continuous increasing or decreasing function on $(0, \infty)$ such that $\psi(Z)Z^\theta$ and Z^θ are integrable for each $\theta > \theta_0(\geqslant 0)$, then

$$E\{\psi(Z)Z^\theta\}/E\{Z^\theta\} \to \lim_{z \to \alpha^*} \psi(z) \quad \text{as } \theta \to \infty,$$

where α^* is the right extremity of the distribution of Z. (One could obviously extend the statement of the lemma further; however, the existing form is sufficient for our purposes.)

Proof of the lemma   There is no loss of generality in assuming ψ to be increasing. Then using the notation $E(Y; A)$ to denote $\int_A Y \, dP$ corresponding to a random variable Y when A is an event, we have, for $z \in (0, \alpha^*)$,

$$\lim_{t \to \alpha^*} \psi(t) \geqslant E\{\psi(Z)Z^\theta\}/E\{Z^\theta\}$$

$$= \frac{E\{\psi(Z)(Z/z)^\theta; Z < z\} + E\{\psi(Z)(Z/z)^\theta; Z \geqslant z\}}{E\{(Z/z)^\theta; Z < z\} + E\{(Z/z)^\theta; Z \geqslant z\}}$$

$$\geqslant \frac{\psi_1(z, \theta) + \psi(z)}{\psi_2(z, \theta) + 1}, \tag{$*$}$$

where

$$\psi_1(z, \theta) = E\{\psi(Z)(Z/z)^\theta; Z < z\}/E\{(Z/z)^\theta; Z \geqslant z\}$$

and

$$\psi_2(z, \theta) = E\{(Z/z)^\theta; Z < z\}/E\{(Z/z)^\theta; Z \geqslant z\}.$$

By the Lebesgue dominated convergence theorem and monotone convergence theorem, it follows that

$$\psi_i(z, \theta) \to 0 \quad \text{as } \theta \to \infty \quad \text{for } i = 1, 2.$$

Consequently, it follows that the extreme right-hand side of $(*)$ tends to $\psi(z)$ as $\theta \to \infty$. Since z is arbitrary, we can then conclude that $(*)$ and the monotonic continuous nature of ψ imply the required result. This concludes the proof of the lemma.

From Lemma 6.3.3 and the fact that $|Q_2(UC)|$ is bounded almost surely, we then get that

$$\frac{d}{d\theta} \log \frac{E\{|Q_2|^\theta\}}{E\{|Q_1|^b|Q_2|^\theta\}} = \frac{E\{(\log|Q_2|)|Q_2|^\theta\}}{E\{|Q_2|^\theta\}} - \frac{E\{(\log|Q_2|)|Q_1|^b|Q_2|^\theta\}}{E\{|Q_1|^b|Q_2|^\theta\}} \quad (6.3.11)$$

$$\rightarrow \log\alpha - \log\alpha \quad \text{as } \theta \rightarrow \infty \quad (6.3.12)$$

with α as the right extremity of the distribution of $|Q_2|$. To establish (6.3.12) we take ψ of the lemma to be such that $\psi(z) = \log z, z \in (0, \infty)$ and Z as respectively $|Q_2|$ and an r.v. Z with d.f.,

$$E\{|Q_1|^b_{\{|Q_2|\leqslant z\}}\}/E\{|Q_1|^b\}.$$

Then it follows that each of the two quantities on the r.h.s. of (6.3.11) tends to $\log\alpha$. From (6.3.10) and (6.3.12), we conclude that

$$\frac{d}{d\theta} \log E\{T^{bk+\theta}\} - \frac{d}{d\theta} \log E\{T^\theta\} \rightarrow 0 \quad \text{as } \theta \rightarrow \infty. \quad (6.3.13)$$

Since $E\{T^\theta\}, \theta \in [0, \infty)$ is the restriction of a moment generating function to $[0, \infty)$, it is obvious that it is log-convex and hence $d\log E\{T^\theta\}/d\theta$ is increasing on $(0, \infty)$. Consequently, from (6.3.13), we conclude that

$$\frac{d}{d\theta} \log E\{T^{\theta+\theta'}\} - \frac{d}{d\theta} \log E\{T^\theta\} \rightarrow 0 \quad \text{as } \theta \rightarrow \infty \quad (6.3.14)$$

for every $\theta' \geqslant 0$. If T and T_1 are two positive r.v.'s for which (6.3.10) is valid, then it follows that $E\{T^\theta\}/E\{(T_1)^\theta\}$ is periodic with period bk for $\theta \geqslant 0$. Since (6.3.14) is valid for T as well as T_1,

$$\frac{d}{d\theta} \log \frac{E\{T^\theta\}}{E\{(T_1)^\theta\}}$$

is independent of θ for $\theta \in (0, \infty)$ or, equivalently,

$$E\{T^\theta\} = \lambda^\theta E\{(T_1)^\theta\}, \quad \theta \in [0, \infty) \quad (6.3.15)$$

for some positive λ. In view of the uniqueness theorem for Mellin transforms, (6.3.15) implies $T \overset{d}{=} \lambda T_1$, which proves the theorem. (Incidentally if both T and T_1 satisfy (6.3.10) we get $\lambda = 1$; however, we have arrived at this situation only because our argument does not take into account scale changes.) ∎

6.3.4   Theorem

Suppose that $|Q_1(X)|$ and $|Q_2(X)|$ are independently distributed, where $X \sim N_n(0, \Sigma) \in \mathcal{E}_n(\Sigma)$. If the independence property holds for any other nondegenerate $Y \sim F \in \mathcal{E}_n(\Sigma)$, then F is the same as $N_n(0, \Sigma)$, except for a change of scale.

Proof   Assume $|Q_1(X)|$ and $|Q_2(X)|$ to be independent if $X \sim N_n(0, \Sigma)$. It is now sufficient if we show that this condition on $|Q_1(X)|$ and $|Q_2(X)|$ with X having nondegenerate distribution in $\mathcal{E}_n(\Sigma)$ implies that X has a distribution of the type $N_n(0, \alpha\Sigma)$ for some α. In view of (6.3.3) and the condition that $P(Q_i(UC) = 0) = 0$, $i = 1, 2$, it follows that the independence of $|Q_1(X)|$ and $|Q_2(X)|$ together with the nondegeneracy of X implies T of (6.3.3) to be > 0 a.s. We then obtain on appealing to the independence of $|Q_1(X)|$ and $|Q_2(X)|$ and condition (6.3.3), that, for s_1, s_2 in some neighborhood of the origin,

$$\frac{E\{T^{i(s_1+s_2)}\}}{E\{T^{is_1}\}E\{T^{is_2}\}} = \frac{E\{|Q_1(UC)|^{is_1}\}E\{|Q_2(UC)|^{is_2}\}}{E\{|Q_1(UC)|^{is_1}|Q_2(UC)|^{is_2}\}} \tag{6.3.16}$$

with both sides well defined. Since the independence of $|Q_1(X)|$ and $|Q_2(X)|$ is assumed to be valid for X with distribution $N_n(0, \Sigma)$, we can claim that (6.3.16) is valid for the T corresponding to $N_n(0, \Sigma)$. Denoting the T for $N_n(0, \Sigma)$ by T_0, we then see that, for s_1, s_2 in some neighborhood of the origin,

$$\psi(s_1 + s_2) = \psi(s_1)\psi(s_2),$$

where

$$\psi(s) = \frac{E\{T^{is}\}}{E\{T_0^{is}\}}.$$

(ψ is indeed well defined for all s; however, we require here only the information that it is well defined in a neighborhood of the origin.) From Lemma 1.3.4 and the fact that $E\{T^{is}\}$ and $E\{T_0^{is}\}$ are characteristic functions, we can then conclude that, for s lying in some neighborhood of the origin,

$$E\{T^{is}\} = \lambda^{is}E\{T_0^{is}\}$$

with $\lambda > 0$. Since the distribution of $\log(\lambda T_0)$ (i.e. the logarithm of the square root of a certain gamma random variable) is determined uniquely by its moments, we have the distribution of $\log T$ to be determined by its moments and

$$\log T \stackrel{\mathrm{d}}{=} \log(\lambda T_0)$$

or, equivalently,

$$T \stackrel{\mathrm{d}}{=} \lambda T_0.$$

Hence we have the required result. ∎

Maxwell (1860), Bartlett (1934), Kac (1939), Hartman and Wintner (1940), Kelker (1970), Nash and Klamkin (1976), Ali (1980), Arnold and Lynch (1982) and several others have characterized the normal distributions in the class $\mathcal{E}_n(I)$, for $n \geqslant 2$, as those having any one of the properties such as:

(i)   The components $X_1, X_2, \ldots, X_n$ of the random vector X are independently distributed.

(ii) $\overline{X} = (1/n)(X_1 + \cdots + X_n)$ and $S^2 = (1/(n-1))[(X_1 - \overline{X})^2 + \cdots + (X_n - \overline{X})^2]$
are independently distributed.

These results follow as corollaries to Theorem 6.3.4. From Theorem 6.3.2, we get
that properties such as

$$E(S^2 | \overline{X}) = c \quad \text{a.s.,} \qquad E(S^2 | |\overline{X}|) = c \quad \text{a.s.,} \qquad E(S | \overline{X}) = c \quad \text{a.s.}$$

$$E(S | |\overline{X}|) = c \quad \text{a.s.,} \qquad E(|\overline{X}| \,|\, S^2) = c \quad \text{a.s.,} \qquad E(\overline{X}^2 | S^2) = c \quad \text{a.s.,}$$

where each of the c's represents a constant, individually characterize the normal
distribution in the class $\mathcal{E}_n(I)$. Theorems 6.3.2 and 6.3.4 also give various versions
of the Darmois–Skitovič theorem under ellipticity or sphericity of the distribution
of $X = (X_1, \ldots, X_n)$.

The following remarks make some further observations on Theorems 6.3.2
and 6.3.4.

6.3.5   Remarks

(i) Theorem 6.3.2 remains valid if $|Q_2(X)|$ in (6.3.7) is replaced by $Q_2(X)$, since
the identity with $Q_2(X)$ in place of $|Q_2(X)|$ yields the original identity (6.3.7). Also
Theorem 6.3.4 remains valid if either $|Q_1(X)|$ is replaced by $Q_1(X)$, or $|Q_1(X)|$
and $|Q_2(X)|$ are replaced respectively by $Q_1(X)$ and $Q_2(X)$; this follows because
the independence of $|Q_1(X)|$ and $Q_2(X)$ implies the independence of $|Q_1(X)|$ and
$|Q_2(X)|$, and also the independence of $Q_1(X)$ and $Q_2(X)$ implies that of $|Q_1(X)|$
and $|Q_2(X)|$.

(ii) A result analogous to that of Theorem 6.3.4 or its variant mentioned in (i) above
remains valid for the uniform distributions on $\{x \in \mathbb{R}^n : \|x\| = \lambda\}$ (with $\lambda > 0$).
(Indeed the proof of Theorem 6.3.4 illustrates that the result is valid if we replace
$N_n(0, \Sigma)$ by a member of $\mathcal{E}_n(\Sigma)$ that has the corresponding $\log T$ well defined
with its distribution determined uniquely by moments; we have not come across
any examples of members of $\mathcal{E}_n(\Sigma)$ other than normal and spherically uniform, for
which some $Q_1(X)$ and $Q_2(X)$ are independent.) It may also be noted that there
exist distributions F which are neither normal nor spherically uniform, for which
a property of the type (6.3.7) is valid. In particular, if X denotes the subvector
containing the first $n(\geqslant 2)$ components of a random vector which is uniformly
distributed on $\{x \in \mathbb{R}^{n+m} : \|x\| = 1\}$, where $m \geqslant 1$, then

$$E\{(n + m - 1)X_1^2 + X_2^2 | X_2^2\} = 1 \quad \text{a.s.;}$$

however, X here is neither normal nor spherically uniform. Clearly the X satisfies
(6.3.3) with $C = I$ and T as the square root of a beta random variable with parameters $n/2$ and $m/2$, respectively. Taking a clue from this example, it is possible to
produce more general examples of the type given below supporting the claim.

Example   Let $X = (X_1, \ldots, X_n)$ be an n-component random vector such that

$$X \stackrel{\mathrm{d}}{=} TU$$

with U as an n-component random vector that is uniformly distributed on $\{x \in \mathbb{R}^n : \|x\| = 1\}$ and T as the square root of a beta random variable with parameters $n/2$ and r, respectively, and distributed independently of U, where r is any fixed positive real number. Observe that

$$E\left\{ (k + 2r) \left(\sum_{i=1}^{k} X_i^2 / k \right) + \sum_{i=k+1}^{n} X_i^2 \,\middle|\, \sum_{i=k+1}^{n} X_i^2 \right\} = 1 \quad \text{a.s.,}$$

where $1 \leqslant k < n$. In the present case, obviously X is neither spherically uniform nor normal; also, if r is not an integral multiple of $\frac{1}{2}$, we cannot have X to be of the type considered in the previous example.

(iii) It is not true that every characterization of the normal distribution based on n independent identically distributed r.v.'s remains valid in the class of s.s.d.s (on $\mathbb{R}^n$). In particular, the characterization of the normal distribution given by Heyde (1970) which is based on the symmetry of a conditional distribution does not, in general, remain valid among s.s.d.s. (A corrected version of Heyde's result appears in Kagan *et al* (1973, p. 418).) On the other hand, from Theorems 6.3.2 and 6.3.4, it follows that there exist characterization properties of normality in the class of s.s.d's such as $E(X_1^2 + \cdots + X_m^2 | X_{m+1}^2 + \cdots + X_n^2) = c$ a.s. or the independence of $X_1^2 + \cdots + X_m^2$ and $X_{m+1}^2 + \cdots + X_n^2$, where $1 \leqslant m < n$, which do not happen to be characterization properties of normality in the class of all probability distributions on $\mathbb{R}^n$ giving the components of X to be independent and identically distributed.

Before closing the section, we may make some further observations that are of relevance to our discussion. In a private communication in 1988, C.G. Khatri conjectured that it was impossible to have a factorization of a gamma random variable (or, in particular, of an exponential random variable) with two independent factors such that one of them is a uniform random variable on (α, β) satisfying $0 < \alpha < \beta < \infty$. That the conjecture is false is shown by the following example appearing in Shanbhag and Kapoor (1993).

Example   Let U be a uniform random variable on $(0, 1)$, λ be a positive real number and Y be a positive random variable independent of U such that its characteristic function is given by

$$\phi(t) = \sum_{r=0}^{\infty} \frac{\lambda}{1 + \lambda} \left(\frac{1}{1 + \lambda} \right)^r \left\{ \frac{(1 + \lambda)^{r+1}}{(1 + \lambda)^{r+1} - it} \right\}^2, \quad t \in \mathbb{R}. \qquad (6.3.17)$$

(Note that the ϕ is a characteristic function of a scale mixture of gamma distributions with index 2.) It is easily seen that we have now $(1 + \lambda U)Y$ to be exponentially distributed with mean 1. We can now see that $(\alpha + (\beta - \alpha)U)((1/\alpha)Y)$, with U

and Y as above having λ replaced by $(\beta - \alpha)/\alpha$, is an exponential random variable with mean 1.

The example given shows, as a by-product, that there exist an s.s.d random vector $X = (X_1, X_2, X_3, X_4)$ and a 4×4 real diagonal matrix A such that XAX' is χ^2 distributed (with 2 degrees of freedom) and the characteristic function of $\|X\|^2$ is given by (6.3.17). As the X obtained here is clearly nonnormal, we have a contradiction to Theorem 1 of Khatri and Mukherjee (1987); see Cacoullos and Khatri (1991) for further details and a correction. It is, however, of interest to note here that the following theorem related to the Khatri–Mukherjee theorem holds.

6.3.6   Theorem

Let $\mathcal{E}_n^*(\Sigma)$ be the subclass of $\mathcal{E}_n(\Sigma)$ with each member of it such that the corresponding T in the representation (6.3.3) satisfies $E(T^\theta) < \infty$ for some $\theta > 0$. Further, let Q_1 be as specified above satisfying the first condition in (6.3.6). Then the distribution of $|Q_1(X)|$ (and hence that of $Q_1(X)$) determines, in $\mathcal{E}_n^*(\Sigma)$, the distribution of X.

Proof   The result is obvious from the uniqueness theorem for Mellin transforms on noting that

$$E(T^\alpha) = \frac{E\{|Q_1(X)|^{\alpha/k}\}}{E\{|Q_1(UC)|^{\alpha/k}\}} \quad \text{for each } \alpha \in [0, \theta_0)$$

for some $\theta_0 > 0$.

6.3.7   Remark

Theorem 6.3.6 is valid under several alternative conditions. We chose the present version of the theorem only for illustrative purposes.

In several of the recent publications, characterizations based on product decompositions of the type met in this section are obtained. Some of these papers implicitly involve versions of the following equation:

$$f(x) = \xi(x) \int_0^x (f(y))^\lambda \eta(y) \, dy \quad \text{for a.a.}[L]\, x \in (0, x_0) \tag{6.3.18}$$

where f, ξ and η are positive Borel measurable functions a.e.[L] on $(0, x_0)$ with ξ and η as given functions, and $x_0 \in (0, \infty]$ and $\lambda \in \mathbb{R}\backslash\{0\}$ are given. We can rewrite (6.3.18) as

$$\frac{(f(x))^\lambda \eta(x)}{\left(\int_0^x (f(y))^\lambda \eta(y) \, dy \right)^\lambda} = (\xi(x))^\lambda \eta(x) \quad \text{for a.a.}[L]\, x \in (0, x_0). \tag{6.3.19}$$

In view of Lemma 1.3.2, it easily follows that (6.3.19) holds if and only if for a.a.$[L]\, x \in (0, x_0)$

$$
f(x) = \begin{cases} \xi(x)\exp\left\{K + \displaystyle\int_c^x \xi(y)\eta(y)\,dy\right\} & \text{if } \lambda = 1, \\[2ex] \xi(x)\left\{K' + (1-\lambda)\displaystyle\int_c^x (\xi(y))^\lambda \eta(y)\,dy\right\}^{(1-\lambda)^{-1}} & \text{if } \lambda \neq 1, \end{cases}
$$

(6.3.20)

with $c \in (0, x_0)$, K and K' as constants, and $f(x)/\xi(x) \to 0$ as $x \to 0$ a.e.$[L]$. (Note that the K' in the second equation is to be taken such that

$$
K' + (1-\lambda)\int_c^x (\xi(y))^\lambda \eta(y)\,dy > 0,
$$

and in both the equations $\int_c^x$ is to be taken as $-\int_x^c$ if $x < c$.) If we take $\eta(y) = \alpha y^{\alpha-1}$ and $\xi(x) = x^{-\alpha}$ (with $\alpha > 0$), then (6.3.20) reduces to

$$
f(x) = \begin{cases} C & \text{if } \lambda = 1 \\ 1 & \text{if } \lambda < 1 \\ (1 + Kx^{\alpha(\lambda-1)})^{(1-\lambda)^{-1}} & \text{if } \lambda > 1 \end{cases}
$$

(6.3.21)

where C is a positive constant and $K > -x_0^{-\alpha(\lambda-1)}$. The function given by (6.3.21) on $(0, x_0)$ is a restriction to $(0, x_0)$ of the Laplace transform of a nondegenerate probability measure with domain of definition having $(0, x_0)$ as a subset (where the definition of a Laplace transform is to be understood as in Chapter 1) if and only if we have the f given by the third expression with $1 < \lambda \leqslant 1 + 2/\alpha$, and K as a nonzero constant such that $K < 0$ if $\alpha(\lambda - 1) > 1$ and $K > 0$ if $\alpha(\lambda - 1) < 1$ to be valid (noting also that the function defined by (6.3.21) requires $K > -x_0^{-\alpha(\lambda-1)}$). The result in question follows on appealing amongst other things to a result of Eaton *et al* (1971) since for $\lambda > 1$ the function f given by the third expression of (6.3.21) on $(0, x_0)$ is such that $\{(f(x/n^{(\alpha(\lambda-1))^{-1}}))^n : n = 1, 2, \ldots\}$ converges to $e^{-K(x^{\alpha(\lambda-1)}/(\lambda-1))}$ for each $x \in (0, x_0)$. In view of the result, it is now an easy exercise to see that the following result holds for any nondegenerate random variable X with moment generating function: for an $\alpha > 0$

$$
X \overset{\mathrm{d}}{=} U^{1/\alpha} Y
$$

(6.3.22)

with U as a uniformly distributed random variable on $(0, 1)$ and Y as a random variable that is distributed independently of U with moment generating function as the λth power of that of X, where $\lambda > 0$, if and only if $1 < \lambda \leqslant 1 + 2\alpha^{-1}$ and

$$
X \overset{\mathrm{d}}{=} V^{(\alpha(\lambda-1))^{-1}} W
$$

(6.3.23)

with W as an extreme strictly stable[†] random variable with exponent $\alpha(\lambda - 1)$ if $\alpha(\lambda - 1) \neq 1$ and a degenerate random variable otherwise, and V as a gamma

---

[†] If W is extreme strictly stable, then $-W$ is extreme strictly stable; also there is no extreme strictly stable law with exponent 1.

random variable with index $(\lambda - 1)^{-1}$ distributed independently of W. Special cases of this latter result have appeared in Kotz and Steutel (1988) and several other places (see, Pakes (1992) for a review).

Using essentially a technique the same as here but involving complex variables, Alamatsaz (1985) has obtained a somewhat different version of the result. This gives that for an $\alpha > 0$, for a nondegenerate random variable X, (6.3.22) with the moment generating function condition on Y replaced by the condition that the restriction of the characteristic function of Y to the largest symmetric interval on which the characteristic function of X does not vanish is given by the λth power of that of X, if and only if $1 < \lambda \leqslant 1 + 2\alpha^{-1}$, and (6.3.21) holds with 'an extreme strictly stable' replaced by 'a strictly stable' and 'strictly stable with exponent 1 or degenerate' in place of 'degenerate'.

The function given by (6.3.21) on $(0, x_0)$ can be written as $G(1 - x)$ with G as a nondegenerate probability generating function if and only if we have it given by the third expression with $1 < \lambda \leqslant 1 + \alpha^{-1}$ and $K > 0$. (Note that the expression cannot be $G(1 - x)$ with G as required for $\alpha(\lambda - 1) > 1$ or $K \leqslant 0$ because it leads to $G'(1-) \leqslant 0$.) This result easily gives a characterization based on a Steutel–van Harn operator product decomposition in the class of random variables that are nondegenerate and nonnegative integer-valued; the distribution characterized in this case is a mixture of discrete stable distributions (introduced by Steutel and van Harn (1979)) of exponent $\alpha(\lambda - 1)$ with the mixing distribution as gamma with index $(\lambda - 1)^{-1}$. Alzaid and Al-Osh (1989) and Sapatinas (1993) among others have considered characterizations of this type.

More recently, Pakes and Khatree (1992) have arrived at characterizations based on a factorization involving length-biasing. With X as a nonnegative real-valued nondegenerate random variable satisfying $E(X^\lambda) = \mu < \infty$, where $\lambda > 0$, they have studied the following property:

$$X \stackrel{\mathrm{d}}{=} V X_\lambda, \qquad (6.3.24)$$

where V is a nonnegative random variable and X_λ is a random variable distributed independently of V with distribution function $(\mu)^{-1} E\{X^\lambda I_{\{X \leqslant x\}}\}$, $x \in \mathbb{R}$. Their Theorem 5.1 characterizing a generalized gamma distribution with some parameter constraints on the basis of (6.3.24) when V has a certain generalized beta distribution turns out to be a simple corollary to Theorem 2.2.2. Also, some of their characterization results based on (6.3.24) happen to be straightforward corollaries to the following general theorem which, in turn, follows via a simpler version of the proof of our Theorem 6.3.2 (since, under the given assumptions, (6.3.24) is equivalent to the condition that $E(X^\theta) = \mu^{-1} E(V^\theta) E(X^{\theta+\lambda})$ for all $\theta \geqslant 0$).

6.3.8   Theorem

If there is an X satisfying the stated conditions for which (6.3.24) holds with the right extremity of the distribution of V as 1, then, given λ, μ and the distribution of V, (6.3.24) determines the distribution of X.

6.4   STABLE DISTRIBUTIONS AND RELATED CHARACTERIZATIONS

Applications of the results given in Chapters 2, 3 and 4, dealing with characterization problems on stable distributions have been considered by many authors including Davies and Shimizu (see, for example, Davies and Shimizu (1976) and Shimizu and Davies (1981)). Several of the results in this connection are discussed in the recent monograph of Ramachandran and Lau (1991). We shall now cover mainly the results that are not touched upon in the cited monograph.

The definitions of univariate and multivariate stable distributions as appearing in Lukacs (1970), Paulauskas (1976) and elsewhere involve several redundant conditions.

According to the definition that is followed in the literature, a nondegenerate d.f. F on $\mathbb{R}^p$ is said to be stable if its characteristic function ϕ satisfies the condition that for every $c_1, c_2 \in (0, \infty)$, there exist $c \in (0, \infty)$ and $a \in \mathbb{R}^p$ for which

$$\phi(c_1 t)\phi(c_2 t) = \exp\{i\langle a, t\rangle\}\phi(ct), \quad t \in \mathbb{R}^p.$$

(A random vector and the characteristic function corresponding to a stable d.f. are also, for convenience, referred to as stable.) Any stable d.f. on $\mathbb{R}^p$ is infinitely divisible, i.e. is such that for every positive integer m, it can be expressed as the m-fold convolution of a d.f. on $\mathbb{R}^p$, or, equivalently, such that its characteristic function ϕ is of the form

$$\phi(t) = \exp\left\{ i\langle a, t\rangle - \tfrac{1}{2}Q(t) + \int_{\mathbb{R}^p} \left(e^{i\langle t, x\rangle} - 1 - \frac{i\langle t, x\rangle}{1 + ||x||^2} \right) v(\mathrm{d}x) \right\},$$

$$t \in \mathbb{R}^p$$

with a as a real vector, Q as a nonnegative definite quadratic form, and v as a measure on $\mathbb{R}^p$ for which $v(\{0\}) = 0$ and

$$\int_{\mathbb{R}^p} \frac{||x||^2}{1 + ||x||^2} v(\mathrm{d}x) < \infty.$$

(The representation for ϕ in the last equation is unique and is referred to as the Lévy canonical or spectral representation, and the measure relative to it as the Lévy measure.) If ϕ is stable, then, corresponding to it, there is a fixed $\alpha \in (0, 2]$ such that each c satisfies the equation, $c^\alpha = c_1^\alpha + c_2^\alpha$; the α in question is referred to as the exponent of the stable law.

The following theorem shows explicitly as to how the definition of a stable distribution referred to above involves many redundancies. Theorem 2.2.2 or Corollary 2.3.2, in addition to a certain result in Hardy (1967), could be used to prove this theorem.

6.4.1   Theorem (Rao and Shanbhag, 1992b; Zeng, 1992).

Let ϕ be the characteristic function of a nondegenerate probability distribution on $\mathbb{R}^p$. Then the following are equivalent:

(i)   For some positive constants c_1, c_2, c_3 with $c_2 \neq 1$, and $a_1, a_2 \in \mathbb{R}^p$,

$$(\phi(t))^2 = \exp\{i \langle a_1, t \rangle\} \phi(c_1 t), t \in \mathbb{R}^p \tag{6.4.1}$$

and

$$\phi(t)\phi(c_2 t) = \exp\{i \langle a_2, t \rangle\} \phi(c_3 t), \quad t \in \mathbb{R}^p. \tag{6.4.2}$$

(The condition implies that $c_1, c_3 \neq 1$, as ϕ is nondegenerate.)

(ii) For some positive constants c_1, c_2, c, and $a \in \mathbb{R}^p$, such that c_1/c and c_2/c are noncommensurable (i.e. such that there are no integers m and n for which $(c_1/c)^m = (c_2/c)^n$),

$$\phi(c_1 t)\phi(c_2 t) = \exp\{i \langle a, t \rangle\} \phi(ct), \quad t \in \mathbb{R}^p. \tag{6.4.3}$$

(iii) ϕ is stable.

Proof   It follows easily that (iii) $\Rightarrow$ (i) and (ii). (To see that (iii) $\Rightarrow$ (ii), note that if ϕ is stable with exponent α, then (ii) holds with c_1, c_2, c_3 such that $c_1^\alpha = x$, $c_2^\alpha = 1 - x$ and $c = 1$, where $x \in (0, 1)$ and is chosen so that $\log x / \log(1 - x)$ is irrational.) Hence it is sufficient if we establish that (i) $\Rightarrow$ (iii) and (ii) $\Rightarrow$ (iii). (6.4.1) implies that ϕ is infinitely divisible and so also does (6.4.3). Assume then that ϕ is infinitely divisible and v is its Lévy canonical measure. To prove the assertion, we can assume without loss of generality that v is concentrated on $\mathbb{R}^p_+$. Then, (6.4.2) implies that

$$v([x, \infty)) + v([c_2^{-1}x, \infty)) = v([c_3^{-1}x, \infty)), \quad x \in \mathbb{R}^p_+ \tag{6.4.4}$$

and (6.4.1) implies that

$$v([x, \infty)) = \tfrac{1}{2} v([c_1^{-1}x, \infty)), \quad x \in \mathbb{R}^p_+, \tag{6.4.5}$$

where ∞ denotes the p-component vector with each component equal to ∞. If v is nonzero, then, in view of Theorem 2.2.2 or Corollary 2.3.2, (6.4.4) and (6.4.5) imply that for some α

$$1 + c_2^\alpha = c_3^\alpha \tag{6.4.6}$$

and

$$2 = c_1^\alpha. \tag{6.4.7}$$

As we cannot have nonzero integers n_1, n_2, n_3, n_4 such that

$$1 + 2^{n_1/n_2} = 2^{n_3/n_4}$$

(cf. Hardy, 1967, p. 37) it is immediate that (6.4.6) and (6.4.7) imply that $\log c_2 / \log c_1$ or $\log c_3 / \log c_1$ is irrational. Hence, in view of Theorem 2.2.2 or Corollary 2.3.2, (6.4.4) and (6.4.5) yield that

$$v([c^{-1}x, \infty)) = c^\alpha v([x, \infty)), \quad x \in \mathbb{R}^p_+, c > 0 \tag{6.4.8}$$

for some $\alpha \in (0, \infty)$. The integration condition

$$\int_{\mathbb{R}^p_+} \frac{\|x\|^2}{1 + \|x\|^2} \nu(dx) < \infty$$

on ν then implies that $\alpha \in (0, 2)$ if $\nu \not\equiv 0$. Note that if $\nu \equiv 0$, we have ϕ to be normal. On the other hand if $\nu \not\equiv 0$ and (6.4.8) is met, we get ϕ to be a non-Gaussian stable characteristic function. (Note that in the latter case, we have $\alpha = (\log 2/\log c_1) \in (0, 2)$, and hence it follows due to (6.4.1) that ϕ cannot have a Gaussian factor.) From what we have seen, it is clear that (i) $\Rightarrow$ (iii).

The implication (ii) $\Rightarrow$ (iii) is easier to get from Theorem 2.2.2 or Corollary 2.3.2 on noting the functional equation satisfied by ν since in this case it is assumed that c_1/c and c_2/c are noncommensurable. ∎

One of the by-products of what we have observed in the proof of Theorem 6.4.1 is that if ϕ is stable, then it is a nonvanishing characteristic function (indeed an infinitely divisible characteristic function) for which for each $c \in (0, \infty)$ there exists an $m_c \in \mathbb{R}^p$ such that

$$\phi(ct) = (\phi(t))^{c^\alpha} \exp\{i \langle m_c, t \rangle\}, \quad t \in \mathbb{R}^p \tag{6.4.9}$$

with $\alpha \in (0, 2]$ (independently of c); the converse of this assertion is trivial. Following Feller (1971), if we call a distribution whose characteristic function ϕ is nonvanishing and satisfying (6.4.9) with each $m_c = 0$, strictly stable with exponent α, then it is now obvious that a p-component random vector has its distribution to be strictly stable with exponent α, on $\mathbb{R}^p$ if and only if each linear combination of the components of the vector has a strictly stable distribution with exponent α, on $\mathbb{R}$.

One may now raise the question as to whether the result just met holds if 'strictly stable' is replaced by 'stable'. The answer to this question is in the negative. Indeed, an example given in Marcus (1983) shows that this modified result does not hold for $\alpha < 1$.

However, for $\alpha \geqslant 1$ we have a different picture and the question is answered in the affirmative, as is shown by the next theorem; the crucial part of the theorem, i.e. the part for $\alpha = 1$, was essentially established recently by Samorodnitsky and Taqqu (1991).

6.4.2   Theorem

Let $\alpha \in [1, 2]$. Then a p-component random vector possesses the property that each linear combination of its components has a stable distribution with exponent α, on $\mathbb{R}$, if and only if its characteristic function ϕ is nonvanishing and is such that for each $c \in (0, \infty)$

$$\phi(ct) = (\phi(t))^{c^\alpha} \exp\{i \gamma_c \langle m, t \rangle\}, \quad t \in \mathbb{R}^p \tag{6.4.10}$$

with $m \in \mathbb{R}^p$ and γ_c such that

$$\gamma_c = \begin{cases} -c \log c & \text{if } \alpha = 1 \\ c - c^\alpha & \text{if } \alpha \in (1, 2]. \end{cases}$$

(In view of what is revealed above, we have then the condition on ϕ to be equivalent to that it is the characteristic function of a stable distribution with exponent α, on $\mathbb{R}^p$.)

Proof   The 'if' part of the assertion is trivial. We shall now prove the 'only if' part of the assertion. Assume then that X is a p-component random vector with the stated property and that ϕ is its characteristic function. As, for every given $t \in \mathbb{R}^p$, we have $\langle t, X \rangle$ to be a stable random variable (with exponent α), it follows that ϕ is nonvanishing and, for each $c \in (0, \infty)$,

$$\phi((cs)t) = (\phi(st))^{c^\alpha} \exp\{i\gamma_c m^*(t)s\}, \quad s \in \mathbb{R}, \quad t \in \mathbb{R}^p \tag{6.4.11}$$

for some function $m^* : \mathbb{R}^p \to \mathbb{R}$ such that

$$m^*(0, \ldots, 0, t_r, 0, \ldots, 0) = t_r m^*(0, \ldots, 0, 1, 0, \ldots 0),$$

with t_r on the left-hand side and 1 on the right-hand side as the rth coordinates, for $r = 1, 2, \ldots p$; this is a corollary to the result for a stable law on $\mathbb{R}$ that is stated on page 9 in Kagan *et al* (1973). (6.4.11), in turn, implies that for $k = 1, 2, \ldots$.

$$\phi(st) = [(\phi(k^{-1/\alpha}st))^k \exp\{i\gamma_k^* s \langle m, t \rangle\}] \exp\{i\gamma_k^* s(m^*(t) - \langle m, t \rangle)\},$$

$$s \in \mathbb{R}, \quad t \in \mathbb{R}^p, \tag{6.4.12}$$

where $\gamma_k^* = k^{-1/\alpha}\gamma_{k^{1/\alpha}}$ and $m = (m^*(1, 0, \ldots, 0), m^*(0, 1, \ldots, 0), \ldots, m^*(0, 0, \ldots, 1))$. The family $\{P_k\}$ of probability measures, where for each k, P_k denotes the probability measure corresponding to the characteristic function

$$\phi_k(t) = (\phi(k^{-1/\alpha}t))^k \exp\{i\gamma_k^* \langle m, t \rangle\}, \quad t \in \mathbb{R}^p$$

is hence tight (in the sense of Billingsley (1968, p. 37)) (because the families of probability measures for the corresponding coordinate mappings are tight in view of (6.4.12)). This, in turn implies that for each $t \in \mathbb{R}^p$, the family of probability measures on $\mathbb{R}$ corresponding to the characteristic functions $\phi_k^{(t)}$, where

$$\phi_k^{(t)}(s) = (\phi(k^{-1/\alpha}st))^k \exp\{i\gamma_k^* s \langle m, t \rangle\}, \quad s \in \mathbb{R}$$

is also tight, and hence (6.4.12) implies that $m^*(t) = \langle m, t \rangle$ for each $t \in \mathbb{R}^p$ (because $\gamma_k^* \to -\infty$ as $k \to \infty$). The result sought then follows from (6.4.11) on taking $s = 1$. ∎

Dudley and Kanter (1974) and Gine and Hahn (1983) have made some further interesting observations on the problem. In particular, Gine and Hahn have shown that if we assume the distribution of the p-component random vector to be infinitely divisible, then the characterization concerning strictly stable distributions also holds in the case of general stable distributions irrespectively of whether or not $\alpha \geqslant 1$. This latter result is easily verified on noting that if X is a random vector having an infinitely divisible distribution, then the condition that each linear combination of the components of X has a stable distribution with exponent α implies that either $\alpha = 2$ and X is Gaussian or $0 < \alpha < 2$ and the characteristic function of X possesses the properties that it does not have a Gaussian factor and the Lévy measure ν corresponding to it is such that for every $c \in (0, \infty)$, $(c_1, \ldots, c_p) \in \mathbb{R}^p$, and Borel set B of $\mathbb{R}$

$$\nu\left(\left\{(x_1, \ldots, x_p) \in \mathbb{R}^p : c \sum_{i=1}^p c_i x_i \in B\right\}\right)$$

$$= c^\alpha \nu\left(\left\{(x_1, \ldots, x_p) \in \mathbb{R}^p : \sum_{i=1}^p c_i x_i \in B\right\}\right)$$

and hence that the distribution of X is stable with exponent α; the converse assertion is obviously trivial.

From Kagan *et al* (1973), assuming $X_1, \ldots, X_n$ to be independent identically distributed (i.i.d.) random variables, regression properties of the type

$$E\{a_1 X_1 + \cdots + a_n X_n | b_1 X_1 + \cdots + b_n X_n\} = 0 \quad \text{a.s.} \tag{6.4.13}$$

follow under some constraints as characteristic properties of stable distributions with characteristic exponent $\alpha > 1$. In view of the characterization for stable distributions with exponent $\alpha \geqslant 1$ mentioned above, we have that these conditions can be modified by taking X_i as p-component random vectors (instead of real random variables) to have characterizations of stable distributions on $\mathbb{R}^p$. (Properties of the type (6.4.13) cannot hold for stable distributions with exponent $\alpha \leqslant 1$, because they assume implicitly the random variables involved to be integrable.) It is also possible to characterize stable distributions under appropriate conditions via factorizations of characteristic functions of the type in Chapter 5 in Kagan *et al* (1973).

Also from Theorem 6.4.1, several other characterization results follow, as observed in Zeng (1992). The following is one such result; according to Zeng (1992), this is an unpublished result of A.K. Gupta, T.T. Nguyen and W.B. Zeng arrived at sometime in 1988.

6.4.3   Corollary

Let X_1, X_2 and X_3 be i.i.d. p-component random vectors. Then X_1 has a stable distribution if and only if there exist $\alpha \in (0, 2]$ and $a_1, a_2 \in \mathbb{R}^p$ such that $2^{1/\alpha} X_1 + a_1 \overset{d}{=} X_1 + X_2$ and $3^{1/\alpha} X_1 + a_2 \overset{d}{=} X_1 + X_2 + X_3$.

This result in turn yields as its corollary, the following result (which is also a result stated in Zeng (1992)).

6.4.4   Corollary

Let X_1, X_2 and X_3 be i.i.d. p-component random vectors. Then

(i) X_1 has a multivariate stable distribution with Cauchy marginals if $2X_1 \overset{d}{=} X_1 + X_2$ and $3X_1 \overset{d}{=} X_1 + X_2 + X_3$;

(ii) X_1 has a multivariate normal distribution with zero mean vector if and only if $\sqrt{2}X_1 \overset{d}{=} X_1 + X_2$ and $\sqrt{3}X_1 \overset{d}{=} X_1 + X_2 + X_3$.

6.4.5   Remarks

(i) From what is seen in the proof of Theorem 6.4.1 or from the validity of (6.4.9) for all $c \in (0, \infty)$, it follows easily that a probability distribution on $\mathbb{R}$ is stable if and only if it is an infinitely divisible distribution that is either Gaussian or is such that its characteristic function has no Gaussian factor and its Lévy measure ν satisfies for some $\alpha \in (0, 2)$

$$\nu(\mathrm{d}x) = \begin{cases} c_1 x^{-\alpha - 1}\,\mathrm{d}x & \text{if } x \in (0, \infty) \\ c_2 |x|^{-\alpha - 1}\,\mathrm{d}x & \text{if } x \in (-\infty, 0) \end{cases}$$

with c_1, $c_2 \geqslant 0$ such that $c_1 + c_2 > 0$. (Note that by definition Gaussian and stable distributions are nondegenerate, but, for simplicity, we have allowed in characterization results in this section degenerate distributions also to be referred to as Gaussian or stable.)

(ii) In view of the form of ν stated in (i) above, we have that m_c in (6.4.9) for $\alpha \in (0, 1)$ is of the form $(c - c^{\alpha})m$; for $\alpha \in [1, 2]$, the form of m_c is obviously as specified by (6.4.10) in the statement of Theorem 6.4.2. (The form of ν also shows easily that m_c is of the form $(c - c^{\alpha})m$ when $\alpha \in (1, 2)$.)

(iii) If $\alpha \in (1, 2]$, then the fact that any stable distribution with characteristic exponent α has a (real) mean vector implies trivially that m_c in (6.4.9) is $(c - c^{\alpha})m$ with m as the mean vector corresponding to the distribution in question. This provides us not only with an alternative argument for that if $\alpha \in (1, 2]$ the form of m_c in (6.4.9) is as stated, but also tells us that in the present case the assertion of Theorem 6.4.2 is a trivial consequence of the result concerning strictly stable distributions that we have met above. This is yet another observation of Gine and Hahn (1983).

From what we have seen so far, it is clear that both Theorem 2.2.2 and Corollary 3.3.2 have applications in the formulation and analysis of stable laws. Recently, Steutel and van Harn (1979) have introduced a new class of distributions on $\mathbb{N}_0$ referred to as the class of discrete (strictly) stable distributions; according to their definition, a probability distribution $\{p_j : j = 0, 1 \ldots\}$ with $0 < p_0 < 1$ is discrete stable with exponent $\alpha > 0$ if and only if its probability generating function P satisfies the equation

$$P(z) = P(1 - c + cz)P(1 - (1 - c^{\alpha})^{1/\alpha}(1 - z)), \quad |z| \leqslant 1, \quad 0 < c < 1. \quad (6.4.14)$$

Note that taking $P^*(x) = -\log P(1 - x)$, $x \in (0, 2]$, we can see that (6.4.14) (even when the condition that c takes all values in $(0,1)$ is replaced by that it is a fixed number in $(0,1)$, provided $(1/\alpha)(\log(1 - c^\alpha)/\log c)$ is irrational) implies, in view of Theorem 2.2.2 (or, on extending the function appropriately, in view of Corollary 2.3.2)

$$P^*(x) \propto x^\lambda, \quad x \in (0, 2]$$

where $\lambda > 0$ is such that $c^\lambda + ((1 - c^\alpha)^{1/\alpha})^\lambda = 1$, i.e. such that $(1 - c^\lambda)^{1/\lambda} = (1 - c^\alpha)^{1/\alpha}$ implying that $\lambda = \alpha$. As P is a probability generating function with $P(0) < 1$, this implies that for some $k > 0$

$$P(z) = e^{-k(1-z)^\alpha}, \quad |z| \leqslant 1, \tag{6.4.15}$$

and $\alpha \leqslant 1$. (Note that $\alpha > 1$ gives $P'(1-) = 0$.) One arrives at the same conclusion, even when we consider the functional equation relative to the restriction to $(0, a)$ of P^*, where $0 < a < 2$, and it is trivial to see that (6.4.15) implies (6.4.14). Steutel and van Harn (1979) have used an alternative approach to prove that a discrete stable distribution has the probability generating function of the form (6.4.15) with $\alpha \leqslant 1$. It is also clear that one could give an alternative definition of a discrete stable distribution involving just one c. Furthermore, if we do not wish to specify a priori the exponent of the stable distribution in the definition, then we can have some alternatives to this as suggested by the next theorem; in the light of the information that is already given including, in particular, the reference of Hardy (1967) in the proof of Theorem 6.4.1, the proof of this theorem could now be considered to be a simple exercise and hence we do not deal with it here.

6.4.6   Theorem

Let P be the generating function of a probability distribution on $\mathbb{N}_0$ such that $0 < P(0) < 1$. Then the following are equivalent:

(i)   For some constants $c_1, c_2, c_3 \in (0, 1)$,

$$(P(z))^2 = P(1 - c_1 + c_1 z), \quad |z| \leqslant 1$$

and

$$P(z)P(1 - c_2 + c_2 z) = P(1 - c_3 + c_3 z), \quad |z| \leqslant 1.$$

(ii)   For some constants $c_1, c_2, c \in (0, 1]$ such that c_1/c and c_2/c are noncommensurable,

$$P(1 - c_1 + c_1 z)P(1 - c_2 + c_2 z) = P(1 - c + cz), \quad |z| \leqslant 1.$$

(iii) P is given by (6.4.15) for some $\alpha \in (0, 1]$. (The theorem also holds if we replace, in the conditions, $|z| \leqslant 1$ by $z \in (0, a)$, where a is a given number in $(0, 1]$.)

6.4.7   Remarks

(i) Following the analogy with the stable distribution that we have dealt with earlier, we could now introduce a (strictly) stable distribution, in the sense of

Steutel and van Harn, on $\mathbb{N}_0^n$ in an obvious way and get easily that a distribution $\{P_{j_1,\dots,j_n}: j_1, \dots, j_n = 0, 1, \dots\}$ with $0 < P_{0,\dots,0} < 1$ is stable on $\mathbb{N}_0^n$ if and only if its probability generating function P satisfies for some $\alpha \in (0, 1]$

$$P(1 - c + cz_1, \dots, 1 - c + cz_n) = (P(z_1, \dots, z_n))^{c^\alpha}, \quad |z_i| \leqslant 1,$$

$$i = 1, 2, \dots n; c \in (0, 1). \tag{6.4.16}$$

The multivariate version of Theorem 6.4.6 could also be seen to be valid. The relation (6.4.16) has an interpretation in terms of the Steutel–van Harn product operator; if we refer to α in (6.4.16) as the characteristic exponent of the stable distribution in question, then with the product operator as $\circ$, it easily follows that a random vector $(X_1, \dots, X_n)$ with nonnegative integer-valued components has a stable distribution with exponent α in the present sense, if and only if for each $c_1, \dots, c_n \in (0, 1)$, the variable $c_1 \circ X_1 + \cdots + c_n \circ X_n$ has a stable distribution with exponent α, on $\mathbb{N}_0$.

(ii) In addition to stable distributions and some others that we have discussed in this monograph, there are many more that have characterization results that are implicitly or explicitly linked with the results of Chapters 2 and 3. Gupta and Richards (1990, 1993) have shown that the Dirichlet distributions and the multivariate Liouville distributions are amongst these. Marshall (1989) gives a construction for a bivariate uniform distribution via a property that is implicitly linked with the Choquet–Deny theorem. Ramachandran and Lau (1991) have dealt with several other applications of these results.

Finally, we make some observations on as to how the results on stable distributions derived in this section lead us to extended versions of the characterizations, based on the factorizations of the type (6.3.22), considered in the previous section. Suppose X is a nondegenerate n-component random vector with a moment generating function. Then, for $\alpha > 0$, (6.3.22), with U and Y as stated but for a modification that Y is now taken as an n-component random vector, holds if and only if (6.3.23) with V and W as stated but for a modification that W is now an n-component random vector (instead of a random variable) holds. (Here by an extreme strictly stable random vector, we mean a strictly stable random vector with the property that the corresponding Lévy canonical measure is concentrated on a set of the type $\prod_{i=1}^n S_i$ with $S_i = \mathbb{R}_+$ or $\mathbb{R}_-$.) Also, the discrete analogue of this result holds. The result in this latter case is that a nondegenerate n-component random vector X with nonnegative integer-valued components satisfies the relation

$$X \overset{d}{=} U^{1/\alpha} \circ Y$$

with $\circ$ as the Steutel–van Harn operator and U and Y possessing the properties of the result stated immediately above, if and only if $1 < \lambda \leqslant 1 + \alpha^{-1}$ and the probability generating function of X is of the form

$$G(\bullet) = E((H_{\alpha(\lambda-1)}(\bullet))^V),$$

where $H_{\alpha(\lambda-1)}(\bullet)$ is the generating function of a stable distribution on $\mathbb{N}_0^n$ with exponent $\alpha(\lambda - 1)$, and V is a gamma random variable with index $(\lambda - 1)^{-1}$. Obviously both the characterizations mentioned here are of mixtures of stable distributions with exponent $\alpha(\lambda - 1)$ relative to gamma mixing distributions with index $(\lambda - 1)^{-1}$. (In the first case, when $\alpha(\lambda - 1) = 1$, we allow the degenerate distributions to be referred to as stable with exponent 1 for convenience.) These results are easy to obtain.

The result of Alamatsaz (1985) also has an obvious extension as discussed below.

Let X be a nondegenerate n-component random vector, and A denote a bounded interval containing the origin on which the characteristic function of X does not vanish but on its closure at least at one point the characteristic function in question vanishes if such an interval exists, and A denote $\mathbb{R}^n$ otherwise. Then

$$X \stackrel{\mathrm{d}}{=} U^{1/\alpha} Y \tag{6.4.17}$$

with U as a uniform random variable on $(0, 1)$ as mentioned before and Y as an n-component random vector independent of U such that the restriction to A of the characteristic function of Y agrees with the λth power of that of X if and only if (6.3.21) with 'random vector' in place of 'random variable', 'a strictly stable' in place of 'an extreme strictly stable', and 'strictly stable with exponent 1 or degenerate' in place of 'degenerate', is valid.

The 'if' part of this last result follows trivially, and we may see the validity of its 'only if' part as follows.

Let (6.4.17) be valid, ϕ be the characteristic function of X and A be as defined above. We have then for $c > 0$ and t such that $ct \in A$

$$\phi(ct) = \alpha c^{-\alpha} \int_0^c (\phi(tz))^\lambda z^{\alpha-1} \, \mathrm{d}z,$$

which implies that

$$\left(c^\alpha h_c(ct) + \alpha c^{\alpha-1}\right) e^{h(ct)} = \alpha \, e^{\lambda h(ct)} c^{\alpha-1}, \tag{6.4.18}$$

where $h(ct) = \log \phi(ct)$ and $h_c(ct) = (\partial/\partial c)h(ct)$. From (6.4.18), we have

$$\frac{\partial}{\partial c}(c^{\alpha(1-\lambda)} e^{(1-\lambda)h(ct)}) = \alpha(1 - \lambda)c^{\alpha(1-\lambda)-1} \quad \text{if } \lambda \neq 1 \tag{6.4.19}$$

and

$$h_c(ct) = 0 \quad \text{if } \lambda = 1. \tag{6.4.20}$$

Since X is nondegenerate, (6.4.20) implies that $\lambda \neq 1$. Now, if $\lambda \neq 1$, we have from (6.4.19)

$$c^{\alpha(1-\lambda)} e^{(1-\lambda)h(ct)} = c^{\alpha(1-\lambda)} + (e^{(1-\lambda)h(t)} - 1) \tag{6.4.21}$$

for all c, t such that t, $ct \in A$. Taking a fixed $t\varepsilon A$ such that $e^{(1-\lambda)h(t)} - 1 \neq 0$ (which exists as X is nondegenerate), we arrive at a contradiction to the fact that

$e^{(1-\lambda)h(ct)}$ is bounded for sufficiently small c, unless, in (6.4.21), $\lambda > 1$. We shall now take $\lambda > 1$ and rewrite (6.4.21) as

$$\exp\{(1 - \lambda)h(ct)\} = 1 + c^{\alpha(\lambda-1)}(\exp\{(1 - \lambda)h(t)\} - 1). \tag{6.4.22}$$

As (6.4.22) holds for c, t such that t and ct lie in A, considering a fixed $c > 1$ and observing that the limit of $|\exp\{(1 - \lambda)h(ct)\}|$ as ct tends to any point in A-closure is finite, we can then conclude that there does not exist a bounded interval A with the stated property. This, in turn, implies that ϕ is nonvanishing on the whole of $\mathbb{R}^n$ and hence that $A = \mathbb{R}^n$; we have then (6.4.22) to be valid for all $c > 0$ and $t \in \mathbb{R}^n$. In view of the continuity theorem for characteristic functions, as (6.4.22) implies that

$$(\exp\{h(ct)\})^{c^{\alpha(1-\lambda)}} \to \exp\{(\exp\{(1 - \lambda)h(t)\} - 1)(1 - \lambda)^{-1}\}$$

as $c \to 0$ with $c^{\alpha(1-\lambda)}$ as an integer, for each $t \in \mathbb{R}^n$ (noting that $\alpha > 0$, $\lambda > 1$), it then follows easily that there exists a nonvanishing characteristic function ϕ^* on $\mathbb{R}^n$ such that

$$\log \phi^*(t) = -K(\exp\{(1 - \lambda)h(t)\} - 1), \quad t \in \mathbb{R}^n$$

where $K = (\lambda-1)^{-1}(> 0)$. The validity of (6.4.22) for all $c > 0$ and $t \in \mathbb{R}^n$ implies hence that $\alpha(\lambda - 1) \leqslant 2$, $\log \phi^*$ or $K^{-1} \log \phi^*$ is the second characteristic (in the sense of Lukacs (1970)) of a strictly stable distribution with exponent $\alpha(\lambda - 1)$ if $\alpha(\lambda - 1) \neq 1$ and strictly stable with exponent 1 or degenerate if $\alpha(\lambda - 1) = 1$, and

$$\phi(t) = (1 - K^{-1} \log \phi^*(t))^{(1-\lambda)^{-1}}, \quad t \in \mathbb{R}^n.$$

The result sought is then obvious.

We have given the proof in this last case only because it is more complicated than the proofs required for the two previous cases.

The multivariate extension to Alamatsaz's result (1985) that we have given subsumes partial results given earlier by Kotz and Steutel (1988) and Alamatsaz (1993); the proof given is adapted from that appearing in Shanbhag (1972b).

CHAPTER 7

Damage Models and Partial Independence

7.1 INTRODUCTION

A damage model can be typified by a random vector (X, Y) with X, Y as scalars (or, more generally, as k-component vectors, where k is a positive integer) satisfying the condition that $0 \leqslant Y \leqslant X$ a.s. (If X and Y are k-component vectors, we take 0 as the k-component vector with all components equal to zero.) We may view Y as the undamaged part of X and $X - Y$ as the damaged part of X, assuming that X is subjected to some destructive process; in that case, we refer to the conditional distribution of Y given X as the survival distribution. Almost all of the research in the area concerns the case of X, Y integer-valued (or of X, Y having integer-valued components when they are vectors) and this is the case we restrict ourselves to herein.

Rao and Rubin (1964) initiated research in the area of damage models, and was followed by many papers, including Srivastava and Srivastava (1970), Talwalker (1970), Shanbhag (1974, 1977), Alzaid et al (1986b) and several others. One of the questions addressed in the cited literature is the following: knowing the survival distribution and that Y and $X - Y$ satisfy a certain partial independence condition, what information can we obtain about the form of the distribution of X? Amongst frequently used partial independence conditions is the Rao–Rubin condition, introduced by Rao (1963), stating that the conditional distribution of Y given that X is undamaged (i.e. given that $X = Y$) equals the (unconditional) distribution of Y. Various modified versions of this have appeared in Patil and Taillie (1980), Talwalker (1980), Rao et al (1980), Alzaid et al (1986b, 1987b, 1988) and others.

To the literature prior to the emergence of Rao and Rubin (1964), a significant contribution came from Moran (1952). From this, it follows essentially that if X and Y are (scalars) such that the survival distribution is binomial with index x for almost all (values) x of X, then, under some mild conditions, Y and $X - Y$ are independent if and only if X is Poisson and the success probability parameter of the survival distribution is independent of x for almost all x. One may view this result as a variant of the main result of Rao and Rubin (1964); the latter result states that if the survival distribution is binomial with parameter vector (x, p) for almost all x, where $0 < p < 1$ and fixed, and $P\{X = 0\} < 1$, then the Rao–Rubin condition is met if and only if X is Poisson. Patil and Seshadri (1964), Menon (1966), Kimeldorf *et al* (1981), Panaretos (1982b) and Alzaid *et al* (1986a) amongst others have extended Moran's result or provided its variants.

We devote the present chapter to discussing various characterizations based on damage models and partial independence, highlighting their salient features and unifying them wherever possible.

7.2 THE RAO–RUBIN AND SHANBHAG THEOREMS AND RELATED RESULTS

Let (X, Y) be a random vector of nonnegative integer-valued components such that

$$P\{X = n, Y = r\} = g_n S(r|n), \quad r = 0, 1, \ldots, n; \quad n = 0, 1, \ldots, \quad (7.2.1)$$

where $\{g_n : n = 0, 1, \ldots\}$ and, for all $n \geqslant 0$, $\{S(r|n) : r = 0, 1, \ldots, n\}$ are discrete probability distributions. (Note that (7.2.1) is equivalent to stating that the marginal distribution of X is $\{g_n\}$ and, for each $n \geqslant 0$ with $g_n > 0$, the conditional distribution of Y given that $X = n$ is $\{S(r|n) : r = 0, 1, \ldots, n\}$.) Recall the definition of a generating function that appeared in Section 1.1. Following it, define G to be the generating function of $\{g_n\}$ and denote by $\mathcal{D}$ its domain of definition. If (7.2.1) is valid with

$$S(r|n) = \binom{n}{r} p^r (1 - p)^{n-r}, \quad r = 0, 1, \ldots, n \quad (7.2.2)$$

for each $n \geqslant 0$, where $0 < p < 1$ and fixed, then the Rao–Rubin condition

$$P\{Y = r\} = P\{Y = r|X = Y\}, \quad r = 0, 1, \ldots \quad (7.2.3)$$

implies that $\mathcal{D} = \mathbb{R}$. This follows because, in the present case, (7.2.3) is equivalent to

$$cg_r = \sum_{n=r}^{\infty} g_n \binom{n}{r} (1 - p)^{n-r}, \quad r = 0, 1, 2, \ldots, \quad (7.2.4)$$

where $c = (P\{X = Y\})^{-1}$, implying that if $z > 0$ and $z \in \mathcal{D}$, then $z + (1 - p) \in \mathcal{D}$.

We now give the following celebrated theorem due to Rao and Rubin (1964).

7.2.1 Theorem

Let (X, Y) be a random vector with X, Y as nonnegative integer-valued satisfying (7.2.1) with $g_0 < 1$, and (7.2.2). Then the Rao–Rubin condition (7.2.3) holds if and only if $\{g_n\}$ is Poisson.

Proof The 'if' part follows easily. To prove the 'only if' part, assume that (7.2.3) holds. We have then that the generating function G of $\{g_n\}$ has its domain of definition to be equal to $\mathbb{R}$ and it is such that

$$G(z + q) = cG(z), \quad z \in \mathbb{R}, \quad (7.2.5)$$

where $q = 1 - p$ and $c = (P\{X = Y\})^{-1} (= (G(p))^{-1})$. From (7.2.5), we get immediately that for each integer k

$$G(z + kq) = c^k G(z), \quad z \in \mathbb{R}. \tag{7.2.6}$$

We may prove the assertion now using either of the following two arguments. These are essentially due to Rao and Rubin (1964) and Shanbhag (1974) respectively.

The first argument   Since G is a probability generating function (with domain of definition $\mathbb{R}$), satisfying (7.2.6), it follows easily that the restriction of G to $(-\infty, 0)$ is absolutely monotonic. Then, essentially from Theorem 3.5.1, we see that

$$G(z) = \int_{\mathbb{R}_+} \exp\{zx\} \nu(dx), \quad z \in (-\infty, 0)$$

with ν as a measure on $\mathbb{R}_+$ such that $\exp\{qx\} = c$ for a.a. $[\nu]x \in \mathbb{R}_+$. Hence, it follows that G is Poisson.

The second argument   Since G is a probability generating function satisfying (7.2.6), we have it to be nonvanishing. Recalling the definition of G, we have then, in view of (7.2.6),

$$\frac{G'(z)}{G(z)} = \frac{G'(z + kq)}{G(z + kq)} = \lim_{k \to \infty} \left\{ \frac{\displaystyle\sum_{j=1}^{\infty} j g_j (z + kq)^{j-1}}{\displaystyle\sum_{j=0}^{\infty} g_j (z + kq)^j} \right\}, \quad z \in \mathbb{R} \tag{7.2.7}$$

and

$$\frac{G''(z)}{G(z)} = \frac{G''(z + kq)}{G(z + kq)} = \lim_{k \to \infty} \left\{ \frac{\displaystyle\sum_{j=1}^{\infty} j(j-1) g_j (z + kq)^{j-2}}{\displaystyle\sum_{j=0}^{\infty} g_j (z + kq)^j} \right\}$$

$$= \lim_{k \to \infty} \left\{ \frac{\displaystyle\sum_{j=1}^{\infty} j^2 g_j (z + kq)^{j-2}}{\displaystyle\sum_{j=0}^{\infty} g_j (z + kq)^j} \right\}, \quad z \in \mathbb{R}, \tag{7.2.8}$$

where G' and G'' are the first and the second derivatives of G. (The existence of G' and G'' and the validity of (7.2.7) and (7.2.8) follow via standard methods; in particular, they are consequences of the Lebesgue dominated convergence theorem.)

From the Cauchy–Schwarz inequality, we have, for $z \in \mathbb{R}$ and integer k such that $z + kq > 0$,

$$\frac{\sum_{j=1}^{\infty} j^2 g_j (z + kq)^{j-2}}{\sum_{j=0}^{\infty} g_j (z + kq)^j} - \left\{\frac{\sum_{j=1}^{\infty} j g_j (z + kq)^{j-1}}{\sum_{j=0}^{\infty} g_j (z + kq)^j}\right\}^2 \geqslant 0,$$

and, hence from (7.2.7) and (7.2.8), we get

$$\frac{G''(z)}{G(z)} - \left(\frac{G'(z)}{G(z)}\right)^2 = \frac{\mathrm{d}}{\mathrm{d}z}\left\{\frac{G'(z)}{G(z)}\right\} \geqslant 0, \quad z \in \mathbb{R}.$$

This implies that $G'(z)/G(z)$ is monotonic increasing in z. As (7.2.6) implies $G'(z)/G(z) = G'(z + kq)/G(z + kq)$ for each real z and integer k, we have then $G'(z)/G(z)$ to be independent of z and G to be Poisson. ∎

The first argument of the proof of the Rao–Rubin theorem links implicitly the integrated Cauchy functional equation to the Rao–Rubin condition. The link in question was shown explicitly and used to get the following general theorem by Shanbhag (1977).

7.2.2   Theorem

Let (X, Y) be a random vector with X, Y nonnegative integer-valued satisfying (7.2.1) with $g_0 < 1$ and, for each n for which $g_n > 0$, $S(\bullet|n)$ given by

$$S(r|n) = \frac{a_r b_{n-r}}{c_n}, \quad r = 0, 1, \ldots, n, \tag{7.2.9}$$

where $\{(a_n, b_n) : n = 0, 1, \ldots\}$ is a sequence of real vectors with $a_n > 0$ for all $n \geqslant 0$, $b_0, b_1 > 0$ and $b_n \geqslant 0$ for all $n \geqslant 2$, and $\{c_n\}$ is the convolution of $\{a_n\}$ and $\{b_n\}$. Then the Rao–Rubin condition (7.2.3) holds if and only if

$$g_n/c_n = (g_0/c_0)\alpha^n, \quad n = 1, 2, \ldots \quad \text{for some } \alpha > 0. \tag{7.2.10}$$

Also, if (7.2.10) holds, then Y and $X - Y$ are independent.

Proof   Observe that (7.2.3) is equivalent to

$$v_m = \sum_{n=0}^{\infty} v_{m+n} w_n, \quad m = 0, 1, \ldots$$

with $v_m = g_m/c_m$, $m = 0, 1, \ldots$ and $w_n = b_n\{\Sigma_{m=0}^{\infty} g_m(a_m/c_m)\}$, $n = 0, 1, \ldots$. Consequently, from Theorem 2.2.1, it follows easily that (7.2.3) holds if and only if (7.2.10) holds. The remainder of the theorem is trivial. ∎

If we consider (X, Y) with X and Y as k-component vectors instead of these as scalars, then, with obvious modifications in (7.2.1) and (7.2.9), one can present a generalization of Theorem 7.2.2. To be precise, let (X, Y) be a random vector such that X and Y be k-component vectors satisfying

$$P\{X = n, Y = r\} = g_n S(r|n), \quad r \in [0, n] \cap \mathbb{N}_0^k, \quad n \in \mathbb{N}_0^k \tag{7.2.11}$$

with $\{g_n : n \in \mathbb{N}_0^k\}$ as a probability distribution and, for each $n \neq 0$ for which $g_n > 0$,

$$S(r|n) = \frac{a_r b_{n-r}}{c_n}, \quad r \in [0, n] \cap \mathbb{N}_0^k, \tag{7.2.12}$$

where $\{a_n : n \in \mathbb{N}_0^k\}$ and $\{b_n : n \in \mathbb{N}_0^k\}$ are respectively positive and nonnegative real sequences with $b_0 > 0$ and $b_n > 0$ if n is of unit length, and $\{c_n : n \in \mathbb{N}_0^k\}$ is the convolution of these two sequences. Then, we have the following theorem.

7.2.3   Theorem

Suppose we have a random vector (X, Y) as stated above. This satisfies the Rao–Rubin condition

$$P\{Y = r\} = P\{Y = r | X = Y\}, \quad r \in \mathbb{N}_0^k \tag{7.2.13}$$

if and only if (in obvious notation)

$$g_n / c_n = \int_{\mathbb{R}_+^k} \left(\prod_{i=1}^k \lambda_i^{n_i} \right) \nu(d\lambda), \quad n \in \mathbb{N}_0^k, \tag{7.2.14}$$

where $0^0 = 1$ and ν is a uniquely determined finite measure on $\mathbb{R}_+^k$ (i.e. by the left-hand side of (7.2.14)) such that it is concentrated for some $\beta > 0$, on $\{\lambda : \Sigma_n b_n \prod_{i=1}^k \lambda_i^{n_i} = \beta\}$.

Proof   The assertion follows essentially via the proof of Theorem 7.2.2 if we use Corollary 3.4.4 in place of Theorem 2.2.1 and recall Remark 3.4.10(v). ∎

7.2.4   Remarks

(i) If we replace the condition that $b_n > 0$ for each n with unit length by that there exists at least one n with $g_n > 0$ for which $S(\bullet|n)$ is nondegenerate (or equivalently that $P\{X = Y\} < 1$) when $g_0 < 1$, then, provided the measure μ defined on $S = \mathbb{N}_0^k$ by

$$\mu(\{n\}) = b_n$$

meets Condition I of Chapter 3, possibly with $\mu(\{0\}) < 1$ excluded, Theorem 7.2.3 still remains valid.

(ii) In view of the information given in Remarks 4.3.11(v) and (x), it follows that a stability version of Theorem 7.2.3 holds. If we take the left-hand side of (7.2.13) to be $(1 - \beta(y))P\{Y = r\}$ in place of $P\{Y = r\}$, where β is as in Corollary 4.3.7 (but with $p = k$), then, for a sufficiently small K, we have

$$g_n/c_n = \int_{\mathbb{R}_+^k} \left(\prod_{i=1}^k \lambda_i^{n_i}\right)(1 + \xi_n(\lambda))\nu(d\lambda), \quad n \in \mathbb{N}_0^k$$

with ν as in (7.2.14) and $\xi.(\lambda)$ satisfying (4.3.23) with μ relative to (7.2.13). One could also give bounds of the type in (4.3.11) (with obviously "$\leqslant -\theta$" deleted) to a version of ξ that appears in the equation here.

The survival distribution (7.2.2) is of the form of (7.2.9) with $a_n = p^n/n!$, $n = 0, 1, \ldots$ and $b_n = (1 - p)^n/n!$, $n = 0, 1, \ldots$. Since, in this case, $c_n = 1/n!$, $n = 0, 1, \ldots$, it is immediate that Theorem 7.2.1 is a corollary to Theorem 7.2.2. Similarly, in the multidimensional case (i.e. the case with X and Y as k-component vectors), if we have (7.2.11) to be valid with

$$S(r|n) = \prod_{i=1}^k \binom{n_i}{r_i} p_i^{r_i}(1 - p_i)^{n_i - r_i}, \quad r \in [0, n] \cap \mathbb{N}_0^k; \quad n \in \mathbb{N}_0^k, \quad (7.2.15)$$

where $0 < p_i < 1$ are fixed (and the notation is standard), then (7.2.12) is met with

$$a_n = \prod_{i=1}^k \frac{p_i^{n_i}}{n_i!}, \quad n \in \mathbb{N}_0^k \quad \text{and} \quad b_n = \prod_{i=1}^k \frac{(1 - p_i)^{n_i}}{n_i!}, \quad n \in \mathbb{N}_0^k$$

(and obviously

$$c_n = \prod_{i=1}^k \frac{1}{n_i!}, \quad n \in \mathbb{N}_0^k).$$

Consequently, we get, in this case, the assertion of Theorem 7.2.3 to be valid with (7.2.14) replaced by

$$g_n = \int_{\mathbb{R}_+^k} \left(\prod_{i=1}^k \frac{\lambda_i^{n_i}}{n_i!}\right)\nu(d\lambda), \quad n \in \mathbb{N}_0^k$$

and $\{\lambda : \Sigma_n b_n \prod_{i=1}^k \lambda_i^{n_i} = \beta\}$ replaced by $\{\lambda : \Sigma_{i=1}^k \lambda_i(1 - p_i) = \log \beta\}$. This latter result follows also via Theorem 3.5.2 on using essentially an extended version of the first argument in the proof of Theorem 7.2.1 given above, and it disproves a conjecture of Srivastava and Srivastava (1970) for $k = 2$.

With the same argument as that used to obtain Theorem 7.2.3, we can also have the theorem given below.

7.2.5 Theorem

Let $n_0 \in \mathbb{N}_0^k$ and (X, Y) be as in Theorem 7.2.3 but for a modification that here we ask for the validity of (7.2.12) only for $n \geqslant n_0$ with $g_n \neq 0$ and we take in place of the condition that $\{a_n : n \in \mathbb{N}_0^k\}$ be a positive sequence the one that it be a nonnegative sequence with $a_n > 0$ for all $n \geqslant n_0$. Then, assuming $P\{X \geqslant n_0\} > 0$, for some constant $\eta > 0$, we have

$$P\{Y = r\} = \eta P\{Y = r | X = Y\}, \quad r \in [n_0, \infty) \cap \mathbb{N}_0^k \qquad (7.2.16)$$

if and only if (7.2.14) is valid with g_n/c_n replaced by g_{n+n_0}/c_{n+n_0}. (In (7.2.16) and some other places in what follows, we have used the notation ∞ to denote the k-component vector with all its components to be equal to infinity.)

We can express (7.2.16) alternatively as

$$P\{Y = r | Y \geqslant n_0\} = P\{Y = r | X = Y, Y \geqslant n_0\}, \quad r \in [n_0, \infty) \cap \mathbb{N}_0^k;$$

consequently Theorem 7.2.5 gives as special cases various characterizations of truncated distributions based on Rao–Rubin type conditions as corollaries. Rao and Rubin (1964), Shanbhag (1977) and Panaretos (1982a) amongst others have implicitly or explicitly addressed the question of characterizing truncated distributions in a damage model set-up. It may be worth pointing out here that Theorem 7.2.5 holds even when we replace the condition that $b_n > 0$ for each n with unit length by Condition I of Chapter 3 with $S = \mathbb{N}_0^k$ and μ such that

$$\mu(\{n\}) = b_n \frac{P\{X = Y, Y \geqslant n_0\}}{b_0 P\{Y \geqslant n_0\}}$$

for each n; in the case of $n_0 = 0$, this reduces effectively to the observation of Remark 7.2.4, (i), although expressed slightly differently.

One may now be interested to know as to how crucial the assumption of $a_n > 0$ for all n is in Theorems 7.2.2 and 7.2.3. In particular, one may raise the following question: suppose (X, Y) is a two-component random vector corresponding to a damage model such that $P\{X \leqslant n_1 + n_2\} = 1$ and for every n with $g_n > 0$, we have

$$S(r|n) = \frac{\binom{n_1}{r}\binom{n_2}{n-r}}{\binom{n_1 + n_2}{n}}, \quad r = 0, 1, \ldots, n \qquad (7.2.17)$$

with n_1 and n_2 as positive integers (and $\binom{k}{m} = 0$ if $m > k$). If X is binomial $(n_1 + n_2, p)$ for some $0 < p < 1$, then (Y and $X - Y$ are independent and hence) the Rao–Rubin condition (7.2.3) is met. Is the converse of this assertion valid? This is a question appearing in Patil and Ratnaparkhi (1975) and it is answered in the negative by the example given below; the example is due to Shanbhag and Panaretos (1979).

Example 1   Let (X, Y) be such that for some positive integer $m (\geqslant 3)$, we have

$$g_0 = g_1 = m^2/(2m^2 + m - 1)$$

and

$$g_m = (m - 1)/(2m^2 + m - 1),$$

and (7.2.17) is satisfied with $n_1 = 1$ and $n_2 = m - 1$. This vector satisfies the Rao–Rubin condition (7.2.3).

Example 1 clearly shows that if we replace in (7.2.9) the condition that $a_n > 0$ for all n by that $a_n \geqslant 0$ for all n but giving $c_n > 0$ for all n, then Theorem 7.2.2, which is obviously a corollary to Theorem 7.2.3, does not remain valid.

In Theorem 7.2.2 (i.e. under the assumptions in the theorem), we found that the Rao–Rubin condition implied the independence of $X - Y$ and Y. However, this was not the case with Theorem 7.2.3 or Remark 7.2.4,(i); indeed, under the assumptions in the theorem or the remark, the Rao–Rubin condition does not imply that $X - Y$ and Y are independent. Also, under the assumptions in Theorem 7.2.5, (7.2.16) does not imply that conditionally upon $y \geqslant n_0$, $X - Y$ and Y are independent. The next two results concern the question of finding additional conditions which together with the Rao–Rubin condition render the independence of $X - Y$ and Y.

7.2.6 Theorem

Under the assumptions in Theorem 7.2.3 or Remark 7.2.4,(i), the Rao–Rubin condition (7.2.13) together with the condition

$$P\{Y = r\} = P\{Y = r | X_{(1)} = Y_{(1)}\} = \cdots = P\{Y = r | X_{(k-1)} = Y_{(k-1)}\}, \quad r \in \mathbb{N}_0^k \tag{7.2.18}$$

where $X_{(i)} = (X_1, \ldots, X_i)$, $Y_{(i)} = (Y_1, \ldots, Y_i)$, $i = 1, 2, \ldots, k$, with X_j and Y_j as components of X and Y respectively, holds if and only if

$$g_n/c_n = (g_0/c_0) \prod_{i=1}^{k} \lambda_i^{n_i}, \quad n \in \mathbb{N}_0^k \tag{7.2.19}$$

for some $(\lambda_1, \ldots, \lambda_k) \in \mathbb{R}_+^k$. If (7.2.19) is valid, then $X - Y$ and Y are independent.

Proof It is sufficient if we prove the 'only if' part of the first assertion, the remainder being easy to see.

To prove the 'only if' part of the first assertion, note that the validity of the Rao–Rubin condition together with (7.2.18) implies in view of Theorem 7.2.3 and Remark 7.2.4,(i), that (7.2.14) holds with measure ν such that (in obvious notation)

$$\sum_{(n_{i+1}, \ldots, n_k)} b_{(0, \ldots, 0, n_{i+1}, \ldots, n_k)} \prod_{i+1}^{k} \lambda_r^{n_r} = \beta_i \quad \text{for a.a. } [\nu]\lambda \in \mathbb{R}_+^k \tag{7.2.20}$$

for $i = 0, 1, \ldots, k - 1$ for some $\beta_0, \ldots, \beta_{k-1} \in (0, \infty)$. (7.2.20) implies that the support of ν is a singleton and hence we have the result. ∎

7.2.7   Theorem

If we have the assumptions in Theorem 7.2.5 (or its modified version implied immediately after its proof) met, then (7.2.15) together with

$$P\{Y = r\} = P\{Y = r | X_{(1)} = Y_{(1)}, \; Y \geqslant n_0\}$$

$$\vdots$$

$$= P\{Y = r | X_{(k-1)} = Y_{(k-1)}, \; Y \geqslant n_0\}, \quad r \in [n_0, \infty) \cap \mathbb{N}_0^k, \quad (7.2.21)$$

where $X_{(i)}$, $Y_{(i)}$ are as defined in Theorem 7.2.6, is valid if and only if

$$g_{n+n_0} \propto c_{n+n_0} \prod_1^k \lambda_i^{n_i}, \quad n \in \mathbb{N}_0^k \quad (7.2.22)$$

for some $(\lambda_1, \ldots, \lambda_k) \in \mathbb{R}_+^k$. Also, if (7.2.22) is valid, then conditionally upon $Y \geqslant n_0$, $X - Y$ and Y are independent.

Proof   The theorem follows via the argument used to establish Theorem 7.2.6, but for obvious alterations. ∎

Theorem 7.2.6 obviously remains valid if we take in place of (7.2.13), (7.2.18), any other equivalent set of conditions. Similarly, Theorem 7.2.7 still holds if we replace (7.2.16), (7.2.21) by any other equivalent set of conditions. Some versions of these theorems were arrived at earlier using merely Theorem 7.2.2 by Shanbhag (1977) and Panaretos (1977). We may also note here that if we take the survival distribution to be that given by (7.2.15), then Theorems 7.2.6 and 7.2.7 hold with

$$c_n = \prod_{i=1}^k \frac{1}{n_i!}, \quad n \in \mathbb{N}_0^k;$$

these results are stronger versions of a result arrived at by Talwalker (1970).

It could be worth noting at this stage that Theorem 7.2.1 does not hold if (7.2.2) is replaced by the condition that $\{S(r|n) : r = 0, 1, \ldots, n\}$ for each $n \geqslant 1$ be binomial, without requiring the success probability parameter to be independent of n. This is shown by the following example of Shanbhag and Panaretos (1979).

Example   Take (X, Y) such that

$$P\{X = 0\} = c \left\{ 1 + \frac{3\sqrt{e} - 1}{8(\sqrt{e} - 1)^2} \right\},$$

$$P\{X = 1\} = c \left\{ 1 + \frac{1}{2(\sqrt{e} - 1)} \right\}, \quad P\{X = 2\} = \frac{9c}{8}$$

and $P\{X = n\} = c/n!$, $n \geqslant 3$, where c is the normalizing constant, and

$$P\{Y = r | X = n\} = \binom{n}{r} \left(\tfrac{1}{2} \right)^n \quad 0 \leqslant r \leqslant n; \quad n \neq 2,$$

and

$$= \binom{n}{r} \left(\tfrac{1}{3}\right)^r \left(\tfrac{2}{3}\right)^{n-r}, \quad r = 0, 1, 2; \quad n = 2.$$

It readily follows that the vector satisfies (7.2.3). However the random variable X considered is not Poisson.

Another relevant question that could be raised now concerning Theorem 7.2.1 is whether, under some mild conditions, (7.2.2) holds given that X is Poisson and (7.2.3) holds. Srivastava and Srivastava (1970) have provided an answer to this question. Considering a family of random vectors $\{(X_\lambda, Y_\lambda) : \lambda \in (a, b)\}$ with $a \geqslant 0$ and $b > a$ of nonnegative integer-valued components (with the vectors not necessarily defined on the same probability space) such that for all values of the parameter

$$P_\lambda\{X_\lambda = n, Y_\lambda = r\} = \exp(-\lambda)\frac{\lambda^n}{n!}S(r|n), \quad r = 0, 1, \ldots, n; \quad n = 0, 1, \ldots,$$

where $\{S(r|n): r = 0, 1, \ldots, n\}$ is a discrete probability distribution independent of λ for each $n \geqslant 0$ and $0 < S(n|n) < 1$ for some n, they have shown that for all values of the parameter λ

$$P_\lambda\{Y_\lambda = r\} = P_\lambda\{Y_\lambda = r | X_\lambda = Y_\lambda\}, \quad r = 0, 1, \ldots$$

if and only if (7.2.2) holds for each $n \in \mathbb{N}_0$ for some $p \in (0, 1)$. Shanbhag and Panaretos (1979) have shown that the Srivastava–Srivastava result does not hold if in place of the family of vectors (X_λ, Y_λ) just one vector (X, Y), with X as Poisson, is taken; Andrews (1988) has recently used an extended version of the Shanbhag–Panaretos argument to show that the situation remains unaltered if two vectors are taken in place of one. As we shall now see, a construction somewhat similar to that given by Moran in a somewhat different context, tells us something more about the problem; in particular, it shows that the Srivastava–Srivastava result does not hold if we restrict ourselves to a family with finite index set.

We give our observations via two separate examples.

Example 1   Let $\{f_r^{(1)} : r = 0, 1, 2, \ldots\}$ and $\{f_r^{(2)} : r = 0, 1, 2, \ldots\}$ be two discrete probability distributions with at least four common support points $0, r_0, r_1, r_2$, assumed to be in ascending order. Further, let

$$0 < \varepsilon \leqslant \min\left\{f_{r_0}^{(1)} f_{r_2}^{(2)}, \; f_{r_1}^{(1)} f_{r_0}^{(2)}, \; f_{r_2}^{(1)} f_{r_1}^{(2)}\right\}.$$

Define now $\{f_{rs} : r, s = 0, 1, \ldots\}$ such that

$$f_{rs} = \begin{cases} f_r^{(1)} f_s^{(2)} + \varepsilon & \text{if } (r, s) = (r_0, r_1) \text{ or } (r_1, r_2) \text{ or } (r_2, r_0) \\ f_r^{(1)} f_s^{(2)} - \varepsilon & \text{if } (r, s) = (r_0, r_2) \text{ or } (r_1, r_0) \text{ or } (r_2, r_1) \\ f_r^{(1)} f_s^{(2)} & \text{otherwise.} \end{cases}$$

If we take (X, Y) to be a random vector such that $(X - Y, Y)$ is distributed with distribution $\{f_{rs}\}$, then it follows that $X - Y$ and Y are dependent random variables satisfying

$$f_r^{(1)} = P\{X - Y = r\} = P\{X - Y = r | Y = 0\}, \quad r = 0, 1, \ldots$$

and

$$f_r^{(2)} = P\{Y = r\} = P\{Y = r | X - Y = 0\}, \quad r = 0, 1, \ldots;$$

we also get here the distribution of X to be the convolution of $\{f_r^{(1)}\}$ and $\{f_r^{(2)}\}$. (Incidentally, if we assume the second moments corresponding to $\{f_r^{(1)}\}$ and $\{f_r^{(2)}\}$ to be finite, we get $X - Y$ and Y to be uncorrelated.) Moreover, if we take here $\{f_r^{(1)}\}$ and $\{f_r^{(2)}\}$ as Poisson, we get X to be Poisson; as $X - Y$ and Y are dependent, we cannot now have (7.2.2) to be valid. (Note that, in this special case, we have both $X - Y$ and Y as Poisson, and that not only the Rao–Rubin condition is met, but also its dual, i.e. the one with the places of Y and $X - Y$ interchanged, holds.)

Example 2   Let k be a positive integer, and $\{f_r^{(1)} : r = 0, 1, \ldots\}$ and $\{f_r^{(2)} : r = 0, 1, \ldots\}$ be probability distributions such that $f_0^{(1)}, \ldots, f_{k+2}^{(1)}, f_0^{(2)}, f_1^{(2)}, f_2^{(2)} > 0$. Let $a_0 + a_1 \lambda + \cdots + a_k \lambda^k$ (with $a_k = 1$) be a polynomial of degree k with zeros as distinct and real positive. Denote by $\lambda_1, \ldots, \lambda_k$ (in ascending order) the zeros of the polynomial in question and define, for a sufficiently small $\varepsilon_k > 0$, $\{f_{rs} : r, s = 0, 1, \ldots\}$ by

$$f_{rs} = \begin{cases} f_r^{(1)} f_s^{(2)} + a_{r-2}\varepsilon_k, & (r, s) = (2, 1), \ldots (k+2, 1) \\ f_r^{(1)} f_s^{(2)} - a_{r-1}\varepsilon_k, & (r, s) = (1, 2), \ldots (k+1, 2) \\ f_r^{(1)} f_s^{(2)} & \text{otherwise.} \end{cases}$$

such that $f_{rs} \geq 0$. Define, for each $\lambda \in \{\lambda_1, \ldots, \lambda_k\}$, a random vector (X_λ, Y_λ) such that

$$P_\lambda\{X_\lambda - Y_\lambda = r, Y_\lambda = s\} \propto f_{rs}\lambda^{r+s}, \quad r, s = 0, 1, \ldots;$$

clearly, we have here $X_\lambda - Y_\lambda$ and Y_λ to be dependent,

$$P_\lambda\{Y_\lambda = r\} = P_\lambda\{Y_\lambda = r | X_\lambda = Y_\lambda\} \propto f_r^{(2)}\lambda^r, \quad r = 0, 1, \ldots$$

and

$$P_\lambda\{X_\lambda = r\} \propto h_r\lambda^r, \quad r = 0, 1, \ldots$$

with $\{h_r\}$ as the convolution of $\{f_r^{(1)}\}$ and $\{f_r^{(2)}\}$. Taking $\{f_r^{(1)}\}$ and $\{f_r^{(2)}\}$ as Poisson with means α and $1 - \alpha$ respectively, where $0 < \alpha < 1$, we can then see that the Srivastava–Srivastava result does not hold if we replace (a, b) by $\{\lambda_1, \ldots, \lambda_k\}$. (That (7.2.2) does not hold here follows because $X_\lambda - Y_\lambda$ and Y_λ are dependent.)

One of the points that we should make now is that in the Srivastava–Srivastava model, we have

$$P_\lambda\{Y_\lambda = r\} = P_\lambda\{Y_\lambda = r | X_\lambda = Y_\lambda\}, \quad r = 0, 1, \ldots \tag{7.2.23}$$

for all λ if and only if $X_\lambda - Y_\lambda$ and Y_λ are independent for all λ. It is an easy exercise to see that this latter result holds even when we replace the condition that X_λ be Poisson by that the distribution of X_λ be power series

$$P_\lambda\{X_\lambda = n\} \propto c_n \lambda^n, \quad n = 0, 1, \ldots$$

with $c_0 > 0$ (see, for example, Alzaid, 1986); incidentally the result in question also holds if we replace the index set (a, b) by a subset of it with a as a cluster point. Also note that if X_λ is as above and $X_\lambda - Y_\lambda$ and Y_λ are independent, then there exist sequences $\{a_n\}$ and $\{b_n\}$ with convolution $\{c_n\}$ such that whenever $c_n > 0$, we have

$$S(r|n) = \frac{a_r b_{n-r}}{c_n}, \quad r = 0, 1, \ldots, n.$$

Hence the following theorem arrived at from Theorem 7.2.2 could be viewed as a variant of the Srivastava–Srivastava result.

7.2.8   Theorem

Under the assumptions in Theorem 7.2.2 together with that $\{g_n\}$ is Poisson, the Rao–Rubin condition (7.2.3) is met if and only if (7.2.2) is valid.

Proof   We need a proof only for the 'only if' part of the theorem (as the other part is trivial and is also implied by Theorem 7.2.2). If $\{g_n\}$ is Poisson and (7.2.3) holds, then Theorem 7.2.2 implies that

$$c_n \propto \frac{\lambda^n}{n!}, \quad n = 0, 1, \ldots$$

for some $\lambda > 0$. From the Raikov theorem, it then follows that

$$a_n \propto \frac{\lambda_1^n}{n!}, \quad n = 0, 1, \ldots$$

and

$$b_n \propto \frac{\lambda_2^n}{n!}, \quad n = 0, 1, \ldots$$

for some $\lambda_1, \lambda_2 > 0$ with $\lambda_1 + \lambda_2 = \lambda$. Hence, we get the required result. ∎

Several variants and extensions of Srivastava–Srivastava–Alzaid results have appeared in the literature. Some of these concern characterizations of discrete infinitely divisible and discrete self-decomposable distributions. We shall not deal with them here.

We conclude the present section with yet another counterexample. This is due to Shanbhag and Panaretos (1979) and it disproves a conjecture of Srivastava and Singh (1975), in the case when the original random variable (i.e. X_λ) has a truncated power-series Poisson distribution.

Example   Let (X_λ, Y_λ) be a random vector, for each $\lambda > 0$, of nonnegative integer-valued components such that $P_\lambda\{Y_\lambda = n\} = (\lambda^n/n!)/(e^\lambda - 1), n = 1, 2, \ldots$

and for each $n \geqslant 1$, $P_\lambda\{Y_\lambda = r | X_\lambda = n\} = \binom{n}{r}\left(\frac{1}{2}\right)^n$ if $r \in \{1, 2, \ldots, n\} \setminus \{n\}$ and $= \left(\frac{1}{2}\right)^{n-1}$ if $r = n$. Observe that

$$P_\lambda\{Y = r | Y \geqslant 1\} = P_\lambda\{Y_\lambda = r | X_\lambda = Y_\lambda\} = P_\lambda\{Y_\lambda = r | X_\lambda > Y_\lambda, Y_\lambda \geqslant 1\},$$

$$r = 1, 2, \ldots$$

but in this case the conditional distribution of Y_λ given that $X_\lambda = n$ and $Y_\lambda \geqslant 1$, is not truncated binomial, and hence is not of the form conjectured by Srivastava and Singh (1975).

7.3   AN EXTENDED SPITZER INTEGRAL REPRESENTATION THEOREM AND MODIFIED RAO–RUBIN CONDITIONS

The material appearing in this section is subdivided into two separate parts.

7.3(a)   *An extended Spitzer integral representation theorem with an application in damage models*

Consider a modified discrete branching process $\{Z_n : n = 0, 1, \ldots\}$ with one-step transition probabilities given by

$$P_{ij} = P\{Z_{n+1} = j | Z_n = i\} = \begin{cases} cp_j^{(i)}, & i = 0, 1, \ldots, j = 1, 2, \ldots \\ 1 - c + cp_0^{(i)}, & i = 0, 1, \ldots, j = 0, \end{cases}$$

$$(7.3.1)$$

where $0 < c \leqslant 1$ and $\{p_j^{(i)} : j = 0, 1, \ldots\}$ is the i-fold convolution of some probability distribution $\{p_j\}$ having $0 < p_0 < 1$ with itself for $i > 0$ and the degenerate distribution at zero for $i = 0$. (It is seen that the process reduces to a Bienaymé–Galton–Watson branching process when $c = 1$.) Define

$$m = \sum_{j=1}^{\infty} jp_j, \qquad m^* = \sum_{j=1}^{\infty} (j \log j)p_j. \qquad (7.3.2)$$

Alzaid *et al* (1987) have established the following theorems on stationary measures of the branching process. (A.N. in the discussion refers to Athreya and Ney (1972).)

7.3.1   Theorem

The g.f. (generating function) $U(s) = \Sigma \eta_j s^j$ of any stationary measure $\{\eta_j\}$ of a modified branching process defined in (7.3.1) is analytic for $|s| < q$, where q is the smallest positive root of the equation $s = f(s)$, and (if normalized so that $U(p_0) = 1$) satisfies the equation

$$cU(f(s)) = c + U(s), \qquad (7.3.3)$$

where f is the g.f. of $\{p_j\}$. Conversely, if $U(s) = \Sigma \eta_j s^j$, $\eta_j \geqslant 0$, $|s| < q$ satisfies (7.3.3), then $\{\eta_j\}$ is a stationary measure.

Proof   If $U(s)$ is analytic for $|s| < s_0$ for some $s_0 > 0$, then

$$U(s) = \sum_{j=1}^{\infty} \eta_j s^j = \sum_{i=1}^{\infty} \eta_i \sum_{j=1}^{\infty} p_{ij} s^j$$

$$= c \sum_{i=1}^{\infty} \eta_i ([f(s)]^i - [f(0)]^i)$$

$$= c(U(f(s)) - U(p_0)), \quad |s| < s_0,$$

which implies that (7.3.3) is valid at least for $|s| < s_0$. Then using the arguments exactly as in A.N. (Theorem 2, page 68), we find that $U(s)$ is analytic for $|s| < q$. The converse assertion easily follows by equating the coefficients of s^j in (7.3.3). ■

7.3.2   Theorem (an extended version of Spitzer's theorem)

If $m < 1$ and $m^* < \infty$ with f as in Theorem 7.3.1, then for every probability measure ν on $[0, 1)$

$$U(s) = K \int_{[0,1)} U(s, t) \nu(dt) \tag{7.3.4}$$

is the g.f. of a stationary measure, where

$$U(s, t) = \sum_{n=-\infty}^{\infty} [\exp\{(\mathfrak{B}(s) - 1)m^{n-t}\} - \exp\{-m^{n-t}\}]c^{n-t},$$

with $\mathfrak{B}(s)$ as the unique p.g.f. among those vanishing at $s = 0$ and satisfying the equation

$$\mathfrak{B}(f(s)) = m\mathfrak{B}(s) + 1 - m, \tag{7.3.5}$$

and K the appropriate normalizing constant. Conversely, every stationary measure has the representation (7.3.4) for some probability measure ν on $[0, 1)$. (For an interpretation of $\mathfrak{B}$ in Bienaymé–Galton–Watson branching processes and a proof of the uniqueness in (7.3.5), see A.N., page 17.)

Proof   The first part of the theorem is easily verified. To prove the converse, it is sufficient to establish that the representation is valid for every $s \in [0, 1)$. Define then for every $s \in [0, 1)$

$$U^*(s) = U(\mathfrak{B}^{-1}(s)),$$

where $\mathfrak{B}$ is as mentioned in the statement of the theorem. In view of Theorem 7.3.1 and (7.3.5) we have

$$c^n U^*(m^n s + 1 - m^n) = U^*(s) + \xi_n, \quad n = 1, 2, \ldots, 0 \leqslant s < 1, \tag{7.3.6}$$

where

$$\xi_n = \begin{cases} c\left(\dfrac{1-c^n}{1-c}\right) & \text{if } c \neq 1, \\ n, & \text{if } c = 1. \end{cases}$$

We can write (7.3.6) also as

$$U^*(s) = c^n U(1 + m^n Q_n(\mathfrak{B}^{-1}(s))) - \xi_n, \quad n = 1, 2, \ldots, \quad 0 \leqslant s < 1,$$

where $Q_n(s) = [f_n(s) - 1]/m^n$ with f_n as the nth iterate of f. Consequently

$$U^*(s) = \lim_{n \to \infty} \{c^n U(1 + m^n Q_n(0) - m^n Q_n(0)\mathfrak{B}_n(\mathfrak{B}^{-1}(s))) - \xi_n\} \qquad (7.3.7)$$

for $0 \leqslant s < 1$, where $\mathfrak{B}_n(s) = [Q_n(0) - Q_n(s)]/Q_n(0)$. If $0 \leqslant s_1 < s < s_2 < 1$, then noting in particular that $\mathfrak{B}_n \to \mathfrak{B}$ pointwise (see A.N., page 47), and $\mathfrak{B}$ and $\mathfrak{B}^{-1}$ are strictly increasing on $[0,1)$, we find that

$$-m^n Q_n(0)\mathfrak{B}_n(\mathfrak{B}^{-1}(s_1)) < -m^n Q_n(0)s = -m^n Q_n(0)\mathfrak{B}(\mathfrak{B}^{-1}(s))$$

$$< -m^n Q_n(0)\mathfrak{B}_n(\mathfrak{B}^{-1}(s_2)) \qquad (7.3.8)$$

for large enough n. Since U is increasing on $[0,1)$, we obtain from (7.3.7) on using (7.3.8)

$$U^*(s_1) \leqslant \liminf_{n \to \infty}\{c^n U(1 + m^n Q_n(0) - m^n Q_n(0)s) - \xi_n\}$$

$$\leqslant \limsup_{n \to \infty}\{c^n U(1 + m^n Q_n(0) - m^n Q_n(0)s) - \xi_n\} \leqslant U^*(s_2). \quad (7.3.9)$$

It is seen that U^* is continuous and $U^*(0) = 0$, and hence from (7.3.9), in particular, we have

$$U^*(s) = \lim_{n \to \infty}\{c^n U(1 + m^n Q_n(0) - m^n Q_n(0)s) - \xi_n\}$$

$$= \lim_{n \to \infty}\{c^n [U(1 + m^n Q_n(0) - m^n Q_n(0)s) - U(1 + m^n Q_n(0))]\} \quad (7.3.10)$$

for $s \in [0, 1)$, with the limits well defined. Since for each $n \geqslant 1$, the expression within the second limit in (7.3.10) can be expressed as the restriction to $[0,1)$ of the g.f. of some nonnegative sequence, Corollary 1.1.13 implies that U^* is the restriction to $[0,1)$ of the g.f. of a nonnegative sequence. Define now a function $\hat{U}$ on $(-\infty, 1)$ such that its restriction to $[0,1)$ is indeed U^* and for $s \in (-\infty, 0)$,

$$\hat{U}(s) = c^n U^*(m^n s + 1 - m^n) - \xi_n,$$

where n is the smallest integer for which $m^n s + 1 - m^n \in [0, 1)$. It is easily seen that

$$c^n \hat{U}(m^n s + 1 - m^n) = \hat{U}(s) + \xi_n, \quad n = 1, 2, \ldots, s \in (-\infty, 1), \qquad (7.3.11)$$

which since $\hat{U} = U^*$ on $[0,1)$, implies that $\hat{U}'$ (i.e., $d\hat{U}(s)/ds$, $s \in (-\infty, 1)$) exists and is absolutely monotonic on $(-\infty, 1)$ with $\lim \hat{U}'(s) = 0$ as $s \to -\infty$. Since $\hat{U}(0) = 0$, we conclude from Theorem 3.5.1 that for some measure μ on $(0, \infty)$

$$\hat{U}(s) = \int_{(0,\infty)} (e^{sx} - 1)\mu(dx), \quad s \in (-\infty, 1). \tag{7.3.12}$$

From (7.3.11) we find that the measure μ is such that, with $\xi_n^* = \xi_n$ if $n \in \mathbb{N}_0$ and equals $-c^n \xi_{|n|}$ if $n \notin \mathbb{N}_0$, we have

$$c^n \int_{(0,\infty)} (e^{sxm^n + (1-m^n)x} - 1)\mu(dx) = \int_{(0,\infty)} (e^{sx} - 1)\mu(dx) + \xi_n^*,$$

$$n \in \mathbb{Z}, s \in (-\infty, 1). \tag{7.3.13}$$

If $s_0 \in (-\infty, 1)$ and $s \in (-\infty, 0]$, then subtracting the identity (7.3.13) from the corresponding identity with s replaced by $s + s_0$, we obtain

$$c^n \int_{(0,\infty)} e^{sxm^n} (e^{s_0 xm^n} - 1)e^{(1-m^n)x}\mu(dx) = \int_{(0,\infty)} e^{sx}(e^{s_0 x} - 1)\mu(dx), \tag{7.3.14}$$

which is valid for all $s \in (-\infty, 0]$. Then, in view of the uniqueness theorem for Laplace–Stieltjes transforms, (7.3.14) implies

$$\int_{[m^n, m^{n-1})} (e^{s_0 x} - 1)\mu(dx) = c^n \int_{[1, m^{-1})} (e^{s_0 xm^n} - 1)e^{(1-m^n)x}\mu(dx) \tag{7.3.15}$$

for $n = 0, \pm 1, \pm 2, \ldots$. Since s_0 in (7.3.15) is arbitrary, we conclude using (7.3.12) that

$$U^*(s) = \hat{U}(s) = \sum_{n=-\infty}^{\infty} \int_{[m^n, m^{n-1})} (e^{sx} - 1)\mu(dx)$$

$$= \int_{[1, m^{-1})} \sum_{n=-\infty}^{\infty} c^n (e^{-(1-s)xm^n} - e^{-xm^n})e^x \mu(dx)$$

$$= K \int_{[0,1)} \left\{ \sum_{n=-\infty}^{\infty} c^{n-t}(e^{-m^{n-t}(1-s)} - e^{-m^{n-t}}) \right\} \nu(dt) \tag{7.3.16}$$

for $s \in [0, 1)$, where K is a positive constant and ν is a probability measure on $[0,1)$ such that for every Borel subset A of $[0,1)$ we have

$$\nu(A) = K^{-1} \int_{S_A} e^x c^{-\log x / \log m} \mu(dx),$$

where $S_A = \{x : (-\log x / \log m) \in A\}$. (The operation of interchanging the order of summation and integration in (7.3.16) is justified by either Fubini's theorem or the monotone convergence theorem.) The required result now follows on observing that $U(s) = U^*(\mathfrak{B}(s))$, $s \in [0, 1)$. ∎

7.3.3   Corollary

If $f(s) = 1 - m + ms$, then every sequence $\{\eta_j\}$ is a stationary measure if and only if it is of the form

$$\eta_j = K \sum_{n=-\infty}^{\infty} \int_{[0,1)} c^{n-t} e^{-m^{n-t}} \frac{m^{(n-t)j}}{j!} \nu(dt), \quad j = 1, 2, \ldots, \tag{7.3.17}$$

where ν is a probability measure on $[0,1)$, and K is a positive constant as in Theorem 7.3.2.

Proof   The corollary is obvious from Theorem 7.3.2 since $\mathfrak{B}(s) = s$. ∎

We now give an application of Corollary 7.3.3 in damage models. Let (X, Y) and (X', Y') be two random vectors with nonnegative integer-valued components such that

$$P\{X = n, Y = r\} = g_n \binom{n}{r} p^r (1 - p)^{n-r}, \quad r = 0, 1, \ldots, n; \quad n = 0, 1, \ldots$$

and

$$P\{X' = n, Y' = r\} = g_n \binom{n}{r} p'^r (1 - p')^{n-r}, \quad r = 0, 1, \ldots, n; \quad n = 0, 1, \ldots,$$

where $\{g_n\}$ is a probability distribution with $g_0 < 1$ and $0 < p, p' < 1$ are fixed numbers. If we have the problem of identifying $\{g_n\}$ for which the modified Rao–Rubin condition

$$P\{Y = r\} = P\{Y' = r \mid X' = Y'\}, \quad r = 0, 1, \ldots \tag{7.3.18}$$

holds, then, for $p = p'$, its solution is immediate from Theorem 7.2.1, and, for $p > p'$ it reduces to a simple exercise with solution that $\{g_n\}$ is binomial with success probability parameter $(p - p')/p(1 - p')$. However, for $p < p'$, we have a totally different picture and the family of $\{g_n\}$ turns out to be somewhat curious and fairly large. Talwalker (1980) and Rao *et al* (1980) have studied the identification problem in this last case. But a most satisfactory answer in this case has been given by Alzaid, *et al* (1987); this is contained in the following theorem, which in turn follows as a consequence of Corollary 7.3.3.

7.3.4   Theorem

Let (X, Y) and (X', Y') be two-component random vectors as considered above with $0 < p < p' < 1$. Then (7.3.18) is valid if and only if

$$g_j = K \sum_{n=-\infty}^{\infty} \int_{[0,1)} c^{n-t} e^{-(p/p')^{n-t}} \frac{(p/p')^{(n-t)j}}{j!} \left[\frac{p' - p}{p'(1 - p)} \right]^j \nu(dt),$$

$$j = 0, 1, \ldots,$$

where ν is a probability measure on $[0,1)$, c is a real number lying in $(0,1)$ and K is a normalizing constant.

Proof   The result follows from Corollary 7.3.3 on noting that (7.3.18) is equivalent to that for some $c \in (0, 1)$,

$$cU(1 - m + ms) = U(s) + cU(1 - m), s \in (-1, 1)$$

with

$$g_0 = \frac{c}{1-c} U(1 - m), \quad m = \frac{p}{p'},$$

$$U(s) = G\left(\frac{(1 - p)p'}{p' - p} s\right) - g_0,$$

where G is the generating function of $\{g_n\}$; observe that (7.3.18) implies that the probability generating function G has the domain of definition

$$\left(-\frac{p'(1 - p)}{p' - p}, \frac{p'(1 - p)}{p' - p}\right). \quad \blacksquare$$

7.3.5   Remarks

(i) A version of Theorem 7.3.2 for $c = 1$ was given by Spitzer (1967) and appears in A.N. The nonuniqueness of a stationary measure in this case was earlier shown by Kingman (1965).

(ii) If the measure ν in Theorem 7.3.4 is taken as the Lebesgue measure on $[0,1)$, then the distribution $\{g_n\}$ in question reduces to a negative binomial distribution.

(iii) We have (7.3.18) to be equivalent to

$$G(1 - p + p's) = G(p's)/G(p'), \quad |s| \leqslant 1, \tag{7.3.19}$$

where G is the generating function of $\{g_n\}$. (We had met (7.3.19) implicitly in the proof of Theorem 7.3.4.) If (7.3.19) is met simultaneously for two pairs (p_i, p'_i), $i = 1, 2$, where $0 < p_i < p'_i < 1$ and $(\log p_1 - \log p'_1)/(\log p_2 - \log p'_2)$ is irrational, then G is the probability generating function of a negative binomial distribution, of the form

$$G(s) = \frac{\{[p'_1(1 - p_1)/(p'_1 - p_1)] - 1\}^\alpha}{\{[p'_1(1 - p_1)/(p'_1 - p_1)] - s\}^\alpha}$$

for some $\alpha > 0$. Since the condition implies G to be well defined also on $(1, s_0)$ where $s_0 = p'_1(1 - p_1)/(p'_1 - p_1)$, the result in question follows as a corollary to the result of Marsaglia and Tubilla (1975) or Theorem 2.2.2, by noting in particular that $f(x) = G(s_0 - s_0 e^{-x})/G(0)$, $x \geqslant 0$ is well defined and satisfies the equation $f(t_i + x) = f(t_i)f(x)$, $x \geqslant 0$, $i = 1, 2$ with $t_i = \log(p'_i/p_i)$, $i = 1, 2$. The result was first established in Rao *et al* (1980) by a different and slightly more involved method.

(iv) Using essentially the argument in A.N. (page 70), one could also give an alternative proof for Theorem 7.3.2 based on the Poisson–Martin integral representation for the stationary measure referred to.

7.3(b)   The Perron–Frobenius theorem and modified Rao–Rubin conditions

Srivastava and Singh (1975) conjectured that Theorem 7.2.1 holds even when the Rao–Rubin condition is replaced by its modified version that

$$P\{Y = r\} = P\{Y = r | X - Y = k\}, \quad r = 0, 1, \ldots, \quad (7.3.20)$$

where k is a fixed positive integer such that $P\{X - Y = k\} > 0$, which we refer to as the RR(k) condition. Patil and Taillie (1979) have shown that this conjecture is false, but its modified version when in place of one RR(k) we take two RR(k)'s, holds. Shanbhag and Taillie (1979), and Alzaid *et al* (1986a, 1988) have extended and unified the results on RR(k)'s. From these studies it has emerged that the main theme of the problem here is linked with either the Perron–Frobenius theory or the Wiener–Hopf factorization met in Chapter 2. We now reproduce some general results illustrating this point.

Let $\{g_n : n = 0, 1, \ldots\}$ and for each $n \geqslant 0$, $\{S(r|n) : r = 0, 1, \ldots, n\}$ be probability distributions, and $k_0 \geqslant 0$ and $k_1 \geqslant 0$ be fixed integers. Define

$$S_{ij}^* = \sup\{S(j + mk_1 | i + k_0 + nk_1) : m \geqslant 0, n \geqslant 0, k_0 \leqslant j + mk_1 \leqslant i + k_0 + nk_1\}.$$

We have the following theorem:

7.3.6   Theorem
Let the random vector (X,Y) be such that

$$P\{X = n, Y = r\} = g_n S(r|n), \quad r = 0, 1, \ldots, r; \quad n = 0, 1, \ldots,$$

where g_n and $S(r|n)$ satisfy the following conditions:

(i) $g_n > 0$ for some $n \geqslant k_0 + k_1$.
(ii) $S(n + rk_1 | n + k_0 + (r + i)k_1) > 0$ for each $n = 0, 1, \ldots k_1 - 1, i = 0, 1$ and $r = 0, 1, \ldots$, and $S(n|n) > 0$ for each $n = 0, 1, \ldots, k_0 - 1$ when $(k_0 > 0)$.
(iii)   The matrix $S^* = (S_{ij}^*)$, $i, j = 0, 1, \ldots, k_1 - 1$, is irreducible.

Then, under the two RR(k) conditions

$$P\{Y = r\} = P\{Y = y | X - Y = k_0\}$$

$$= P\{Y = r | X - Y = k_0 + k_1\}, \quad y = 0, 1, \ldots \quad (7.3.21)$$

and for a fixed value $\lambda = P\{X - Y = k_0 + k_1\}/P\{X - Y = k_0\}$, the family of distributions

$$\{S(r|n) : \quad r = 0, 1, \ldots, n; \quad n = 0, 1, \ldots\}$$

determines the $\{g_n : n = 0, 1, \ldots\}$ uniquely and $g_n > 0$ for $n \geqslant k_0$.

Proof   Using the equations $P\{Y = r | X - Y = k_0\} = P\{Y = r | X - Y = k_0 + k_1\}$ of the condition (7.3.21) and assumption (ii) of the theorem, we have

$$g_{r+bk_1+k_0} = \lambda g_{r+(b-1)k_1+k_0} \frac{S(r + (b-1)k_1 | r + (b-1)k_1 + k_0)}{S(r + (b-1)k_1 | r + bk_1 + k_0)}$$

$$r = 0, 1, \ldots, k_1 - 1, \quad b = 1, 2, \ldots .$$

Hence

$$g_{r+bk_1+k_0} = \lambda^b g_{r+k_0} \prod_{m=0}^{b-1} \frac{S(r + mk_1 | r + k_0 + mk_1)}{S(r + mk_1 | r + k_0 + (m+1)k_1)}$$

$$r = 0, 1, \ldots, k_1 - 1, \quad b = 1, 2, \ldots, \tag{7.3.22}$$

which, in view of assumptions (i) and (ii) implies that

$$g_{k_0} + \cdots + g_{k_0+k_1-1} \neq 0.$$

The system of equations $P\{Y = r\} = P\{Y = r | X - Y = k_0\}, r = 0, 1, \ldots$ implies that

$$\frac{g_{r+mk_1+k_0}}{P(X - Y = k_0)} S(r + mk_1 | r + mk_1 + k_0) = \sum_{n=r+mk_1}^{\infty} g_n S(r + mk_1 | n), r, m \geq 0. \tag{7.3.23}$$

Summing (7.3.23) over m such that $r + mk_1 \geq k_0$ and using (7.3.22), we find

$$\lambda^* g_{r+k_0} = \sum_{s=0}^{k_1-1} g_{s+k_0} q_{sr}, \quad r = 0, 1, \ldots, k_1 - 1, \tag{7.3.24}$$

where $\lambda^* = [P\{X - Y = k_0\}]^{-1}$ and q_{sr} are certain uniquely determined functions of $\{\{S(r|n)\}\}$. The irreducibility of S^* (assumption (iii)) implies that the matrix (q_{sr}) arrived at in (7.3.24) is irreducible. The fact that $S(r + mk_1 | r + mk_1 + k_0) > 0$ for all $m \geq 0$ and $r = 0, 1, \ldots, k_1 - 1$ implies that $q_{rr} > 0$ in (7.3.24) for all $r = 0, 1, \ldots, k_1 - 1$. Consequently, the matrix $Q = (q_{sr})$ is primitive. Then, from the Perron–Frobenius theorem (see section 1.3.10), it follows that λ^* is the unique eigenvalue of Q having the largest absolute value and

$$(g_{k_0}, \ldots, g_{k_0+k_1-1})$$

is the eigenvector of Q associated with λ^* having all its components positive. Using (7.3.22) and (7.3.23) and the fact that $S(r|r) > 0$ for $r < k_0$, we can express $g_{k_0-1}, \ldots, g_0$ as linear combinations of $g_{k_0}, \ldots, g_{k_0+k_1-1}$ with coefficients which are uniquely determined by $\{\{S(r|n)\}\}$. Using the fact that $\Sigma_{n=0}^{\infty} g_n = 1$ or that $P\{X - Y = k_0\} = (\lambda^*)^{-1}$, we see that $\{g_n\}$ is uniquely determined by $\{\{S(r|n) : r = 0, 1, \ldots, n\} : n \geq 0\}$. The remainder of the theorem easily follows. ∎

7.3.7   Remarks

(i) With appropriate modifications in assumptions (ii) and (iii) of the theorem, such as replacing $S^* = (S_{ij}^*)$ in (ii) by $S^{**} = (S_{ij}^{**})$ where

$$S_{ij}^{**} = \sup\{S(j + mk_1|i + k_0 + nk_1) :$$
$$m \geqslant 0, n \geqslant 0, k_0 \leqslant j + mk_1 \leqslant \min(i + k_0 + nk_1, k_0 + k_1 - 1)\}$$

it is possible to prove the validity of the above theorem with (7.3.21) replaced by a somewhat weaker assumption of the type

$$P\{Y = r|X - Y = k_0\} = P\{Y = r|X - Y = k_0 + k_1\}, \quad r = 0, 1, \ldots,$$
$$P\{Y = r\} = P\{Y = r|X - Y = k_0\}, \quad r = 0, 1, \ldots, k_0 + k_1 - 1. \quad (7.3.25)$$

Several other possibilities exist.

(ii) If the matrix S^* is such that $S_{i,i+1}^* > 0$, $i = 0, 1, \ldots, k_1 - 2$, and $S_{k_1-1,0}^* > 0$, then clearly it is irreducible. Also, if it is such that $S_{i,i-1}^* > 0$, $i = 1, \ldots, k_1 - 1$ and $S_{0,k_1-1}^* > 0$, then it is irreducible. Using this information one could give slightly weaker but at the same time simpler versions of the theorem.

7.3.8   Corollary

Let $\{(a_n, b_n) : n = 0, 1, \ldots\}$ be a sequence of vectors with nonnegative real components such that $a_n > 0$ for all n and $b_0 > 0$. Let (X, Y) be a random vector with nonnegative integer-valued components such that for each n with $P\{X = n\} > 0$, we have

$$P\{Y = r|X = n\} = a_r b_{n-r}/c_n, \quad r = 0, 1, \ldots, n,$$

where $\{c_n\}$ is the convolution of $\{a_n\}$ and $\{b_n\}$. Assume that $P\{X - Y = k_0\} > 0$ and $P\{X - Y = k_0 + k_1\} > 0$ with k_0, k_1 as considered earlier. Then the following conditions are equivalent:

(i) Y and $X - Y$ are independent.

(ii) $P\{Y = r\} = P\{Y = r|X - Y = k_0\} = P\{Y = r|X - Y = k_0 + k_1\}, r = 0, 1, \ldots$.

(iii)   For some $\theta > 0$ and some periodic sequence $\{q_n : n = 0, 1, \ldots\}$ with the largest common divisor (l.c.d.) of the n for which $b_n > 0$ as one of its periods

$$P\{X = n\} = q_n c_n \theta^n, \quad n = 1, 2, \ldots .$$

Proof   We prove the corollary by showing (i) $\Rightarrow$ (ii) $\Rightarrow$ (iii) $\Rightarrow$ (i). Except for the implication (ii) $\Rightarrow$ (iii), all the others are easy to establish. To show (ii) $\Rightarrow$ (iii), let us assume that (ii) is valid. Let τ denote the largest common divisor of those n for which $b_n > 0$. Define $\{(X_i^*, Y_i^*) : i = 0, 1, \ldots, \tau - 1\}$ to be a sequence of random vectors such that the joint distribution of X_i^* and Y_i^* is the same as the conditional distribution of $(X - i)/\tau$ and $(Y - i)/\tau$ given that $X \in \{i, i + \tau, i + 2\tau, \ldots\}$ for each i. Assuming without loss of generality, $P\{X \in \{i, i + \tau, i + 2\tau, \ldots\}\} > 0$, for each $i = 0, 1, \ldots, \tau - 1$, it follows that for each $i \in \{0, 1, \ldots, \tau - 1\}$

$$P\{X_i^* = n, Y_i^* = r\} = g_n^*(i)a_r^*(i)b_{n-r}^*(i)/c_n^*(i), \quad r = 0, 1, \ldots, n; \ n = 0, 1, \ldots$$

with $a_r^*(i) = a_{i+r\tau}, b_r^*(i) = b_{r\tau}, r = 0, 1, \ldots$, and $\{c_r^*(i)\}$ as the convolution of $\{a_r^*(i)\}$ and $\{b_r^*(i)\}$, and

$$g_n^*(i) \propto P\{X = i + \tau n\}, \quad n = 0, 1, \ldots .$$

Observe that $c_n^*(i) = c_{i+\tau n}; i = 0, 1, \ldots, \tau - 1; n = 0, 1, \ldots$, and that (ii) is valid for X_i^* and Y_i^* with k_0 and k_1 replaced by k_0/τ and k_1/τ respectively. The required result follows if it is established that for some $\theta > 0$

$$g_n^*(i) \propto c_n^*(i)\theta_i^n, \quad n = 0, 1, \ldots$$

since the form of $\{g_n^*(i)\}$ implies in view of (ii) that for some r

$$
\begin{aligned}
\theta_i &= \left[\frac{b_{k_0} P\{X_i^* - Y_i^* = (k_0 + k_1)/\tau, Y_i^* = r\}}{b_{k_0+k_1} P\{X_i^* - Y_i^* = k_0/\tau, Y_i^* = r\}}\right]^{\tau/k_1} \\
&= \left[\frac{b_{k_0} P\{X - Y = k_0 + k_1, Y = i + r\tau\}}{b_{k_0+k_1} P\{X - Y = k_0, Y = i + r\tau\}}\right]^{\tau/k_1} \\
&= \left[\frac{b_{k_0} P\{X - Y = k_0 + k_1\}}{b_{k_0+k_1} P\{X - Y = k_0\}}\right]^{\tau/k_1}, \quad i = 0, 1, \ldots, \tau - 1,
\end{aligned}
$$

which is clearly independent of i. Consequently, the required result follows if it is established when $\tau = 1$; this is so because (X_i^*, Y_i^*) satisfies the requirement of (X, Y) with $\tau = 1$. Suppose

$$\lambda = P\{X - Y = k_0 + k_1\}/P\{X - Y = k_0\}.$$

Clearly in that case, the positivity of g_n for $n \geqslant k_0$ together with the straightforward relation (7.3.22) appearing in the proof of Theorem 7.3.6 implies that

$$\sum_{n=0}^{\infty} c_n(\lambda^*)^{n/k_1} < \infty$$

where $\lambda^* = \lambda b_{k_0}/b_{k_0+k_1}$.
  The distribution

$$g_n \propto c_n(\lambda^*)^{n/k_1}, \quad n = 0, 1, \ldots \tag{7.3.26}$$

together with the distributions $\{S(r|n)\}$ satisfies the requirement of the theorem and gives $P\{X - Y = k_0 + k_1\}/P\{X - Y = k_0\} = \lambda$. The unique determination of $\{P(X = n)\}$ here implies that it is given by $\{g_n\}$ of (7.3.26). Consequently, we can conclude that (iii) is valid when $\tau = 1$ and hence for all τ. ∎

7.3.9 Remarks

(i) Corollary 2.3.9 gives immediately that Corollary 7.3.8 with (ii) replaced by

(ii)' $P\{Y = r\} = P\{Y = r|X - Y = k_0\}, \quad r = 0, 1, \ldots;$

$$P\{Y = r|X - Y = k_0\} \propto P\{Y = r|X - Y = k_0 + k_0\}, \quad r = 0, 1, \ldots, k_0 + \tau - 1,$$

where τ is as defined in the proof of the corollary, holds.

(ii) In view of Corollary 2.2.3, it follows that Corollary 7.3.8 or its modified version given in (i) above holds with (iv) given below included in its statement.

'(iv) $P\{Y = r\} = P\{Y = r|X = Y\}, \quad r = 0, 1, \ldots.$'

(iii) In view of Corollary 2.3.10 it is seen that Theorem 4 of Krishnaji (1974) is not valid. This also follows from the counterexample given by Patil and Taillie. The error in Krishnaji's argument appears in the last sentence of the proof in which it is claimed that since $X = \Lambda \exp\{\Lambda(\theta - 1)\}$ is degenerate, Λ has to be degenerate. We may, however, point out here that Krishnaji's theorem with the portion '$G(t)$ is nonnegative for all real t' in it replaced by '$G(t)$ is infinitely divisible' is valid.

(iv) Corollary 2.3.10 not only shows Krishnaji's Theorem 4 and hence the Srivastava–Singh conjecture to be void, but also identifies, under the assumptions in Corollary 7.3.8, the class of distributions relative to which (7.3.20) holds. In particular, the corollary implies under the stated assumptions that (7.3.20) with $k = 1$ holds if and only if either $g_n/c_n \propto \theta\lambda_1^{n-1} + (1-\theta)\lambda_2^{n-1}$ for some $\theta < 1$ and $0 < \lambda_1 \leqslant \lambda_2$ satisfying $\sum_{n=0}^{\infty}b_n\lambda_1^{n-1} = \sum_{n=0}^{\infty}b_n\lambda_2^{n-1}$, or $g_n/c_n \propto \{\theta+(1-\theta)n\}\lambda^n$ for some $\theta \in [0, 1)$ and $\lambda > 0$ satisfying $\sum_{n=0}^{\infty}(n-1)b_n\lambda^n = 0$, where $g_n = P\{X = n\}$.

(v) Theorem 7.3.6 does not hold if we replace (7.3.21) by the original Rao–Rubin condition (7.2.3) (with λ as $P\{X - Y = 0\}$) even when we assume $S(r|n) > 0$ for all $0 \leqslant r \leqslant n, n \geqslant 0$, as is shown by the following example based on Theorem 7.3.4.

Example Let $0 < p < p' < 1$ and $c > 1$ be fixed. Theorem 7.3.4 implies that there exist infinitely many distributions $\{g_n^*\}$ with probability generating function G^* satisfying

$$G^*(1 - p + pz) = cG^*(p'z), \quad |z| \leqslant 1.$$

Define

$$S(r|n) = \begin{cases} \beta_n \binom{n}{r} p^r(1-p)^{n-r}, & r = 0, 1, \ldots, n-1; \ n \geqslant 1 \\ \beta_n \{(p')^n - c^{-1}p^n\}, & r = n; \ n \geqslant 0, \end{cases}$$

where $\beta_n = c/\{c(1 - p^n) + c(p')^n - p^n\}, n \geqslant 0$, and

$$g_n \propto \beta_n^{-1}g_n^*, \quad n = 0, 1, \ldots.$$

If we take $\{g_n\}$ and $\{S(r|n) : r = 0, 1, \ldots, n; n = 0, 1, \ldots\}$ as above, then

$$P\{Y = r\} = P\{Y = r | X - Y = 0\}, \quad r = 0, 1, \ldots$$

with

$$P\{X - Y = 0\} = 1/(1 + c)$$

which is fixed.

7.4   SOME CHARACTERIZATIONS BASED ON PARTIAL INDEPENDENCE

We briefly touch upon some characterization properties based on partial independence. These properties are linked with those met in Sections 7.2 and 7.3 corresponding to damage models and hence one could view these as their variants. However, the techniques required to deal with these latter properties are of more elementary nature and could be, in many cases, dismissed as those necessary for solving only simple exercises. As we believe that the results that we are to discuss in this section tell us something more about the structure and scope of a damage model, we have decided to cover them here. However, the discussion in this section is of very restrictive nature, addressing only the questions that we found appealing; the results given here are taken either from Alzaid *et al* (1986b) or Shanbhag and Kapoor (1993).

Let $\{(a_n, b_n, h_n) : n = 0, 1, \ldots\}$ be a sequence of real vectors such that $a_n > 0, b_n \geqslant 0, h_n > 0$ for all $n \geqslant 0$ with b_0, b_1 and $b_2 > 0$. Define the family of discrete distributions

$$F(r|n) = c_n^{-1} h_n^r a_r h_n^{-(n-r)} b_{n-r}, \quad r = 0, 1, \ldots, n : n \geqslant 0, \tag{7.4.1}$$

where c_n are such that $F(0|n) + \cdots + F(n|n) = 1, n \geqslant 0$. We have then the following theorem.

7.4.1   Theorem

Let (X, Y) be a random vector with nonnegative integer components, and denote $P\{X = n\} = g_n$ and $P\{Y = r | X = n\} = S(r|n)$. Assume that $P\{X = 0\} = g_0 < 1$ and $S(r|n) = F(r|n)$ as defined in (7.4.1). Then the following are equivalent:

(i) $P(Y = r | X = Y) = P(Y = r | X = Y + 1) = P(Y = r | X = Y + 2)$, $r = 0, 1, \ldots$

(ii) $h_n = A\alpha^n$ and $g_n = (g_0/c_0)c_n\beta^n, n = 1, 2, \ldots$, for some A, α and $\beta > 0$.

(iii) Y is independent of $X - Y$.

Proof   It is easy to check (ii) $\Rightarrow$ (iii) $\Rightarrow$ (i). So, it is sufficient to show that (i) $\Rightarrow$ (ii) to prove the theorem. Suppose that (i) is valid. Then, in view of the assumptions imposed on the sequence $\{(a_n, b_n, h_n) : n \geqslant 0\}$, we get

$$\frac{g_r S(r|r)}{P(X = Y)} = \frac{g_{r+1} S(r|r + 1)}{P(X = Y + 1)} = \frac{g_{r+2} S(r|r + 2)}{P(X = Y + 2)}, \quad r = 0, 1, \ldots . \quad (7.4.2)$$

(with the quantities involved well defined). (7.4.2) implies that $g_r > 0$ for all $r \geqslant 0$. Consequently

$$\frac{g_{r+2}}{g_{r+1}} \frac{c_{r+1}}{c_{r+2}} = \frac{P(X = Y + 1)}{P(X = Y)} \frac{h_{r+1}^{r+1}}{h_{r+2}^{r}} \frac{b_0}{b_1}$$

$$= \frac{P(X = Y + 2)}{P(X = Y + 1)} \frac{h_{r+1}^{r-1}}{h_{r+2}^{r-2}} \frac{b_1}{b_2}, \quad r = 0, 1, \ldots \quad (7.4.3)$$

and hence

$$\frac{h_{r+2}}{h_{r+1}} = \left[\frac{P^2(X = Y + 1)}{P(X = Y)P(X = Y + 2)} \frac{b_0 b_2}{b_1^2}\right]^{1/2} = \alpha(\text{say}), \quad r = 0, 1, \ldots,$$

which implies that

$$h_r = A\alpha^r, \quad r = 1, 2, \ldots \quad (7.4.4)$$

for some positive constant A. Substituting (7.4.4) in (7.4.3) leads to

$$\frac{g_{r+2}}{c_{r+2}} = \beta \frac{g_{r+1}}{c_{r+1}}, \quad r = 0, 1, \ldots,$$

where $\beta = b_0 A\alpha P(X = Y + 1)/b_1 P(X = Y)$. Hence, using (7.4.2) with $r = 0$

$$\frac{g_r}{c_r} = \frac{g_0}{c_0} \beta^r, \quad r = 1, 2, \ldots,$$

which completes the proof. ∎

The following corollary is an improved version of Moran's result.

7.4.2 Corollary

Let (X, Y) be as in Theorem 7.4.1 with

$$S(r|n) = \binom{n}{r} p_n^r (1 - p_n)^{n-r}, \quad r = 0, 1, \ldots, n; n \geqslant 0$$

for some fixed sequence $\{p_n : p_n \in (0, 1), n = 0, 1, \ldots\}$. Then Theorem 7.4.1 holds with (ii) replaced by:

(ii)′ p_n is independent of n for $n \geqslant 1$ and $\{g_n\}$ is a Poisson distribution.

Proof   Note that $S(r|n)$ here are the same as $F(r|n)$ given by (7.4.1) with $a_n = b_n = (n!)^{-1}$,

$$h_n = \left(\frac{p_n}{1 - p_n}\right)^{1/2} \quad \text{and} \quad c_n = \frac{[p_n(1 - p_n)]^{-n/2}}{n!}.$$

It is sufficient if (ii) of Theorem 7.4.1 implies (ii)′. Assume the (ii) in question. Then

$$[p_n/(1 - p_n)] = a\gamma^n, \quad n = 1, 2, \ldots,$$

for some positive constants a and γ. Hence

$$c_n = (\gamma^{-n/2} + a\gamma^{n/2})^n a^{-n/2}/n!, \quad n = 1, 2, \ldots,$$

which implies that, for appropriate B and β,

$$g_n = B\beta^n(\gamma^{-n/2} + a\gamma^{n/2})^n/n!, \quad n = 0, 1, \ldots .$$

Since $\Sigma_{n=0}^\infty g_n = 1$, we must have $\gamma = 1$ in which case $\{g_n\}$ is a Poisson distribution and p_n is independent of n for $n \geq 1$. (Note that $\gamma \neq 1$ contradicts $\Sigma_{n=0}^\infty g_n < \infty$.) This completes the proof. ∎

7.4.3   Remarks

(i) If the condition (i) in Theorem 7.4.1 is replaced by

$$P\{Y = r\} = P\{Y = r | X = Y + j\}, \quad j = 2, 3, \ldots; \quad r \geq 0, \tag{7.4.5}$$

then the conclusion of the theorem does not remain valid even when $b_j > 0$ for all j as the following example shows. Let

$$g_n = \begin{cases} e^{-\lambda}\lambda^n/n!, & n = 2, 3, \ldots \\ \alpha e^{-\lambda}, & n = 1 \\ e^{-\lambda}(1 + \overline{1 - \alpha}\lambda), & n = 0 \end{cases}$$

and

$$S(r|n) = \begin{cases} \binom{n}{r}(\alpha p)^r(1 - \alpha p)^{n-r}, & r = 0, 1, \ldots, n : n \geq 2 \\ \binom{1}{r}p^r(1 - p)^{1-r}, & r = 0, 1; \ n = 1, \end{cases}$$

where $0 < \alpha, p < 1$ and $\lambda > 0$. It is easy to see that if (X, Y) is the corresponding random vector, then it satisfies (7.4.5), but the distribution of X here is not of the form given in (ii) of Theorem 7.4.1.

(ii) A multivariate version of Theorem 7.4.1 is given by Gerber (1980).

We now give two theorems extending respectively a result of Janardan and Rao (1982) and a result of Kourouklis (1986); the result of Kourouklis referred to here subsumes a result of Kimeldorf et al. (1981).

7.4.4   Theorem

Let $t \geqslant 0$, $c \geqslant -t$ and $\{(a_n, b_n) : n = 1, 2, \ldots\}$ be a sequence of real vectors with positive components such that $a_n + b_n = A$ for all $n \geqslant 1$, where A is fixed. Further, let (X, Y) be a random vector with nonnegative integer-valued components such that $g_0 < 1$, where $g_n = P\{X = n\}$, and the conditional distribution $S(\bullet|n)$ is GMPD (n, a_n, b_n, c, t) i.e. generalized Markov–Polya distribution with parameter vector as stated.[†] Then the following conditions are equivalent:

(i) $P\{Y = r|X = Y\} = P\{Y = r|X = Y + 1\}$;
   $P\{X - Y = r|Y = 0\} = P\{X - Y = r|Y = 1\}, r = 0, 1, \ldots$.

(ii) X has GPED (A, t, c, λ), i.e. generalized Polya–Eggenberger distribution with parameter vector as stated, [‡] for some $\lambda > 0$, and a_n is independent of n.

(iii) Y and $X - Y$ are independent.

Proof   It is easy to check (ii) $\Rightarrow$ (iii) $\Rightarrow$ (i). We now show that (i) $\Rightarrow$ (ii). Clearly (i) implies, in view of the fact that $a_r, b_r > 0$ for all $r \geqslant 1$ and $g_0 < 1$, that

$$\left. \begin{array}{l} 0 < S(r|r)g_r/P(X = Y) = S(r|r + 1)g_{r+1}/P(X = Y + 1) \\ 0 < S(0|r)g_r/P(Y = 0) = S(1|r + 1)g_{r+1}/P(Y = 1) \end{array} \right\} \quad r = 0, 1, \ldots .$$

$$(7.4.6)$$

Once it is shown that a_r is independent of r, then, from either of the two identities in (7.4.6), it follows that $X \sim \text{GPED}(A, t, c, \lambda)$ for some $\lambda > 0$. Hence, it is sufficient if we show that a_r is independent of r.
   From (7.4.6), we get

$$[S(r|r)/S(0|r)] = \alpha S(r|r + 1)/S(1|r + 1), \quad r = 0, 1, 2, \ldots, \tag{7.4.7}$$

where $\alpha = P(Y = 1)P(X = Y)/P(X = Y + 1)P(Y = 0)$. Recalling the form of GMPD (n, a_n, b_n, c, t), we can rewrite (7.4.7) as

$$\frac{a_r (a_r + rt + c)^{(r-1,c)}}{b_r (b_r + rt + c)^{(r-1,c)}} = \alpha \frac{(a_{r+1} + rt + c)^{(r-1,c)}}{(b_{r+1} + rt + c)^{(r-1,c)}}, \quad r = 1, 2, \ldots . \tag{7.4.8}$$

[†] This is to mean that

$$S(r|n) = \binom{n}{r} \frac{a_n b_n (A + nt)(a_n + rt)^{(r,c)}(b_n + (n - r)t)^{(n-r,c)}}{A(a_n + rt)(b_n + (n - r)t)(A + nt)^{(n,c)}}, \quad r = 0, 1, \ldots, n,$$

where $x^{(k,c)} = x(x + c) \cdots (x + (k - 1)c)$ if $k \geqslant 1$ and $x^{(0,c)} = 1$.
[‡] This is to mean that

$$P\{X = n\} \propto \frac{A(A + nt)^{(n,c)}}{(A + nt)n!} \lambda^n, n = 0, 1, 2, \ldots,$$

where the notation used is as in the previous footnote.

Note that if we assume $(a_k/b_k) > \alpha$ for some k, (7.4.8) implies that $\{(a_r/b_r) : r = k, k+1, \ldots\}$ strictly increasing and hence

$$\lim_{r \to \infty} \frac{a_r}{b_r} > \alpha; \tag{7.4.9}$$

note also that in that case $\{a_r\}$ and $\{b_r\}$ converge to finite limits. If we take $(a_k/b_k) < \alpha$, then (7.4.9) holds with $>$ replaced by $<$ but the remark following (7.4.9) still holds.

Consider now the case of $c + t > 0, t > 0$ and assume that $(a_k/b_k) > \alpha$; (7.4.8) then implies in view of the observation following (7.4.9) that

$$\frac{a_r}{b_r} \prod_{i=1}^{r-1} \left\{ 1 + o_i \left(\frac{1}{r} \right) \right\} = \alpha \prod_{i=1}^{r-1} \left\{ 1 + o_i^* \left(\frac{1}{r} \right) \right\},$$

where o_i and o_i^* are smaller order functions of $(1/r)$ uniformly for $i = 1, 2, \ldots, r-1$. The above equation implies that $\{a_r/b_r\}$ converges to α and hence we have a contradiction to (7.4.9). Hence it follows that we cannot have a k such that $(a_k/b_k) > \alpha$ and, by symmetry, also a k such that $(a_k/b_k) < \alpha$. Consequently, we have here $(a_r/b_r) = \alpha$ for all r or equivalently all a_r to be equal.

Now noting that the case $c + t > 0, t = 0$ can be reduced to the case $c + t = 0, t > 0$, we claim that we need a proof only for $c + t = 0$. Assume that $a_k \neq a_1$; then clearly the observation following (7.4.9) is valid. If $c + t = 0, t > 0$, we get from (7.4.8)

$$\frac{a_r}{b_r} \frac{\Gamma\left(\dfrac{a_r}{t} + r\right)}{\Gamma\left(\dfrac{a_{r+1}}{t} + r\right)} = \alpha \frac{\Gamma\left(\dfrac{b_r}{t} + r\right)}{\Gamma\left(\dfrac{b_{r+1}}{t} + r\right)} \left\{ \frac{\Gamma\left(\dfrac{a_r}{t} + 1\right)}{\Gamma\left(\dfrac{a_{r+1}}{t} + 1\right)} \frac{\Gamma\left(\dfrac{b_{r+1}}{t} + 1\right)}{\Gamma\left(\dfrac{b_r}{t} + 1\right)} \right\},$$

$$r = 1, 2, \ldots . \tag{7.4.10}$$

Using Stirling's formula in (7.4.10), we see that (a_r/b_r), $r = k, k+1, \ldots$, converges to α. Hence we have a contradiction to an earlier observation that the limit of the sequence is not equal to α. This, in turn, implies that we cannot have a k such that $a_k \neq a_1$ in (7.4.8). In that case either of the two equations in (i) of the theorem implies the validity of (ii) of the theorem. Thus, we have the assertion for $c + t = 0, t > 0$. It now remains to prove the result when $c = t = 0$ in which case (7.4.8) reduces to

$$\left(\frac{a_r}{b_r} \right)^r = \alpha \left(\frac{a_{r+1}}{b_{r+1}} \right)^{r-1}, \qquad r = 1, 2, \ldots,$$

which yields easily that

$$\frac{a_{r+1}}{b_{r+1}} = \left(\frac{a_2}{b_2} \right)^r \alpha^{1-r}, \qquad r = 0, 1, \ldots . \tag{7.4.11}$$

Either of the two equations in (i) of the theorem implies then that for some $\lambda > 0$

$$g_r \propto \lambda^r \beta^{-r(r-1)/2}(1 + \alpha\beta^{r-1})^r/r!, \quad r = 0, 1, \ldots, \tag{7.4.12}$$

where $\beta = a_2/(b_2\alpha)$. (7.4.12) contradicts (as in the proof of Corollary 7.4.12) the requirement that $\Sigma_{r=0}^{\infty} g_r < \infty$ (or indeed that $\Sigma_{r=0}^{\infty} g_r = 1$), unless $\beta = 1$ or equivalently, in view of (7.4.11) all a_r's are equal. The validity of the required assertion is hence obvious. ∎

7.4.5   Theorem

Let $B_1, B_2, \ldots$ be a sequence of nondegenerate independent (0–1 valued) Bernoulli random variables. Further let (X, Y) be a random vector of nonnegative integer-valued components such that $P\{X \geqslant 1\} > 0$ and for almost all $n \geqslant 0$

$$P\{Y = r | X = n\} = P\left(\sum_{i=1}^{n} B_i = r\right), \quad r = 0, 1, \ldots, n.$$

Then the following are equivalent:

(i) $P\{Y = r | X - Y = 0\} = P\{Y = r | X - Y = 1\}$,
      $P\{X - Y = r | Y = 0\} = P\{X - Y = r | Y = 1\}, r = 0, 1, \ldots$

(ii) Conditionally upon $X - Y \in \{0, 1, 2\}$, Y and $X - Y$ are independent.

(iii) Conditionally upon $Y \in \{0, 1, 2\}$, Y and $X - Y$ are independent.

(iv) The random variable X is such that

$$P\{X = n\} = \begin{cases} K\lambda^n \displaystyle\prod_{m=1}^{n} \left(\dfrac{1 + \alpha\beta^m}{1 + \beta + \cdots + \beta^{m-1}} \right), & n = 1, 2 \ldots \\ K, & n = 0 \end{cases}$$

and B_i's satisfy

$$P\{B_i = 1\} = \frac{\alpha\beta^i}{1 + \alpha\beta^i}, \quad i = 1, 2, \ldots,$$

where K, λ, α and β are appropriate positive constants (with β not necessarily equal to 1). (If $\beta = 1$, we have X as Poisson and B_i's as identically distributed.)

(v) $X - Y$ and Y are independent.

Proof   Using essentially the argument leading to (7.4.6) and (7.4.7) in the proof of Theorem 7.4.4, we can see that each of (i), (ii) and (iii) implies individually that (iv) is valid. We shall now see that (iv) implies (v).

Denote $P\{Y = r, X - Y = n - r\}$ by $C(r, n)$. We have then

$$C(r, n) = \frac{\lambda}{1 + \beta + \cdots + \beta^{n-1}} C(r, n - 1) + \frac{\lambda\alpha\beta^n}{1 + \beta + \cdots + \beta^{n-1}} C(r - 1, n - 1),$$

$$r = 0, 1, \ldots, n; \quad n \geqslant 1$$

where $C(-1, n - 1) = 0$, $C(n, n - 1) = 0$ and $C(0, 0) = K$. Consequently, by induction on n, it is easily seen that

$$C(r, n) = K\lambda^n \left(\prod_{m=1}^{r} \frac{\alpha\beta^m}{1 + \beta + \cdots + \beta^{m-1}} \right) \left(\prod_{m=1}^{n-r} \frac{1}{1 + \beta + \cdots + \beta^{m-1}} \right),$$

$$r = 0, 1, \ldots, n; \quad n = 0, 1, \ldots,$$

where $\Pi_{m=1}^0 = 1$. In view of the form of C, it is now immediate that Y and $X - Y$ are independent, i.e. (v) holds.

As (v) implies (i), (ii) and (iii), we have then the theorem. ∎

7.4.6 Remarks

(i) Theorem 7.4.4, but for some trivial situations, is stronger than the result of Moran (1952). As mentioned before, the result is also stronger than a result of Janardan and Rao (1982). Both these cited results concern characterizations based on the independence of Y and $X - Y$.

(ii) If we take in Theorem 7.4.4, in place of the GMPD (n, a_n, b_n, c, t) its extended version of the form

$$\left\{ \frac{\xi_r^{(1)} \xi_{n-r}^{(2)} (a_n + rt)^{(r,c)} (b_n + (n - r)t)^{(n-r,c)}}{\xi_n^{(3)} (a_n + rt)(b_n + (n - r)t)} : r = 0, 1, \ldots, n \right\}$$

with $\xi_j^{(i)}$'s positive, then our modified theorem still holds provided (ii) is replaced by (ii)' that either $g_n \propto \xi_n^{(3)} \lambda^n$, $n = 0, 1, \ldots$ for some $\lambda > 0$ and a_n is independent of n, or $c = t = 0$, $g_n \propto \xi_n^{(3)} (a_n b_n)^{(2-n)/2} \lambda^n$, $n = 0, 1, \ldots$, for some $\lambda > 0$ and a_n/b_n is of the form $\alpha\beta^n$ for all n.

(iii) As GMPD includes several well known distributions such as binomial and negative hypergeometric as special cases, and GPED distributions such as Poisson, Lagrangian Poisson and negative binomial as special cases, we can obtain via Theorem 7.4.4 characterizations of various specialized distributions.

(iv) According to the definitions in Benkherouf and Bather (1988), the distributions of Y and $X - Y$ in (iv) of Theorem 7.4.5 are respectively Heine and Euler distributions if $\beta < 1$, and vice versa if $\beta > 1$. Consequently, we get the distribution of X in the stipulation in question to be the convolution of Heine and Euler distributions if $\beta \neq 1$; we can easily obtain the p.g.f. of X here using the results in Benkherouf and Bather (1988) or directly using the expression for $P\{X = n\}$ given. In particular, if we use the expression for $P\{X = n\}$, then denoting the p.g.f. of X by G, we get, for $\beta < 1$,

$$G(z) = \left(\frac{1 + \lambda\alpha\beta(1 - \beta)z}{1 - \lambda(1 - \beta)z} \right) G(\beta z)$$

$$= K \prod_{r=1}^{\infty} \left\{ \frac{1 + \lambda\alpha\beta^r(1 - \beta)z}{1 - \lambda\beta^{r-1}(1 - \beta)z} \right\}, \quad |z| \leqslant 1$$

(because $\lim_{n \to \infty} G(\beta^n z) = K$); as the p.g.f. of X in the case of $\beta > 1$ can easily be reduced to that for $\beta < 1$ with appropriate notational changes, the result in this latter case is now obvious.

Another result that could be worth pointing out at this stage is the following theorem due to Alzaid *et al* (1986b). This theorem extends some of the earlier results given by Patil and Seshadri (1964) and Panaretos (1982b) under more restrictive conditions.

7.4.7   Theorem

Let (X, Y) be a two-component random vector, as in a damage model, with nonnegative integer-valued components satisfying (7.2.1), and there exist integers $n_0 = 0 < n_1 < \cdots < n_r$ (with $r \geqslant 1$) and $\xi \geqslant n_i - n_{i-1}, i = 1, 2, \ldots, r$ and a nonempty subset A of $\{1, 2, \ldots, r\}$ such that $P\{X - Y = n_i\} > 0, i = 0, 1, \ldots, r$, and the largest common divisor of $n_i - n_{i-1}, i \in A$ equals 1 and

$$\min\{n_i - n_{i-1} : i \in A\} + \max\{n_i - n_{i-1} : i \in A\} \leqslant \max\{2, \xi\}.$$

Assume that $S(m|m + n_i) > 0$ for $i = 0, 1, \ldots, r$ and $m = 0, 1, \ldots, \xi$. Then the condition that for some positive constants $k_i, i = 1, \ldots, r$,

$$P\{Y = m|X - Y = 0\} = k_i P\{Y = m|X - Y = n_i\},$$

$$m = 0, 1, \ldots, \xi; \quad i = 1, \ldots, r \quad (7.4.13)$$

implies that either $P\{X \leqslant n_r + \xi\} = 0$ or $P\{X = n\} = P\{X = 0\}c_n\lambda^n$, $n = 1, \ldots, n_r + \xi$ for some (positive) sequence $\{c_n\}$ determined uniquely by $\{\{S(m|n)\}\}$ and some positive λ (together with n_i's and ξ).

A proof of the theorem appears in Alzaid *et al* (1986); one could also prove the result via an argument based on the Perron–Frobenius theorem. We shall not discuss the details of these here.

CHAPTER 8

Order Statistics, Record Values and Properties in Applied Probability

8.1 INTRODUCTION

The strong memoryless properties of exponential and geometric distributions and the corresponding characterizations that we met in Chapter 5 play crucial roles in applied probability in topics such as reliability, queueing networks, point processes and biological models. Apart from these, various other properties based on order statistics and record values lead us to important results either in distribution theory or applied probability. Many of these results follow via the results on integrated Cauchy functional equations, appearing in Chapters 2 and 3; the necessary details in this connection are provided in the present chapter. We also demonstrate herein as to how several of the other results in applied probability follow from the cited results of earlier chapters.

Galambos and Kotz (1978) and Rao and Shanbhag (1986) amongst others have reviewed some major characterizations based either on order statistics or on record values. Çinlar (1972), Daley (1976), Kelly (1978), Rao and Shanbhag (1988), Alzaid *et al* (1988), and others have discussed many results in applied probability that have links with the results given in Chapter 2. Blackwell (1948, 1953, 1955), Kendall (1951, 1953), Lindley (1952) and Wishart (1956) contain some of these results; Feller (1966) also includes some. Amongst the problems that we are to revisit in this chapter, there are many of these.

Some of the derivations given in this discussion are taken from Rao and Shanbhag (1986, 1989a). We also use herein arguments of other authors that are of relevance, whenever necessary.

8.2 CHARACTERIZATIONS BASED ON ORDER STATISTICS AND RECORD VALUES

Ferguson (1964, 1965) and Crawford (1966) were among the earliest authors who characterized geometric and exponential distributions via properties of order statistics. They showed that if X and Y are independent nondegenerate random variables, then $\min\{X, Y\}$ is independent of $X - Y$ if and only if for some $\alpha > 0$ and $\beta \in \mathbb{R}$, we have $\alpha(X - \beta)$ and $\alpha(Y - \beta)$ to be either both exponential or both geometric (in the usual sense). In what follows, using implicitly Theorem 2.2.2, we shall prove a somewhat stronger version of this result. Indeed, it should be clear from the proof in question that the Ferguson–Crawford result as well as its extension given here are linked with the strong memoryless property characterization of the geometric and exponential distributions. The extension is given by the following theorem.

8.2.1   Theorem

Let X and Y be as in the Ferguson–Crawford result and y_0 be a point such that there are at least two support points of the distribution of $\min\{X, Y\}$ in $(-\infty, y_0]$. Then $X - Y$ and $\min\{X, Y\}I_{\{\min\{X,Y\}\leqslant y_0\}}$ are independent if and only if for some $\alpha \in (0, \infty)$ and $\beta \in \mathbb{R}$, $\alpha(X - \beta)$ and $\alpha(Y - \beta)$ are both exponential, or geometric on $\mathbb{N}_0$, in which case $X - Y$ and $\min\{X, Y\}$ are independent.

Proof   The 'if' part and the result that $\min\{X, Y\}$ and $X - Y$ are independent if X and Y are as in the 'if' part are trivial. We shall now prove the 'only if' part. Assume hence that $X - Y$ and $\min\{X, Y\}I_{\{\min\{X,Y\}\leqslant y_0\}}$ are independent. We have at least one of $P\{X \geqslant Y\}$ and $P\{Y \geqslant X\}$ to be positive. Without loss of generality, we shall assume that the first one of these is positive. Then, for any $x \in \mathbb{R}_+$, we have

$$P\{X - Y \geqslant x \mid \min\{X, Y\} \leqslant y\}/P\{X - Y \geqslant 0 \mid \min\{X, Y\} \leqslant y\}$$

to be independent of y for $y \in (l, y_0]$, where l is the left extremity of the distribution of Y. Note that

$$P\{X \geqslant Y, Y \leqslant y\}(= P\{X \geqslant Y\}P\{\min\{X, Y\} \leqslant y\}) > 0$$

for each $y \in (l, y_0]$. In view of what we have established, we have further that for each $x \in \mathbb{R}_+$,

$$P\{X \geqslant Y + x, Y \leqslant y\}/P\{X \geqslant Y, Y \leqslant y\}$$

is independent of y if $y \in (l, y_0]$. This last ratio equals $P\{X \geqslant Y + x \mid X \geqslant Y, Y \leqslant y\}$. If it is assumed that $l = -\infty$, we get for any $x \in \mathbb{R}_+$ and $y \in (-\infty, y_0]$

$$P\{X \geqslant Y + x \mid X \geqslant Y, Y \leqslant y\} = \lim_{y' \to -\infty} P\{X \geqslant Y + x \mid X \geqslant Y, Y \leqslant y'\} = 1$$

which leads us to a contradiction because for every $y \in (l, y_0]$,

$$\lim_{x \to \infty} P\{X \geqslant Y + x \mid X \geqslant Y, Y \leqslant y\} = 0.$$

This, in turn, implies that we have $l > -\infty$, and for every $x \in \mathbb{R}_+$,

$$P\{X \geqslant Y + x \mid X \geqslant Y, Y \leqslant y_0\} = \lim_{y' \to l+} P\{X \geqslant Y + x \mid X \geqslant Y, Y \leqslant y'\}$$

$$= P\{X \overset{*}{\geqslant} l + x \mid X \overset{*}{\geqslant} l\}$$

where '$\overset{*}{\geqslant}$' denotes '$\geqslant$' if l is a discontinuity point of the distribution of Y, and '$>$' otherwise. As

$$P\{Y = l | X \geqslant Y, Y \leqslant y_0\} = P\{\min\{X, Y\} = l\}/P\{\min\{X, Y\} \leqslant y_0\} < 1$$

(where we have used the independence stated in the assertion), we have from Theorem 2.2.2 that the conditional distribution of $X - l$ given that $X \geqslant l$ is exponential if the conditional distribution of $Y - l$ given that $Y \leqslant y_0$ is nonarithmetic, and that of $\lambda[(x - l)/\lambda]$ given that $X \geqslant l$, where $[\bullet]$ denotes the integral part, is geometric on $\{0, \lambda, 2\lambda, \ldots\}$ if the conditional distribution of $Y - l$ given that $Y \leqslant y_0$ is arithmetic with span λ. Note that this implies that the left extremity of the distribution of X is less than or equal to that of Y and that $P\{Y \geqslant X\} > 0$. Hence, by symmetry, we get a further result with the places of X and Y interchanged (and the obvious notational change in l). When this latter result is used in conjunction with the earlier result, we get that the left extremity of the distribution of X is also l and that both $X - l$ and $Y - l$ are either exponential, or geometric on $\{0, \lambda, 2\lambda, \ldots\}$ for some $\lambda > 0$. Hence, we have the theorem. ∎

The following two obvious corollaries of the theorem are worth noting at this stage; the latter of these two extends essentially a result of Fisz (1958).

8.2.2   Corollary

If in Theorem 8.2.1, X and Y are additionally assumed to be identically distributed, then the assertion of the theorem holds with $|X - Y|$ in place of $X - Y$.

Proof   The corollary follows on noting that, under the assumptions, for any $y \in \mathbb{R}$, $|X - Y|$ and $\min\{X, Y\}I_{\{\min\{X,Y\}\leqslant y\}}$ are independent if and only if $X - Y$ and $\min\{X, Y\}I_{\{\min\{X,Y\}\leqslant y\}}$ are independent. ∎

8.2.3   Corollary

Let X and Y be two i.i.d. nondegenerate positive random variables and y_0 be as defined in Theorem 8.2.1. Then $\min\{X, Y\}/\max\{X, Y\}$ and $\min\{X, Y\}I_{\{\min\{X,Y\}\leqslant y_0\}}$ are independent if and only if for some $\alpha > 0$ and $\beta \in \mathbb{R}$, $\alpha(\log X - \beta)$ is either geometric or exponential.

Proof   Define $X^* = \log X$, $Y^* = \log Y$ and $y_0^* = \log y_0$. (Note that y_0 here has to be positive.) Noting that $-\log(\min\{X, Y\}/\max\{X, Y\}) = |X^* - Y^*|$ and

$$(\log(\min\{X, Y\}))I_{\{\log(\min\{X,Y\})\leqslant \log y_0\}} = \min\{X^*, Y^*\}I_{\{\min\{X^*,Y^*\}\leqslant y_0^*\}},$$

we can hence get the result from Corollary 8.2.2. ∎

8.2.4   Remark

If we replace in Corollary 8.2.3 the condition on the existence of y_0 by that there exists a point y_0' such that there are at least two support points of the distribution of $\max\{X, Y\}$ in $[y_0', \infty)$, then the assertion of the corollary with $\min\{X, Y\}I_{\{\min\{X,Y\} \leqslant y_0\}}$ replaced by $\max\{X, Y\}I_{\{\max\{X,Y\} \geqslant y_0'\}}$ and $\log X$ replaced by $-\log X$ holds. This follows because $\min\{X^{-1}, Y^{-1}\} = (\max\{X, Y\})^{-1}$ and $\max\{X^{-1}, Y^{-1}\} = (\min\{X, Y\})^{-1}$. The present version of Corollary 8.2.3 is indeed a direct extension of Fisz's (1958) result. (Fisz characterizes the distribution in question via the independence of $\max\{X, Y\}/\min\{X, Y\}$ and $\max\{X, Y\}$.)

There is an interesting variant of Theorem 8.2.1. This is due to Rossberg (1972), Ramachandran (1980) and (for the special case of $n = 2$ in a somewhat restricted form) Puri and Rubin (1970). This is given by the next theorem; that this follows as a corollary to Theorem 2.2.2 is essentially an observation in Rao (1983). (See also Lau and Rao (1982).)

8.2.5   Theorem

Let $n \geqslant 2$ and $X_1, \ldots, X_n$ be i.i.d. random variables with d.f. F that is not concentrated on $\{0\}$. Further, let $X_{1:n} \leqslant \cdots \leqslant X_{n:n}$ denote the corresponding order statistics. Then, for some $1 \leqslant i < n$,

$$X_{i+1:n} - X_{i:n} \overset{\mathrm{d}}{=} X_{1:n-i}, \tag{8.2.1}$$

where $X_{1:n-i} = \min\{X_1, \ldots, X_{n-i}\}$, if and only if one of the following two conditions holds:

(i) F is exponential.

(ii) F is concentrated on some semilattice of the form $\{0, \lambda, 2\lambda, \ldots\}$ with $F(0) = \alpha$ and $F(j\lambda) - F((j-1)\lambda) = (1-\alpha)(1-\beta)\beta^{j-1}$ for $j = 1, 2, \ldots$ for some $\alpha \in (0, \binom{n}{i}^{-1/i}]$ and $\beta \in [0, 1)$ such that $P\{X_{i+1:n} > X_{i:n}\} = (1-\alpha)^{n-i}$ (which holds with $\alpha = \binom{n}{i}^{-1/i}$ or $\beta = 0$ if and only if

$$F(0) - F(0-) = \binom{n}{i}^{-1/i}$$

and

$$F(\lambda) - F(\lambda-) = 1 - \binom{n}{i}^{-1/i}$$

for some $\lambda > 0$). (The existence of cases $\beta > 0$ can easily be verified.)

Proof   We shall first establish the 'only if' part of the assertion. (8.2.1) implies that $P\{X_{1:n-i} \geqslant 0\} = (1 - F(0-))^{n-i} = 1$ and hence that F is concentrated on $\mathbb{R}_+$. Hence, we can conclude from (8.2.1) that

$$\binom{n}{i} \int_{\mathbb{R}_+} (\overline{\overline{F}}(x + y))^{n-i} F^i(\mathrm{d}y) = (\overline{\overline{F}}(x))^{n-i}, \quad x \in \mathbb{R}_+, \tag{8.2.2}$$

where $\overline{\overline{F}}(x) = 1 - F(x)$, $x \in \mathbb{R}_+$. As $P\{X_1 = 0\} < 1$, we have then $\overline{\overline{F}}(0) > 0$ and (8.2.2) implies that $\binom{n}{i}(F(0))^i \leqslant 1$ (where obviously we have used the property of

F that $F(0-) = 0$). If F is a nonarithmetic distribution (concentrated on $\mathbb{R}_+$) with $\binom{n}{i} F^i(0) < 1$, then (8.2.2) implies, in view of Theorem 2.2.2, that (i) holds. On the other hand, if F is arithmetic with span λ with $\binom{n}{i}(F(0))^i < 1$, then the equation in question implies, in view of the cited theorem, that F is such that $F(0) = \alpha$ and

$$F(j\lambda) - F((j-1)\lambda) = (1-\alpha)(1-\beta)\beta^{j-1}$$

for $j = 1, 2, \ldots$, for some $\alpha \in \left(0, \binom{n}{i}^{-1/i}\right)$ and $\beta \in (0, 1)$ such that $P\{X_{i+1:n} > X_{i:n}\} = (1-\alpha)^{n-i}$. Finally, if $\binom{n}{i}(F(0))^i = 1$, then (8.2.2) is not met if F is nonarithmetic, or arithmetic with some span λ with at least two positive support points, but is met by any F concentrated on $\{0, \lambda\}$ for some $\lambda(> 0)$. This completes the proof of the 'only if' part of the assertion.

It is easily seen that (8.2.1) is equivalent to the condition that (8.2.2) holds with $\overline{\overline{F}}(0-) = 1$. For all the distributions that we have arrived at in the proof of the 'only if' part, we have $\overline{\overline{F}}(0-) = 1$ and (8.2.2) met. Hence, we have the 'if' part of the assertion, and consequently the theorem. ∎

There are other variations of Theorem 8.2.5. A special case for $n = 2$ of one of these was dealt with by Arnold and Ghosh (1976) and this latter result was later extended by Arnold (1980) via Theorem 2.2.1. (There is a minor error in the proof of the Arnold–Ghosh result in the cited paper, see Fosam, *et al* (1993) for a correction.) We shall give this variation below. However, as this result follows essentially via the same argument as that used to prove Theorem 8.2.5 (but for a minor simplification), we shall refrain from producing a proof here.

8.2.6   Theorem

Let $n \geqslant 2$ and $X_1, \ldots, X_n$ be nondegenerate i.i.d. random variables with d.f. F. Also, let $X_{1:n}, \ldots, X_{n:n}$ be order statistics as in Theorem 8.2.5. Then, for some $i \geqslant 1$, the conditional distribution of $X_{i+1:n} - X_{i:n}$ given that $X_{i+1:n} - X_{i:n} > 0$ is the same as the distribution of $X_{1:n-i}$, where $X_{1:n-i}$ is as defined in Theorem 8.2.5 if and only if F is either exponential, or, for some $\lambda > 0$, geometric on $\{\lambda, 2\lambda, \ldots\}$.

Zijlstra (1983) also gives a specialized version of Theorem 8.2.6 and makes some interesting observations on order statistics of geometric distributions.

The next two theorems are extended versions of certain results appearing in Beg and Kirmani (1979) and Kirmani and Alam (1980) respectively.

8.2.7   Theorem

Let F be continuous and, as before, let $X_{1:n}, \ldots, X_{n:n}$ for $n \geqslant 2$ be n ordered observations based on a random sample of size n from F. Further, let i be a fixed positive integer less than n and ϕ be a nonarithmetic (or nonlattice) real monotonic function on $\mathbb{R}_+$ such that $E(|\phi(X_{i+1:n} - X_{i:n})|) < \infty$. Then, for some constant $c \neq \phi(0+)$,

$$E(\phi(X_{i+1:n} - X_{i:n})|X_{i:n}) = c \quad \text{a.s} \tag{8.2.3}$$

if and only if F is exponential, within a shift.

Proof   Using the fact that F is continuous, we can see that (8.2.3) is equivalent to

$$\int_{[x,\infty)} \phi(y-x) F_x^{(n-i)}(\mathrm{d}y) = c \quad \text{for a.a. } [F] \, x \in \mathbb{R}, \tag{8.2.4}$$

where $F_x^{(n-i)}(y) = 1 - (\overline{F}(y)/\overline{F}(x))^{n-i}$, $y \in [x,\infty)$ with $\overline{F}(z) = 1 - F(z-)(= 1 - F(z))$, $z \in \mathbb{R}$. We can take without loss of generality ϕ to be increasing and, as F is continuous, also to be right continuous. We can then see in view of Fubini's theorem, that (8.2.4) is equivalent to

$$\int_{(0,\infty)} (\overline{F}(y+x))^{n-i} \phi(\mathrm{d}y) = (c - \phi(0))(\overline{F}(x))^{n-i}, \quad x \in \text{supp}[F]. \tag{8.2.5}$$

As $\overline{F}$ is monotonic, if there exist points x_1 and x_2 such that $\overline{F}(x_1) = \overline{F}(x_2)$ and the equation in (8.2.5) is met at $x = x_1$ and $x = x_2$, then it follows that the equation is met at all x lying in $[x_1, x_2]$. Using this, we can easily see that (8.2.5) is equivalent to

$$\int_{(0,\infty)} (\overline{F}(y+x))^{n-i} \phi(\mathrm{d}y) = (c - \phi(0))(\overline{F}(x))^{n-i}, \quad x \in (l, \infty) \tag{8.2.6}$$

where l is the left extremity of the d.f. F. In view of Theorem 2.2.2, we hence get that (8.2.6) holds, for some c, if and only if F is exponential, within a shift. ∎

8.2.8   Theorem

Let X_1 and X_2 be i.i.d. nondegenerate integer-valued random variables with support of the type $I \cap \mathbb{Z}$ with I as an interval and $\phi : \mathbb{N}_0 \to \mathbb{R}$ a function such that $E(|\phi(|X_1 - X_2|)|) < \infty$, $\phi(1) > \phi(0)$ and $\phi(n+2) - 2\phi(n+1) + \phi(n) \geq 0$ for all $n \in \mathbb{N}_0$ (i.e. the second differences of ϕ are nonnegative on $\mathbb{N}_0$). Then, for some c,

$$E(\phi(|X_1 - X_2|) \mid \min\{X_1, X_2\}) = c \text{ a.s.} \tag{8.2.7}$$

if and only if X_1 is geometric, but for a shift.

Proof   The 'if' part follows trivially because, as we have seen before, $X_1 - X_2$ and $\min\{X_1, X_2\}$ are independent if X_1 is geometric. To prove the 'only if' part, note that (8.2.7) implies

$$2 \sum_{y=1}^{\infty} \phi(y) P_{y+x} P_x + \phi(0) P_x^2 = c\{2\overline{F}(x+1) P_x + P_x^2\} \quad \text{for all } x \in \mathbb{Z} \tag{8.2.8}$$

where $\{P_x\}$ denotes the distribution of X_1 and $\overline{F}(x) = P\{X_1 \geq x\}$, $x \in \mathbb{R}$. From (8.2.8), we get (in view of Fubini's theorem) after a minor manipulation

$$2 \sum_{y=1}^{\infty} \overline{F}(y+x) \phi^*(y) = (c - \phi(0))\{\overline{F}(x) + \overline{F}(x+1)\} \quad \text{for all } x \text{ with } P_x > 0$$

$$\tag{8.2.9}$$

where $\phi^*(y) = \phi(y) - \phi(y-1)$. Hence the assumption on the support of the X_i's (i.e., on the support of $P\{X_i \leqslant x\}, x \in \mathbb{R}$) gives that

$$2\sum_{y=1}^{\infty} \overline{F}(y+x)\phi^*(y) = (c - \phi(0))\{\overline{F}(x) + \overline{F}(x+1)\}, \quad x > l^* \qquad (8.2.10)$$

where l^* is such that $l^* + 1$ is the left extremity of the distribution of X_1, (8.2.10), in turn, gives that $c - \phi(0) > 0$ and (on using Fubini's theorem) that

$$2\sum_{y=2}^{\infty} \overline{F}(y+x)\phi^{**}(y) + 2\overline{F}(x+1)\phi^*(1) + (c - \phi(0))\overline{F}(x+2)$$

$$= (c - \phi(0))\overline{F}(x), \quad x > l^*, \qquad (8.2.11)$$

where $\phi^{**}(y) = \phi^*(y) - \phi^*(y-1)$. In view of Theorem 2.2.1, we then get the required result. ∎

A somewhat different version of Theorems 8.2.7 and 8.2.8 is given by the following theorem.

8.2.9   Theorem

Let $X_{1:1}, \ldots, X_{1:n}$ be ordered observations based on a random sample of size $n (\geqslant 2)$ from a nondegenerate distribution with d.f. F. Let $1 \leqslant i \leqslant n - 1$ be a given integer and ϕ be a monotonic real left continuous nonconstant function on $\mathbb{R}_+$ such that $E(|\phi(X_{i+1:n} - X_{i:n})|) < \infty$. Then, for some constant $c \neq \phi(0+)$,

$$E(\phi(X_{i+1:n} - X_{i:n}) \mid X_{i+1:n} > X_{i:n}, X_{i:n}) = c \quad \text{a.s} \qquad (8.2.12)$$

if and only if the left extremity, l, of F is finite, and either ϕ is nonarithmetic (or nonlattice) and the conditional distribution of X_1 given that $X_1 > l$ is exponential, within a shift, or for some $\lambda > 0$, ϕ is arithmetic (or lattice) with span λ and the conditional survivor function, $\overline{F}_l$, of X_1 given that $X_1 > l$ satisfies for some $\beta \in (0, 1)$

$$\overline{F}_l(x + n\lambda) = \beta^n \overline{F}_l(x), \quad x > l, n = 0, 1, 2, \ldots . \qquad (8.2.13)$$

Proof   The assertion follows essentially via the argument used in the proof of Theorem 8.2.7 because (8.2.12) is equivalent to

$$\int_{(x,\infty)} \phi(y-x) F_x^{(n-i)}(dy) = c \quad \text{for a.a. } [F]\, x \in \mathbb{R},$$

where

$$F_x^{(n-i)}(y) = 1 - \left(\frac{\overline{\overline{F}}(y)}{\overline{\overline{F}}(x)}\right)^{n-i}, \quad y \in (x, \infty) \text{ with } \overline{\overline{F}}(z) = 1 - F(z), \quad z \in \mathbb{R}. \blacksquare$$

202 ORDER STATISTICS, RECORD VALUES AND PROPERTIES IN PROBABILITY

8.2.10 Remarks

(i) Theorem 8.2.9 gives Theorem 8.2.7 as a corollary. However, because of its more appealing nature, we have dealt with Theorem 8.2.7 separately.

(ii) If we have in addition to the assumptions in Theorem 8.2.9 the assumption that both ϕ and F be arithmetic with the span of F to be greater than or equal to the span of ϕ, then as a corollary of Theorem 8.2.9, we have that for some $c \neq \phi(0+)$, (8.2.12) holds if and only if we have, for some integer n_0, F to be a mixture of the degenerate distribution at $n_0\lambda$ and the distribution of $\lambda(Y + n_0 + 1)$, where λ is the span of ϕ and Y is a geometric random variable.

(iii) Some specialized versions of Theorems 8.2.8 and 8.2.9 have appeared in Rao and Shanbhag (1986) and elsewhere. One of the commonly used ϕ in such characterizations is that given by

$$\phi(x) = \min\{x, \alpha\}, \quad x \in \mathbb{R}_+,$$

where α is an appropriate positive constant.

(iv) As a corollary to Theorem 8.2.9, it follows that the distributions characterized in the theorem are also characterized by the independence of $X_{i+1:n} - X_{i:n}$ and $X_{i:n}$ conditionally upon $\{X_{i+1:n} > X_{i:n}\}$, provided $X_{1:n}, \ldots, X_{n:n}$ are as stated in the theorem. This latter result, in turn, gives as its corollary Rogers's (1963) extension of Fisz's result; we have met Fisz's result in Remark 8.2.4.

(v) One could arrive at further characterization results based on order statistics via Theorems 2.2.1 or 2.2.2 or their extensions or variants in Chapter 2. Arnold (1983), Zijlstra (1983), Srivastava (1986) and several others have given some results using the methods that we have implied; Aly (1988) has reviewed the literature in this connection. However, as an illustration, we could mention one such result here. Let $X_{1:1} \leqslant \cdots \leqslant X_{n:n}$ denote the n ordered observations in a random sample of size $n(\geqslant 2)$ from a nondegenerate d.f. F concentrated on $\mathbb{N}_0$. Arnold (1980) effectively raised the question as to whether the independence of $X_{2:n} - X_{1:n}$ and the event $\{X_{1:n} \doteq m\}$ for a fixed $m \geqslant 1$ implies F to be geometric (possibly within a shift or a change of scale). The question has not as yet been answered completely. However, the result of Corollary 2.3.9 shows that, under mild conditions assuring certain points to be atoms of F, the independence of $X_{2:n} - X_{1:n}$ and $\{X_{1:n} = m\}$ together with the condition

$$P\{X_{2:n} - X_{1:n} > j | X_{1:n} = m\} \propto P\{X_{2:n} - X_{1:n} > j | X_{1:n} = m + m'\},$$

$$j = 0, 1, \ldots, m,$$

for some fixed integer $m' > 0$ characterizes F as a geometric distribution. This extends a result of Sreehari (1983) showing that the independence of $X_{2:n} - X_{1:n}$ and $\{X_{1:n} = m\}$ and the independence of $X_{2:n} - X_{1:n}$ and $\{X_{1:n} = m + m'\}$ for some fixed integer $m' > 0$ characterizes, under some mild conditions, F as a geometric distribution.

We may conclude our discussion of order statistics by presenting two further characterization results on them. These have features somewhat different from those of the results that we have dealt with earlier. The results are respectively due to Shimizu, but in a somewhat modified form suggested by Deny's theorem (i.e. Corollary 2.3.2), and Ferguson.

8.2.11   Theorem

Let $X_1, \ldots, X_n$, $n \geqslant 2$, be i.i.d. positive random variables and $a_1, \ldots, a_n$ be positive real numbers such that the smallest closed subgroup of $\mathbb{R}$ containing $\log a_1, \ldots, \log a_n$ equals $\mathbb{R}$ itself. Then

$$\min\{X_1/a_1, \ldots, X_n/a_n\} \overset{\mathrm{d}}{=} X_1 \qquad (8.2.14)$$

if and only if the survivor function of X_1 is of the form

$$\overline{F}(x) = \exp\{-\lambda_1 x^{\alpha_1} - \lambda_2 x^{\alpha_2}\}, \qquad x \in \mathbb{R}_+ \qquad (8.2.15)$$

with $\lambda_1, \lambda_2 \geqslant 0$, $\lambda_1 + \lambda_2 > 0$ and $\alpha_r (r = 1, 2)$ as positive numbers such that $\Sigma_{i=1}^n a_i^{-\alpha_r} = 1$. (If $\alpha_1 = \alpha_2$, the distribution corresponding to (8.2.15) is Weibull.)

Proof   Denoting the survivor function of the distribution of X_1 (once again) by $\overline{F}$, we see that (8.2.14) is equivalent to

$$\prod_{i=1}^n \overline{F}(x/a_i) = \overline{F}(x), \qquad x \in \mathbb{R}_+. \qquad (8.2.16)$$

In view of the assumption in the statement of the theorem, it is impossible that $a_1 = a_2 = \cdots a_n = 1$. Consequently, in view of (8.2.16), the fact that $\overline{F}(x) > 0$ for some $x > 0$ implies that $\overline{F}(x) > 0$ for all $x \in \mathbb{R}_+$. Hence, it follows that (8.2.16) holds if and only if $\overline{F}(x) > 0$ for all $x > 0$ and

$$G(y - \log a_1) + \cdots + G(y - \log a_n) = G(y), \qquad y \in \mathbb{R}$$

where $G(y) = -\log \overline{F}(e^y)$. In view of Corollary 2.3.2, we have then the theorem; note that $G(y) \to 0$ as $y \to -\infty$. ∎

8.2.12   Theorem

Let F be a continuous d.f. with finite mean and $X_{1:n}, \ldots, X_{n:n}$ be the ordered observations based on a random sample of size $n(\geqslant 2)$ from it. Then, for some $1 \leqslant i < n$

$$E(X_{i:n}|X_{i+1:n} = x) = ax - b \quad \text{for a.a. } [F] x \in \mathbb{R}, \qquad (8.2.17)$$

only if $a > 0$ and the d.f. has the following form, to within a shift and a change of scale:

(i) $F(x) = e^x$ for $x \leqslant 0$ if $a = 1$.

(ii) $F(x) = x^\theta$ for $x \in [0, 1]$ if $a \in (0, 1)$.

(iii) $F(x) = (-x)^\theta$ for $x \leqslant -1$ if $a > 1$, where $\theta = a/[i(1 - a)]$.

Proof   (8.2.17) gives that

$$
\frac{\int_{-\infty}^{x} y F^i (\mathrm{d}y)}{(F(x))^i} = ax - b \qquad \text{for a.a. } [F] \, x \in \mathbb{R}. \tag{8.2.18}
$$

If s_1 and s_2 are support points of F such that $F(s_1) > F(s_2) > 0$ (implying that $s_1 > s_2$), then it is easily seen that the left-hand side of (8.2.18) is larger at s_1 than at s_2. This implies that $a > 0$. (8.2.18) also implies that we cannot have support points s_1 and s_2 of F such that $s_1 > s_2$ and $F(s_1) = F(s_2) > 0$. (If it is assumed that these points exist, then we get on $[s_2, s_1]$, the left-hand side of the equation in (8.2.18) to be constant and the right-hand side of the equation in question to be strictly increasing, leading us to a contradiction.) Consequently, (8.2.18) gives

$$
\frac{\int_{-\infty}^{x} (x - y) F^i (\mathrm{d}y)}{(F(x))^i} = (1 - a)x + b, \qquad \text{for all } x > l, \tag{8.2.19}
$$

where l is the left extremity of F. As the identity (8.2.19) can be translated into the identity corresponding to the mean residual life for the random variable $-X_{i:i}$ (in obvious notation), we have then, in view of Theorem 5.2.1, the assertion of the present theorem to be valid. ∎

8.2.13   Remarks

(i) In view of the proof based on Theorem 5.2.1 for Theorem 8.2.12, it is clear that this latter result is implicitly linked with the characterization of the class of distributions for which the mean residual life function is linear; the latter result has appeared in Hall and Wellner (1981) and other places.

(ii) Appealing to Theorem 5.2.1, we could also see, under the assumptions in Theorem 8.2.12, that if $E\{X_{i:n}|X_{i+1:n} = x\}$ has a continuous strictly increasing version, then it determines F. Similarly, it follows that if $E\{X_{i+1:n}|X_{i:n} = x\}$ has a continuous strictly increasing version, then it also determines F. Aly (1988) has given some results of this type.

We shall now deal with some important characterization results based on record values. Before, doing so, let us recall the definition of record values.

8.2.14   Definition

Let $\{X_n : n = 1, 2, \ldots\}$ be a sequence of i.i.d. random variables such that the right extremity of the common distribution of X_j's is not one of its discontinuity points.

Then, if we define $\{L(i) : i = 1, 2, \ldots\}$ such that $L(1) = 1$ and for $i > 1$

$$L(i) = \inf\{j : j > L(i-1), X_j > X_{L(i-1)}\},$$

the sequence $\{X_{L(i)} : i = 1, 2, \ldots\}$ is called the sequence of record values and the sequence $\{L(i) : i = 1, 2 \ldots\}$ as that of record times.

If X_n's have a continuous distribution, then it is easily seen that

$$P\{R_i \leqslant x\} = \int_{-\infty}^{x} \frac{(F^*(y))^{i-1}}{(i-1)!} F(dy), \quad i = 1, 2, \ldots; \quad x \in \mathbb{R}, \tag{8.2.20}$$

and

$$P\{R_i \leqslant x, R_j \leqslant y\}$$

$$= \int_{-\infty}^{x} \left(\int_{x_1}^{y} \frac{(F^*(x_1))^{i-1}}{(i-1)!} \frac{(F^*(x_2) - F^*(x_1))^{j-i-1}}{(j-i-1)!} F^*(dx_2) \right) F(dx_1),$$

$$i < j = 1, 2, \ldots; x, y \in \mathbb{R}, \tag{8.2.21}$$

where $\int_{x_1}^{y} = 0$ if $x_1 \geqslant y$, F is the d.f. of X_n, R_k is the kth record value of $\{X_n\}$ and $F^*(x) = -\log(1 - F(x))$ for a.a. $[F]x \in \mathbb{R}$; Karlin (1966, pp. 267–268) has essentially proved these results.

The following are some of the major characterization results based on record values, that are corollaries of the results in Chapter 2.

8.2.15   Theorem

Let $\{R_i : i = 1, 2, \ldots\}$ be a sequence of record values corresponding to a d.f. F (with its right extremity satisfying the condition in Definition 8.2.14). For some $k \geqslant 1$, $R_{k+1} - R_k \overset{d}{=} X_1$ where X_1 is a random variable with d.f. F, if and only if X_1 is exponential or, for some $a > 0$, X_1 is geometric on $\{a, 2a, \ldots\}$ (i.e. $a^{-1}X_1 - 1$ is geometric in the usual sense).

Proof   Clearly the condition that $R_{k+1} - R_k \overset{d}{=} X_1$ is equivalent to that $X_1 > 0$ a.s. (i.e. $F(0) = 0$) and

$$\int_{(0,\infty)} \frac{\overline{\overline{F}}(x+y)}{\overline{\overline{F}}(y)} F_k(dy) = \overline{\overline{F}}(x), \quad x \in (0, \infty), \tag{8.2.22}$$

where F_k is the d.f. of R_k and $\overline{F}(\bullet) = 1 - F(\bullet)$. Note that if (8.2.22) holds with $F(0) = 0$ then given any point $s_0 \in \text{supp}[F]$ there exists a point $s_1 \in \text{supp}[F_k]$ such that $s_0 + s_1 \in \text{supp}[F]$ and hence $\in \text{supp}[F_k]$. Consequently, from the condition, we get that the smallest closed subgroup of $\mathbb{R}$ containing $\text{supp}[F_k]$ equals that containing $\text{supp}[F]$. In view of Theorem 2.2.2 we have then immediately that if (8.2.22) holds with $F(0) = 0$, then either X_1 is exponential or for some $a > 0$,

$a^{-1}X_1 - 1$ is geometric (in the usual sense). The converse of the assertion is trivial and hence we have the theorem. ∎

8.2.16   Theorem

Let $\{R_i : i = 1, 2, \ldots\}$ be as in Theorem 8.2.15, k be a positive integer, and ϕ be a nonconstant real monotonic left continuous function on $\mathbb{R}_+$ such that $E\{|\phi(R_{k+1} - R_k)|\} < \infty$. Then, the following assertions hold:

(i)   If F is continuous or has its left extremity as one of continuity points and ϕ is nonarithmetic, then, for some $c \neq \phi(0+)$,

$$E\{\phi(R_{k+1} - R_k)|R_k\} = c \quad \text{a.s.} \tag{8.2.23}$$

   if and only if F is exponential, within a shift.

(ii)   If ϕ is arithmetic with span a and F is arithmetic with span greater than or equal to a, then, for some $c \neq \phi(0+)$, (8.2.23) is valid if and only if F has a finite left extremity and the conditional distribution of the residual value of X_1 over the kth support point of F given that this is positive is geometric on $\{a, 2a, \ldots\}$, where X_1 is a random variable with d.f. F.

Proof   We have (8.2.23) to be equivalent to

$$\int_{(x,\infty)} \phi(y - x)\frac{F(\mathrm{d}y)}{\overline{\overline{F}}(x)} = c \quad \text{for a.a. } [F_k]\, x \in \mathbb{R}, \tag{8.2.24}$$

where $\overline{\overline{F}}(x) = 1 - F(x)$ and F_k is the d.f. of R_k. Once again appealing to Fubini's theorem, we get that (8.2.24) is equivalent to

$$\int_{(0,\infty)} \overline{\overline{F}}(x + y)\phi(\mathrm{d}y) = (c - \phi(0+))\overline{\overline{F}}(x) \quad \text{for some a.a. } [F_k]\, x \in \mathbb{R}. \tag{8.2.25}$$

Following the relevant argument in the proofs of Theorems 8.2.7 and 8.2.9, we can see that (8.2.25) is equivalent to the condition obtained from it by replacing 'for a.a. $[F_k]\, x \in \mathbb{R}$' by '$x \geq l_k$' if $l_k > -\infty$ and by '$x \in \mathbb{R}$' otherwise, where l_k is the left extremity of F_k. Theorem 2.2.2 hence gives the two assertions of the theorem. ∎

8.2.17   Theorem

Let $\{R_i\}$ be as defined in the two previous theorems, but with F continuous. Let $k_2 > k_1 \geq 1$ be fixed integers. Then, on some interval of the type $(-\infty, a]$, with $a >$ the left extremity of the distribution of R_{k_1}, the conditional distribution of $R_{k_2} - R_{k_1}$ given $R_{k_1} = x$ is independent of x for almost all x if and only if F is exponential, within a shift.

Proof   Let for any $c < b$, the right extremity of the d.f. F, $\{R_i^{(c)} : i = 1, 2, \ldots\}$ be a sequence of record values from the distribution with d.f.

$$F_c(x) = \begin{cases} \dfrac{F(x) - F(c)}{1 - F(c)} & \text{if } x > c, \\ 0 & \text{otherwise.} \end{cases}$$

Note that the independence condition in the assertion holds if and only if for a.a. $[F]c$ in an interval of the type mentioned in the assertion, we have $R^{(c)}_{k_2-k_1} - c$ to be independent of c. The distribution of $R^{(c)}_{k_2-k_1} - c$ is computed as

$$P\{R^{(c)}_{k_2-k_1} - c \leqslant x\} = \begin{cases} [(k_2 - k_1 - 1)!]^{-1} \displaystyle\int_{\alpha_{x,c}}^1 (-\log y)^{k_2-k_1-1} \, dy & \text{if } c \leqslant c + x < b \\ 1 & \text{if } c + x \geqslant b \\ 0 & \text{if } x < 0, \end{cases}$$

where $\alpha_{x,c} = [1 - F(c + x)]/[1 - F(c)]$. Consequently, it follows that the independence condition is valid if and only if $b = \infty$ and for a.a. $[F]c$ in an interval of the type stated, $(1 - F(c + x))/(1 - F(c))$, $x \in \mathbb{R}_+$ is independent of c. As F is continuous, the latter condition is seen to be equivalent to the condition that the left extremity, l, of F is finite and for some $a > l$

$$(1 - F(c + x)) = (1 - F(l + x))(1 - F(c)), \quad c \in (-\infty, a) \cap \text{supp}[F], \quad x \in \mathbb{R}_+.$$

(Note that the last equation implies that $b = \infty$). In view of the Marsaglia–Tubilla result, the assertion of the theorem then follows. ∎

8.2.18   Remarks

(i) Various versions of Theorem 8.2.15 have appeared in Lau and Rao (1982), Rao and Shanbhag (1986) and elsewhere. The present version is due to Witte (1988). Downton's (1969) result (with the correction as in Fosam et al (1993)) may be viewed as a specialized version of the theorem for $k = 1$. If we assume F to be concentrated on $\{0, 1, 2, \ldots\}$ with $F(1) - F(1-) > 0$ (and the right extremity condition met), then as a corollary to the theorem it follows that, for some $k \geqslant 1$, $R_{k+1} - R_k \overset{d}{=} X_1 + 1$ if and only if F is geometric.

(ii) The proof of Theorem 8.2.16 indicates that a version stronger than that produced by the theorem holds. Rao and Shanbhag (1986) have given two specialized results, essentially containing the main theme of the present theorem. Gupta (1984) and Huang and Li (1993) contain some specialized versions of the theorem. (In particular, Gupta considers the case of F continuous with $\phi(x) = x^r$, $x \in \mathbb{R}_+$, $r > 0$; for the details of his result and a correction, see Rao and Shanbhag (1986).) We may also note here that Theorem 8.2.16 gives as a corollary the characterization for the distributions in question by the independence of $R_{k+1} - R_k$ and R_k; note further that the specialized versions of Theorems 8.2.9 and 8.2.16 when F is continous hold without the requirement that ϕ be left continous.

(iii) Theorem 8.2.17 is an improved version of a result of Dallas (1981), (see also Nayak, 1981). With obvious alterations in the proof of the theorem, it follows that on assuming $n \geqslant k_2 > k_1 \geqslant 1$, the theorem with $X_{k_1:n}$ in place of R_{k_1} and $X_{k_2:n}$ in place of R_{k_2} (where $X_{i:n}$ is the ith order statistic as mentioned earlier) holds. A variant of this latter result appears in Gather (1989); the result in this case is that if F is a continuous d.f. with support equal to $\mathbb{R}_+$ (which is equivalent to stating that F is continuous with $F(0) = 0$ and F is strictly increasing on $\mathbb{R}_+$), then F is exponential if and only if

$$X_{j_r:n} - X_{i:n} \overset{\mathrm{d}}{=} X_{j_r-i:n-i}, \quad r = 1, 2$$

holds for fixed $1 \leqslant i < j_1 < j_2 \leqslant n$ and $n \geqslant 3$.

We have touched upon in this section various characterization results based on order statistics or record values that follow as corollaries to either Theorem 2.2.1 or 2.2.2. However, there are many other characterizations in the area that we have not discussed. Galambos and Kotz (1978), Gupta (1984), Witte (1993), Huang *et al* (1993) and Huang and Li (1993) among others provide fine reviews of the literature in the area, and the reader could explore the possibility of using the techniques given in this monograph to deal with characterization problems not studied by us.

8.3   CHARACTERIZATIONS OF POINT PROCESSES

We have already met some characterization problems that are related to stochastic processes, in earlier discussion. The lack of a memory property of exponential distributions met in Chapter 5 is implicitly related to Poisson and birth and death processes and plays a crucial role in their mechanism. Spitzer's integral representation theorem met in Chapter 7 deals with an important characterization problem in branching processes. Characterizations of probability distributions on the basis of properties of record values, studied in the previous section, provide us with further illustrations in this connection.

In the present section, we shall concentrate on studying some characterization results on point processes. We need some notation for this purpose.

We define a renewal process as usual. Let F be a d.f.[†] (possibly a defective one) on $\mathbb{R}_+$ with $F(0) = 0$ and $0 < F(\infty)(= \lim_{x \to \infty} F(x)) \leqslant 1$, and with mean μ which may be infinite. Let the class of all such F be $\mathcal{F}$ and $\{X_n : n = 1, 2, \ldots\}$ be a sequence of independent random variables with d.f. F. Define $\xi_0 = 0$, $\xi_n = X_1 + \cdots + X_n$, $n \geqslant 1$ and denote by $N(x)$ the number of renewals in $(0, x]$, i.e. $N(x) = \sup\{n : \xi_n \leqslant x\}$, for all $x \geqslant 0$ and by $H(x)$ the renewal function $E(N(x))$ on $\mathbb{R}_+$. (It is clear that H is an increasing and right continuous real function with $H(0) = 0$, and it uniquely determines F.) By the forward recurrence time from t for the process, we mean $\xi_{N(t)+1} - t$; in the case when the starting point t for the forward recurrence time is chosen randomly with d.f. Φ we denote the d.f. of the forward recurrence time by V_Φ.

[†] For convenience, we take this to be nondegenerate, even when some of the results given do not require this assumption.

A Poisson process is, of course, a renewal process for which F is exponential. Its arithmetic equivalent, when F is the d.f. of a constant multiple of a geometric random variable, we refer to the renewal process as a binomial process. For both of these processes, we obviously have $\mu < \infty$.

We need the key renewal theorem in what follows and its proof provides one with an important application of the Choquet–Deny theorem. In view of this, we reproduce these now; the proof of the renewal theorem given here is essentially as in Feller (1966, pp. 350–351):

8.3.1   The key renewal theorem

If F is nonarithmetic, then as $y \to \infty$

$$H(x+y) - H(y) \to \frac{x}{\mu} \qquad (8.3.1)$$

for every fixed $x > 0$. Furthermore, if F is arithmetic with span λ, the same is true when x is an integral multiple of λ.

Proof   As mentioned before, we have H to be an increasing and right continuous real function on $\mathbb{R}_+$ with $H(0) = 0$. (That it is a real function follows on noting that there exists, for each $x \in \mathbb{R}_+$, a positive integer k_x such that $F^{k_x *}(x) < 1$ and $H(x) \leqslant k_x(1 - F^{k_x *}(x))^{-1}$; with a slight extension of this argument, it even follows that $\Sigma_{n=0}^{\infty} F^{n*}(x)z^n$ is an entire function of z for each $x \in \mathbb{R}_+$. On noting that for each $x \in \mathbb{R}_+$

$$\int_{[0,x]} (1 - F(x - y)) H(\mathrm{d}y) = 1,$$

we see that $(1 - F(\alpha))(H(x) - H(x - \alpha)) \leqslant 1$ for each $x \geqslant \alpha$ and $\alpha > 0$. As $F(\alpha) < 1$ for a sufficiently small α, we have $H(\alpha + y) - H(y)$ for a sufficiently small α and hence for any $\alpha > 0$, to be bounded in y. We shall prove the theorem by showing that the assumption that it is not valid leads us to a contradiction. Assume then that for an F, the theorem does not hold. This implies that there exists a sequence $\{y_n : n = 1, 2, \ldots\}$ of positive numbers converging to infinity and an $x > 0$, with y_n's and x as integral multiples of λ if F is arithmetic with span λ, for which $\{H(x + y_n) - H(y_n)\}$ converges and has a limit different from x/μ. In view of the Helly selection theorem, it follows that there exists a subsequence $\{y_n^* : n = 1, 2, \ldots\}$ of $\{y_n : n = 1, 2, \ldots\}$ such that $\{H(\bullet + y_n^*) - H(y_n^*)\}$ (defining for convenience, $H(y) = 0$ for $y < 0$) converges weakly to a Lebesgue–Stieltjes distribution function (referred to also as a Lebesgue–Stieltjes measure function) H^* on $\mathbb{R}$.

Now if z is a continuous function vanishing outside a finite interval in $\mathbb{R}_+$ then, for each $x \in \mathbb{R}$,

$$\int_{\mathbb{R}_+} z(x - s) H(y_n^* + \mathrm{d}s) \to \int_{\mathbb{R}} z(x - s) H^*(\mathrm{d}s) \qquad (8.3.2)$$

as $n \to \infty$. Denoting the limit in (8.3.2) by $\zeta(x)$ and recalling that $H = F^{0*} + F^* H$, we get easily that ζ is a bounded continuous function satisfying

$$\zeta(x) = \int_{\mathbb{R}_+} \zeta(x - y) F(\mathrm{d}y), \quad x \in \mathbb{R}. \tag{8.3.3}$$

The specialized versions of the Choquet–Deny theorem for the real line and the set of integers then imply that $\zeta \equiv \zeta(0)$ if F is nonarithmetic and it is periodic with period λ if F is arithmetic with span λ. (This also follows from a special case of Theorem 2.2.2.) This, in turn, implies that the measure determined by H^* is a constant multiple of the Lebesgue measure if F is nonarithmetic, and it is a constant multiple of a measure that is concentrated on the lattice $\{n\lambda : n = 0, \pm 1, \ldots\}$ with its restriction to the lattice as λ times the counting measure (on the lattice) if F is arithmetic with span λ. Denote the constant in each case by c.

As $H(\alpha + y) - H(y)$ is bounded in y for any $\alpha > 0$ and

$$\mu = \int_{\mathbb{R}_+} (1 - F(x)) \, \mathrm{d}x,$$

it follows that if $\mu < \infty$, (8.3.2) holds with z replaced by z^*, where $z^*(x) = 1 - F(x)$ if $x \geqslant 0$, and $= 0$ otherwise. As in that case, the left-hand side of (8.3.2) equals 1, and its right-hand side equals $c\mu$, we have that $c = \mu^{-1}$. If $\mu = \infty$, then, noting that for any $\beta > 0$, (8.3.2) holds with z replaced by z_β^*, where

$$z_\beta^*(x) = \begin{cases} 1 - F(x) & \text{if } x \in [0, \beta] \\ 0 & \text{otherwise,} \end{cases}$$

we see that, for any $\beta > 0$,

$$1 \geqslant c \int_{[0,\beta]} (1 - F(x)) \, \mathrm{d}x,$$

implying that $1 \geqslant c\mu$ and hence $c = 0$. As $\{y_n^*\}$ is a subsequence of $\{y_n\}$, we have a contradiction to the existence of $\{y_n^*\}$ with stated properties. This, in turn, implies that the theorem holds. ∎

Feller (1966) includes several other interesting results in renewal theory that are based implicitly or explicitly on the Choquet–Deny theorem.

We are now in a position to give the first characterization theorem of the section. The theorem is due to Isham *et al* (1975), and it extends, amongst other things, certain results given earlier by Chung (1972), Çinlar and Jagers (1973), and Kotz and Johnson (1974).

8.3.2   Theorem

(i) A renewal process with nonarithmetic distribution $F \in \mathcal{F}$ is a Poisson process if $V_\Phi = F$ for some Φ not concentrated at zero. Conversely, if the process is Poisson, then $V_\Phi = F$ for any Φ.

(ii) A renewal process with arithmetic distribution $F \in \mathcal{F}$ with span λ is a binomial process if $V_\Phi = F$ for some Φ not concentrated at zero with support in $\{0, \lambda, 2\lambda, \ldots\}$. Conversely, if the process is binomial, then $V_\Phi = F$ for any Φ with support in $\{0, \lambda, 2\lambda, \ldots\}$, where λ is the span of F.

Proof The converses in both (i) and (ii) are easy to verify. We shall now establish the first assertions of both (i) and (ii). Let T be a random variable with d.f. Φ, independent of $\{X_i\}$; if F is arithmetic, we assume that T takes values in $\{0, \lambda, 2\lambda, \ldots\}$, where λ is the span of F. Then the hypothesis $V_\Phi = F$ implies that the forward recurrence time from T has the same distribution as a typical interval, and hence that $N(T + x) - N(T)$ and $N(x)$ are identically distributed. Taking expectations of these variables and using Fubini's theorem, we find that

$$E\{H(T + x) - H(T)\} = H(x), \quad x \geqslant 0. \tag{8.3.4}$$

Thus, it suffices to prove that the only admissible solutions of (8.3.4) are linear (on $\mathbb{R}_+$ if F is nonarithmetic and on the semilattice $\{0, \lambda, 2\lambda, \ldots\}$ otherwise). To do this, we may use either of the following two arguments.

The first argument Let $\{T_i\}$ be a sequence of independent random variables, identically distributed as T, independent of $\{X_i\}$, with $S_n = \Sigma_{i=1}^n T_i$. It is easy to see from (8.3.4) that

$$E\{H(T + x + y) - H(T + y)\} = H(x + y) - H(y) \quad x, y \geqslant 0. \tag{8.3.5}$$

By successively setting y equal to $S_{n-1}, S_{n-2}, \ldots S_0 = 0$, T equal to $T_n, T_{n-1}, \ldots T_1$ and then taking expectations in (8.3.5), we deduce that, for any $x \geqslant 0$,

$$E\{H(S_n + x) - H(S_n)\} = E\{H(S_{n-1} + x) - H(S_{n-1})\} = \cdots = H(x). \tag{8.3.6}$$

Now $S_n \to \infty$ almost surely, since T is nonnegative with $P\{T = 0\} < 1$, and by Theorem 8.3.1

$$H(x + y) - H(y) \to \frac{x}{\mu} \quad \text{as } y \to \infty, \quad x \geqslant 0;$$

in the arithmetic case x must be a multiple of the span λ of F. So from a simple variant of Slutsky's theorem

$$H(S_n + x) - H(S_n) \to \frac{x}{\mu} \quad \text{a.s. as } n \to \infty, \quad x \geqslant 0. \tag{8.3.7}$$

In the proof of Theorem 8.3.1, we have seen that for each fixed $x \geqslant 0$, $H(x + y) - H(y)$ is bounded in y. (This also follows trivially from Theorem 8.3.1 as $H(y) < \infty$ for all $y \in \mathbb{R}_+$.) So applying the dominated convergence theorem to (8.3.6), we get

$$H(x) = \frac{x}{\mu} \tag{8.3.8}$$

As $H \not\equiv 0$, (8.3.8) implies that $\mu < \infty$ and hence we have the required result.

The second argument   Before going through the details of this argument, let us make a general remark.

8.3.3   Remark

Theorem 2.2.5 still holds if the equation in (2.2.8) is replaced by that $\int_{\mathbb{R}_+} (H(x + y) - H(y))\mu(\mathrm{d}y) = H(x)$.

In view of Theorem 2.2.5, with the modification stated in Remark 8.3.3, we get from (8.3.4) that $H(x) = \beta x$, $x \in \mathbb{R}_+$ if Φ is nonarithmetic, and $H(x + n\lambda^*) = H(x) + \beta'n$, $x \in \mathbb{R}_+$, $n = 0, 1, 2, \ldots$ if Φ is arithmetic with span λ^*, where β and β' are positive constants. (Note that $H \not\equiv 0$, and it is increasing and right continuous with $H(0) = 0$.) It now remains to get the required result only when Φ is arithmetic. In this case, taking λ^* as before, we can define

$$\hat{H}(x) = \lim_{n \to \infty} \{H(x + n\lambda^*) - H(n\lambda^*)\}, \quad x \in \mathbb{R}, \tag{8.3.9}$$

and see from the renewal equation that, for any fixed $y \in \mathbb{R}_+$, the function H_y defined by $\hat{H}(y + \bullet) - \hat{H}(\bullet)$ to be such that $H_y = F^*H_y$. Also, $\hat{H}(x) = H(x)$ if $x \in \mathbb{R}_+$. From Theorem 2.2.2, we can hence claim that for each $y \in \mathbb{R}_+$

$$H(y + x) - H(x) = H(y), \quad x \in \mathbb{R}_+ \tag{8.3.10}$$

if F is nonarithmetic, and

$$H(y + n\lambda) = H(n\lambda) + H(y), \quad n = 0, 1, \ldots \tag{8.3.11}$$

if F is arithmetic with span λ. (The result required here follows more easily than Theorem 2.2.2 because of the constraints on F and H_y.) (8.3.10) gives that $H(x) = \beta''x$, $x \in \mathbb{R}_+$ if F is nonarithmetic and (8.3.11) gives $H(x + n\lambda) = H(x) + \beta'''n$, $x \in \mathbb{R}_+$, $n = 0, 1, \ldots$ if F is arithmetic with span $\lambda(\beta''$ and β''' being appropriate constants). Hence, we have the required result. ∎

In both the arguments used to prove the first assertions in (i) and (ii) of Theorem 8.3.2, specialized versions of Theorem 2.2.2 with H bounded and μ as a probability or a subprobability measure are involved directly or indirectly. If we are prepared to use a more general version of Theorem 2.2.2 for getting the results, then we could give an argument for the results that is much shorter than the above two. This is as follows.

The relation $V_\Phi = F$ implies that

$$1 - F(x) = \int_{\mathbb{R}_+} (1 - F(x + y))\mu(\mathrm{d}y), \quad x \in \mathbb{R}_+ \tag{8.3.12}$$

for some σ-finite measure μ such that $\mu(\{0\}) < 1$ and either supp$[\mu]$ has 0 as a cluster point or there exist positive points y_0 and y_1 satisfying the conditions that $y_0 \leqslant y_1$, $y_0 \in$ supp$[\mu]$ and y_1 is the smallest support point relative to F. In view of Theorem 2.2.2, (8.3.12) hence gives that we have either F to be exponential, or it to be geometric on $\{y_0, 2y_0, \ldots\}$. The required results are then immediate. ∎

The next theorem gives further characterization properties of Poisson and binomial processes based on the forward recurrence times. The theorem is essentially due to Rao and Shanbhag (1989) and it gives, amongst other things, certain results of Gupta and Gupta (1986) and Huang et al (1993) as corollaries. In their arguments, the latter authors appeal to a result of Shimizu (1978) instead of its stronger version due to Lau and Rao (1982), which appears as Theorem 2.2.2 in this monograph, and hence end up requiring an unnecessary integral condition that goes with it.

8.3.4 Theorem

Let the renewal process be such that $F(\infty) = 1$ (i.e. such that F is a nondefective d.f.) and for each $t \in \mathbb{R}_+$, let R_t denote the forward recurrence time t of the process. Also, let v be a measure on $\mathbb{R}_+$. Then, if v is nonarithmetic, we have for all $t \in \mathbb{R}_+$

$$E(v((0, R_t])) = c \tag{8.3.13}$$

with c as a positive constant if and only if for some $\alpha > 0$

$$\int_{(0,\infty)} e^{-\alpha x} v(\mathrm{d}x) = c \tag{8.3.14}$$

and the renewal process is Poisson with intensity α. Furthermore if v is arithmetic with span λ such that $v(\{\lambda\}) < c$ and F is concentrated on $\{0, \lambda, 2\lambda, \ldots\}$, then (8.3.13) holds for $t \in \{0, \lambda, 2\lambda, \ldots\}$ if and only if (8.3.14) holds and the renewal process is binomial with the span and the mean of F equal to λ and $\lambda(1 - e^{-\alpha\lambda})^{-1}$ respectively.

Proof   As the 'if' parts of the two assertions are easy to see, it is sufficient if we establish their 'only if' parts. (8.3.13) for $t = 0$ implies that

$$\int_{\mathbb{R}_+} v((0, x]) F(\mathrm{d}x) = c,$$

which in turn gives that $v((0, x]) < \infty$ for each x such that $\overline{F}(x)(= 1 - F(x-)) > 0$. If we now define a measure v^* on $\mathbb{R}_+$ such that its restriction to $\{x : x > 0$ and $\overline{F}(x) > 0\}$ agrees with that of v and $v^*(\{x : x = 0$ or $\overline{F}(x) = 0\}) = 0$, then it follows that v^* is σ-finite. Clearly, if (8.3.13) is valid for all $t \in \mathbb{R}_+$ or all $t \in \{0, \lambda, 2\lambda, \ldots\}$, then the corresponding equation with v^* in place of v is also valid for the t's in question. Suppose now that we have the case with v nonarithmetic. In this case, we can assume (8.3.13) with v^* in place of v to be valid for all $t \in \mathbb{R}_+$. In view of Fubini's theorem, we hence get

$$\int_{(0,\infty)} P\{R_t \geqslant x\} v^*(\mathrm{d}x) = c, \quad t \in \mathbb{R}_+. \tag{8.3.15}$$

From the total probability law, we have for each $t \in \mathbb{R}_+$ and $x \in (0, \infty)$,

$$P\{R_t \geqslant x\} = \overline{F}(t + x) + \int_{[0,t]} P\{R_{t-y} \geqslant x\} F(\mathrm{d}y)$$

and hence on using (8.3.15), in view of Fubini's theorem, we get that

$$c = \int_{(0,\infty)} \overline{F}(t + x)v^*(\mathrm{d}x) + cF(t), \quad t \in \mathbb{R}_+. \tag{8.3.16}$$

From (8.3.16), we get

$$c\overline{F}(t+) = \int_{(0,\infty)} \overline{F}(t + x)v^*(\mathrm{d}x), \quad t \in \mathbb{R}_+. \tag{8.3.17}$$

As a corollary to Theorem 2.2.2, we can now see immediately that the right extremity of F is equal to ∞ and hence that the restriction to $(0, \infty)$ of v^* agrees with that of v. Appealing to Theorem 2.2.2 once more (and taking into account the conditions that $F(0) = 0$ and that $\overline{F}$ is left continuous), we see that for some $\alpha > 0$, (8.3.14) holds and F is exponential with mean α^{-1}. (From (8.3.17), it follows, because of the conditions that $F(0) = 0$ and $\overline{F}$ is left continuous, without any reference to Theorem 2.2.2 that F is continuous. However, we do not need this information in our argument.) Hence we have the 'only if' part of the first assertion. The 'only if' part of the second assertion follows essentially via the same argument. ■

8.3.5   Remarks

(i) With a slightly more stringent a priori requirement, one could express Theorem 8.3.4 also in terms of an increasing or decreasing right continuous function G on $\mathbb{R}_+$ with $G(0) = 0$. Gupta and Gupta (1986), Rao and Shanbhag (1989) and Huang *et al* (1993) have essentially used this formulation in their results.

(ii) As a corollary to the first assertion of Theorem 8.3.4 we have that for the renewal process, $E(R_t) = c$ for all $t \in \mathbb{R}_+$ where c is a positive real number, if and only if the process is Poisson with intensity $\alpha = c^{-1}$. This latter result was established by Holmes (1974).

Shanbhag (1972c) gave a characterization of Poisson processes in the class of renewal processes via the property that the renewal (or counting) variable $N(x)$ be power series with x as the parameter taking all values in $(0, \infty)$. To be more precise, the characterization that he gave was based on the property that

$$P\{N(x) = n\} = \frac{c_n(\xi(x))^n}{C(\xi(x))}, \quad n \in \mathbb{N}_0, \quad x \in (0, \infty) \tag{8.3.18}$$

for some function $\xi : (0, \infty) \rightarrow \mathbb{R}_+$, with $C(\xi(\bullet))$ as a positive normalizing function. As for any fixed n, the left-hand side of (8.3.18) is positive for some

$x > 0$, it follows that, in (8.3.18), $c_n > 0$ for all $n \in \mathbb{N}_0$; also, by taking the Laplace transforms of the two sides of the equation in (8.3.18), we see that for some $k > 0$, there exists a moment sequence $\{\mu_n^* : n \in \mathbb{N}_0\}$ such that

$$\mu_n^* = kc_n^{-1}, \quad n \in \mathbb{N}_0. \tag{8.3.19}$$

In essence, the result of Shanbhag (1972c) here could be expressed by the theorem that follows.

8.3.6   Theorem

In a renewal process, (8.3.18) holds with $\{\mu_n^*\}$ of (8.3.19) determining the corresponding distribution if and only if the process is Poisson.

Proof   As the 'if' part is standard, it is sufficient if we establish the 'only if' part. To do this, assume that (8.3.18) holds with $\{\mu_n^*\}$ of (8.3.19), and the d.f. F is as defined before. From (8.3.18) and (8.3.19), we get

$$\int_{(0,\infty)} (\exp\{-\theta x\}) \frac{(\xi(x))^n}{C(\xi(x))} \, dx = \left(\frac{1 - \Phi(\theta)}{\theta} \right) (\Phi(\theta))^n k^{-1} \mu_n^*,$$

$$\theta \in (0, \infty), \quad n \in \mathbb{N}_0, \tag{8.3.20}$$

where

$$\Phi(\theta) = \int_{\mathbb{R}_+} (\exp\{-\theta x\}) F(dx), \quad \theta \in (0, \infty). \tag{8.3.21}$$

In view of the property of $\{\mu_n^*\}$, (8.3.21) implies that

$$\int_{\{x:\xi(x) \leqslant y\}} (\exp\{-\theta x\})(C(\xi(x)))^{-1} \, dx$$

$$= \left(\frac{1 - \Phi(\theta)}{k\theta} \right) \hat{F}((\Phi(\theta))^{-1} y), \quad \theta \in (0, \infty), \quad y \in \mathbb{R}_+, \tag{8.3.22}$$

where $\hat{F}$ is the d.f. corresponding to the moment sequence $\{\mu_n^*\}$. (8.3.22) implies that $\hat{F}((\Phi(\theta))^{-1} y)$ is differentiable any number of times with respect to θ, given $y \in (0, \infty)$. This, in turn, implies that the restriction to $(0, \infty)$ of $\hat{F}$ is differentiable any number of times. In view of (8.3.18) with $n = 0$, we have that ξ is monotonic increasing. From (8.3.18), we have also that the left extremity of F is zero or equivalently $\xi(x) > 0$ for each $x \in (0, \infty)$. The latter observation implies that d.f. $\hat{F}$ in (8.3.22) to be such that

$$\hat{F}(0) = 0. \tag{8.3.23}$$

On differentiating both sides of the equation in (8.3.22) with respect to y (as the left-hand side is differentiable since the right-hand side is so), we now get that for some functions $\xi^* : (0, \infty) \to (0, \infty)$ and $f^* : (0, \infty) \to \mathbb{R}_+$ with ξ^* monotonic and such that $\xi^*(x) \to 0$ as $x \to 0$,

$$\hat{f}((\Phi(\theta))^{-1}y) = \left(\frac{k\theta\Phi(\theta)}{1 - \Phi(\theta)} \right) \exp\{-\theta\xi^*(y)\}f^*(y), \theta, y \in (0, \infty), \quad (8.3.24)$$

where $\hat{f}(x) = \mathrm{d}\hat{F}(x)/\mathrm{d}x$, $x \in (0, \infty)$. In view of (8.3.23), (8.3.24) implies that both $\hat{f}$ and f^* are positive functions on $(0, \infty)$. (Note that if it is assumed that $\hat{f}$ or f^* vanishes at some point of $(0, \infty)$, then (8.3.24) implies that $\hat{f}$ and f^* vanish everywhere on $(0, \infty)$, contradicting (8.3.23).) On letting $\theta \to 0$, we get from (8.3.24) that $\mu < \infty$ (and f^* is a constant multiple of $\hat{f}$). We have also that $\hat{f}$ is differentiable (i.e. on $(0, \infty)$) and hence (8.3.24) implies, on differentiating the logarithms of both sides with respect to θ and carrying out a minor manipulation, that for some real constants β_0, β_1 and β_2 with $\beta_2 > 0$,

$$\left(\frac{\Phi'(\theta)}{\Phi(\theta)} \right) \left(\beta_0 + \beta_1\xi^*(\beta_2 y(\Phi(\theta))^{-1}) \right) \cdot$$

$$= \frac{\mathrm{d}}{\mathrm{d}\theta} \left(\log \left(\frac{\mu\theta\Phi(\theta)}{1 - \Phi(\theta)} \right) \right) - \xi^*(y), \theta, y \in (0, \infty). \quad (8.3.25)$$

Since $\xi^*(x) \to 0$ as $x \to 0$, it easily follows from (8.3.25) that $\mu\theta(\Phi(\theta))^{1-\beta_0}/(1 - \Phi(\theta)) = 1$, $\theta > 0$, and for some $\gamma > 0$, $-\Phi'(\theta)/(\Phi(\theta))^{\gamma+1} = \mu$, $\theta > 0$. The latter (is a consequence of a version of the Cauchy equation and) gives that F is gamma; as we have also $(1 - \Phi(\theta))/\mu\theta = (\Phi(\theta))^{1-\beta_0}$, $\theta > 0$, it then follows that $\beta_0 = 0$ and hence F is exponential. Thus, we have the required result. ∎

8.3.7   Corollary

A renewal process with $\limsup_{\delta \to 0+}(F(\delta)/\delta) < \infty$ has the counting variable infinitely divisible (in the sense of Feller (1968)) for all t if and only if it is a Poisson process.

Proof   As the 'if' part is standard, it is sufficient if we establish the 'only if' part. Assume then that we have a renewal process and that $\limsup_{\delta \to 0+}(F(\delta)/\delta) = \lambda < \infty$. Then, for a sufficiently small δ, we have that if $t \in (0, \delta)$,

$$P\{N(t) = n\} \leqslant \frac{\left(\lambda + \frac{1}{2} \right)^n t^n}{n!}, \quad n \in \mathbb{N}_0. \quad (8.3.26)$$

This follows because

$$P\{N(t) = n\} = F_n(t) - F_{n+1}(t)$$

$$\leqslant F_n(t)$$

$$= \int_{(0,t]} F(t - y)F_{n-1}(\mathrm{d}y), \quad n \in \{1, 2, \ldots\}, \quad t \in (0, \infty), \quad (8.3.27)$$

where $F_m(t)$ for $m \geqslant 1$ denotes the m-fold convolution of F with itself and F_0 is the degenerate d.f. at zero. (8.3.26) clearly holds for $n = 0, 1$. Assuming now that it holds for $n = 0, 1, \ldots, k$, where $k \geqslant 1$, we get for $n = k + 1$, from (8.3.27) and Fubini's theorem, that

$$P\{N(t) = n\} \leqslant \int_{(0,t]} \left(\lambda + \tfrac{1}{2}\right)(t - y) F_{n-1}(dy)$$

$$= \left(\lambda + \tfrac{1}{2}\right) \int_{(0,t]} \left(\int_{[y,t]} dx\right) F_{n-1}(dy)$$

$$= \left(\lambda + \tfrac{1}{2}\right) \int_{(0,t]} F_{n-1}(x)\, dx$$

$$\leqslant \left(\left(\lambda + \tfrac{1}{2}\right)^n / (n-1)!\right) \int_{(0,t]} x^{n-1}\, dx$$

$$= \left(\lambda + \tfrac{1}{2}\right)^n t^n / n!, \quad t \in (0, \infty). \tag{8.3.28}$$

If $P\{N(t) \neq 0\} = 0$ for some $t \in (0, \infty)$, then it follows that the left extremity of F is positive, which in turn implies that there exists a point t at which $N(t)$ is a nondegenerate bounded random variable and hence a noninfinitely divisible random variable. Hence, we have to assume $P\{N(t) \neq 0\} > 0$ for all $t \in (0, \infty)$. If it is then $N(t)$ is infinitely divisible for all t, we get that unless all $N(t)$ for $t \in (0, \delta)$ are Poisson, we have a contradiction (because the existence of a non-Poisson infinitely divisible $N(t)$ for some $t \in (0, \delta)$ implies that $E\{z^{N(t)}\} / \exp\{\left(\lambda + \tfrac{1}{2}\right) z\} \to \infty$ as $z \to \infty$, contradicting the above inequality). It is easily seen that if $z \geqslant 1$ and k is a positive integer

$$E(z^{N(t)}) \leqslant z^{k-1} (E(z^{N(t/k)}))^k,$$

which implies that if $N(t)$ is infinitely divisible for all t, then, as $N(t)$ for $t \in (0, \delta)$ are Poisson, we have $N(t)$ to be Poisson for all t (because, for a sufficiently large k, the ratio of the left-hand side to the right-hand side of this last inequality tends to infinity otherwise). Hence the corollary follows from the theorem. ∎

8.3.8 Remarks

(i) Versions of Corollary 8.3.7 have appeared in Haberland (1975) and Alamatsaz (1983). The reader could explore the possibility of investigating whether or not under weaker assumptions Theorem 8.3.6 or Corollary 8.3.7 or both hold.

(ii) There are several characterizations of the Poisson processes that we have not dealt with which have appeared in a nice review given by Çinlar (1972). Some of the results that he has mentioned as well as a characterization given more recently by Samuels (1974) are linked implicitly or explicitly with the Choquet–Deny or the Deny theorems or other tools discussed in this monograph.

(iii) Theorem 8.3.6 could have possible applications in 'Inference in Stochastic Processes' because (8.3.18) is obviously a discrete exponential family assuming that ξ is positive.

Finally in this section we provide a characterization theorem related to the Yule process. The result in the form that we have produced was given by Rao and Shanbhag (1989a); an earlier version of this result had appeared in Shanbhag (1973b).

Suppose that $\{X_n : n = 1, 2 \ldots\}$ is a sequence of positive random variables. Define $N(t) = \sup\{n : X_1 + \cdots + X_n \leqslant t\}$, $t > 0$. Let $n_0 (\geqslant 2)$ be a fixed positive integer and let $\{X_n : n = 1, 2, \ldots\}$ be such that it satisfies additionally $P\{N(t) = n\} > 0$ for $n = 1, 2, \ldots n_0$ and all $t > 0$. We have then the following theorem.

8.3.9 Theorem

The conditional distribution of $N(y)$ given $N(t) = n$ for each $0 < y < t, t > 0$ and $n = 1, 2, \ldots n_0$ is nondegenerate binomial with index n and success probability parameter independent of n if and only if for some $\lambda_0 > 0$ and $\lambda \neq 0$ such that $\lambda_0 + n_0\lambda > 0$,

$$P\{X_i > x\} = \exp\{-(\lambda_0 + (i-1)\lambda)x\}, \quad x \in \mathbb{R}_+, \quad i = 1, 2, \ldots, n_0 + 1 \quad (8.3.29)$$

Proof The 'if' part is easy to prove, and is essentially a corollary to what appears in Neuts and Resnick (1971). We shall now prove the 'only if' part. Assume then that the conditional distribution stated is binomial $(n, p(y, t))$ with $0 < p(y, t) < 1$ for each $0 < y < t, t > 0$ and $n = 1, 2, \ldots n_0$. It follows that unless the distribution of X_1 has full support on $\mathbb{R}_+$, we have a contradiction to the implicit assumption that $0 < p(y, t) < 1$. We have for $0 < y < t < \infty$

$$\frac{1}{\Psi_n(t)} \int_0^y [B_{n-1}(t-x) - B_n(t-x)] A(\mathrm{d}x) = 1 - \{q(y, t)\}^n, \quad n = 1, \ldots, n_0 \quad (8.3.30)$$

where $q(y, t) = 1 - p(y, t)$, $\Psi_n(t) = P\{N(t) = n\}$, $B_r(t) = P\{X_2 + \cdots + X_{r+1} \leqslant t\}$, $B_0(t) = 1$, and A is the d.f. of X_1. If $y \in (0, t)$, then it follows from (8.3.30) that

$$\frac{B_{n-1}((t-y)-) - B_n((t-y)-)}{\Psi_n(t)} = n(q(y, t))^{n-1} q^*(y, t), \quad n = 1, 2, \ldots, n_0, \quad (8.3.31)$$

where

$$q^*(y, t) = \lim_{m \to \infty} \left(\frac{q(y, t) - q(y + m^{-1}, t)}{A(y + m^{-1}) - A(y)} \right)$$

(which is easily seen to exist). Also, it follows from (8.3.30) with $n = 1$ that

$$A(y)(1 - B_1(t - y)) \geqslant p(y, t)\Psi_1(t) > 0, \quad (8.3.32)$$

implying that $(1 - B_1(t - y)) > 0$ and hence $1 - B_1((t - y)-) > 0$ for all $0 < y < t < \infty$. Dividing (8.3.31) by its special case for $n = 1$ we can then conclude that $q(y, t)$ is of the form $\Psi^*(t - y)/\Psi(t)$. Since, given $t > 0$, each continuity point y of A in $(0, t)$ is, in view of (8.3.30), a continuity point of $q(y, t)$, it follows that Ψ^* is continuous on $(0, \infty)$. Furthermore, in view of the fact that $\lim_{y \to 0+} q(y, t) = 1$ (with obviously $q(y, t) \in (0, 1)$ for each $y \in (0, t)$ with $q(t-, t) = 0$) for each t, we can then immediately conclude that $q(y, t)$ is of the form $\Psi(t - y)/\Psi(t)$ with Ψ continuous and strictly increasing on $(0, \infty)$ with $\Psi(0+) = 0$. We have now for $0 < y < t < \infty$

$$\frac{\int_0^y (1 - C_{n+1}(t - x)) A_n(dx)}{\Psi_n(t)(\Psi(t))^{-n}} = (\Psi(t) - \Psi(t - y))^n, \quad n = 1, 2, \ldots, n_0, \quad (8.3.33)$$

where A_n and C_{n+1} are d.f.'s of $X_1 + \cdots + X_n$ and X_{n+1} respectively, which implies, in view of the properties of Ψ, that each A_n is continuous with support $\mathbb{R}_+$. From the observation immediately below (8.3.32), we get, in view of (8.3.31), that $q^*(y, t) \neq 0$ for each y, t. Consequently, (8.3.33) implies on appealing to (8.3.31) that for some function ξ_n

$$\xi_n(t - y) a_n^*(y) = \Psi(t) - \Psi(t - y), \quad 0 < y < t < \infty, \quad (8.3.34)$$

where $a_n^*(y) = \lim_{m \to \infty} \{(A_n(y + m^{-1}) - A_n(y))/(A(y + m^{-1}) - A(y))\}^{1/(n-1)}$ (which is clearly seen to exist). In (8.3.34) we should have clearly ξ_n and a_n^* to be positive and continuous on $(0, \infty)$ and indeed $a_n^*(y) \propto \Psi(y)$. Writing in (8.3.34) for y and t respectively t and $y + t$, and dividing both sides by $\xi_n(y)$, we get

$$K\Psi(t) = \frac{\Psi(t + y) - \Psi(y)}{\xi_n(y)}, \quad 0 < y, t < \infty, \quad (8.3.35)$$

where K is a positive constant. From Theorem 2.2.5, or directly from (8.3.34) noting (first essentially as in the proof of Lemma 1.3.4 that Ψ is differentiable and) that Ψ', the derivative of Ψ, satisfies for some nonzero constant K^*

$$\Psi'(t + y) = K^* \Psi'(t) \Psi'(y), \quad y, t \in (0, \infty),$$

we see that $\Psi(t)$ is either proportional to t or to $|1 - e^{-\lambda t}|$ for some $\lambda \neq 0$. In either of the two situations, we have in (8.3.33), A_n and Ψ to be differentiable on $(0, \infty)$. (To see the differentiability of A_n, first choose t such that $t - y$ is a continuity point of $C_{n+1}(t - y)$ given y.) Differentiating both sides of (8.3.33) with respect to y, we get

$$\frac{(1 - C_{n+1}(t - y)) a_n(y)}{\Psi_n(t)(\Psi(t))^{-n}} = n(\Psi(t) - \Psi(t - y))^{n-1} \Psi'(t - y), \quad 0 < y < t < \infty,$$

where $a_n(y) = dA_n(y)/dy$. Substituting the form of Ψ, we see by induction on n here (or somewhat more easily using in the equation an observation based on (8.3.31) that $\Psi_n(t)(\Psi(t))^{-n} \propto a_1(t)$ or exclusively from (8.3.31) that $X_1, \ldots, X_{n_0+1}$ are exponential random variables of the form in the theorem. (It may be mentioned here that the situation $\lambda = 0$ arises when Ψ is linear, otherwise, we get $\lambda \neq 0$.) Hence, we have the assertion. ∎

8.3.10   Corollary

If we assume $P\{N(t) = n\} > 0$ for every $n \geqslant 1$ and every $t > 0$, then the conditional distribution of $N(y)$ given $N(t) = n$ is nondegenerate binomial with parameters as stated in Theorem 8.3.9 for every $0 < y < t < \infty$ and every $n \geqslant 1$ if and only if the process $\{N(t)\}$ is Yule. (The process constructed with intervals such that $P\{X_n > x\} = e^{-\{\lambda_0 + (n-1)\lambda\}x}$, $x \in \mathbb{R}_+$, with $\lambda_0 > 0$ and $\lambda \geqslant 0$, is referred to as Yule.)

8.3.11   Remarks

(i) It is implicit from the proof of Theorem 8.3.9 that the parameter $p(y, t)$ of the binomial distribution appearing in it equals $(e^{\lambda y} - 1)/(e^{\lambda t} - 1)$ if $\lambda \neq 0$ and y/t if $\lambda = 0$.

(ii) A version of Liberman's (1985) characterization of a Poisson process in the class of renewal processes is a special case of Corollary 8.3.10.

(iii) Suppose we have a renewal process, but for a modification that for a fixed positive integer m_0, its first m_0 intervals are not necessarily identically distributed or distributed as other intervals. Then, a slightly modified version of the proof of Theorem 8.3.6 implies that the counting variables corresponding to this process satisfy (8.3.18) with $\{c_n\}$ so that (8.3.19) is met with $\{\mu_n^*\}$ as a moment sequence corresponding to a distribution that is determined by its moments if and only if the process is Poisson, provided we take the left extremity of the distribution of the first interval of the process to be zero. It is also now easily seen that Corollary 8.3.7 still holds if we take in place of 'a renewal process', 'a modified renewal process', where a modified renewal process is defined as above, provided the left extremity of the distribution of the first interval of the process is taken as zero (once again) and F is taken as the common d.f. of the intervals following the m_0th of the process. Even when the a priori conditions that $P\{N(t) = n\} > 0$ for $n = 1, 2, \ldots, n_0$ and all $t > 0$ in Theorem 8.3.9 and that $P\{n(t) = n\}$ for every $n \geqslant 1$ and every $t > 0$ in Corollary 8.3.10 are not assumed, the respective results still hold, provided we understand by the conditional distributions their versions selected such that they are as stated whenever $P\{N(t) = n\} = 0$. These assertions are easy to establish.

8.4   FUNCTIONAL EQUATIONS IN APPLIED STOCHASTIC PROCESSES AND RELATED RESULTS

The tools developed in this monograph or minor variants of these could be used to solve functional equations in various other stochastic processes in applied probability. In what follows, we provide some illustrations of this.

8.4(a)   A Problem in Epidemiology

In a problem of deterministic spread of a simple epidemic, Daniels (1975) derived the following equation, with f nonnegative, not identically zero and everywhere differentiable,

$$-cf'(z) = \int_{-\infty}^{\infty} f(z - u) F(du), \quad z \in \mathbb{R}, \tag{8.4.1}$$

where F is a probability distribution function. There is no loss of generality in assuming that $c > 0$ in equation (8.4.1) and hence we consider this to be so. Then, in view of the nonnegativity of f, the equation implies that f is decreasing with $\lim f(z) = 0$ as $z \to \infty$. In that case, integrating both sides of (8.4.1) from x to ∞ and dividing by c, we have using Fubini's theorem

$$f(x) = \int_{-\infty}^{\infty} f(x - u) c^{-1} (1 - F(u)) du \quad x \in \mathbb{R}. \tag{8.4.2}$$

From Deny's theorem for the real line or a specialized version of Corollary 2.3.2, we have that there exist either two real roots θ_1, θ_2 of

$$\int_{-\infty}^{\infty} c^{-1} e^{\theta u} (1 - F(u)) \, du = 1 \tag{8.4.3}$$

(or equivalently of $\int_{-\infty}^{\infty} e^{\theta u} F(du) = c\theta$) giving the solution for f as

$$f(x) = p_1 \exp(-\theta_1 x) + p_2 \exp(-\theta_2 x), \quad x \in \mathbb{R}$$

with $p_1, p_2 \geqslant 0$ and at least one $p_i > 0$, or there exists just one real root θ_0 of (8.4.3) giving

$$f(x) = p \exp(-\theta_0 x), \quad x \in \mathbb{R}$$

with $p > 0$. Daniels arrived at this result in an intuitive way assuming a priori the existence of the moment generating function F on some interval which does not appear to be necessary.

The observations that we have made here on Daniels's equation are taken from Rao and Shanbhag (1989a).

8.4(b)   *A functional equation in a branching random walk with a barrier*

We have already come across in our extension of the Spitzer integral representation theorem, relative to stationary measures of branching processes, in Section 7.3a, a functional equation that is valid for certain branching processes. We shall now touch upon a functional equation that is met in a branching random walk with a barrier; this latter equation was, amongst other things, studied in a recent article of Biggins *et al* (1991).

Before dealing with the equation in question, we may make some observations on Theorem 2.3.1 that are of relevance to the equation. Suppose we have (2.1.2) with a modification that $\int_{\mathbb{R}} H(x + y)\mu(dy)$ on its right-hand side is now replaced by the integral plus c, where c is a nonnegative constant. With trivial alterations in the proof of Theorem 2.3.1, we can then see that the theorem in question holds subject to a modification that (2.3.3) is now replaced by

$$H(x) = \int_{(-\infty, 0)} H(x + y)\rho(dy) + c \left(1 + \sum_{n=1}^{\infty} \mu_{1n}(\mathbb{R}_+) \right) + \xi(x) \exp\{\eta x\}$$

$$\text{for a.a. } [L] \, x \in \mathbb{R}_+. \tag{8.4.4}$$

Suppose we have a point δ such that $\mu^*(\delta) = 1$, $\int_{\mathbb{R}} x(\exp\{\delta x\})\mu(dx) < 0$ and $(\exp\{-\delta x\})H(x)$ is bounded for $x \in \mathbb{R}_+$, where the notation is as in Chapter 2. As in the proof of Theorem 2.3.8, we have from Feller (1966) that $\rho^*(\delta) = 1$, $-\infty < \int_{(-\infty,0)} x(\exp\{\delta x\})\rho(dx) < 0$ and $\tau^*(\delta) < 1$; from the last condition it follows that $\delta < \eta$ should η as stated exist, and hence that (8.4.4) with $\xi \equiv 0$ holds. If we have $c \neq 0$, then it easily follows that we should also have $\delta > 0$. Hence, if we have additionally the restriction of H to $(-\infty, 0)$ to be identically equal to zero and μ to be nonarithmetic, we get in this case in view of Theorem 2 on page 349 in Feller (1966), which in turn is a corollary to our Theorem 8.3.1, that as $x \to \infty$,

$$H(x)e^{-\delta x} \to c\left(1 + \sum_{n=1}^{\infty} \mu_{1n}(\mathbb{R}_+)\right) \bigg/ \left(-\delta \int_{(-\infty,0)} ye^{\delta y}\rho(dy)\right), \qquad (8.4.5)$$

except, possibly, for the points x lying in a set of Lebesgue measure zero. (If we take μ to be arithmetic instead of nonarithmetic, then the corresponding result, with obvious alterations, also holds.)

Biggins et al (1991) showed that, under some mild conditions, the function f, where $f(x)$ is the mean total family size in a branching random walk with a single initial ancestor and the initial prowess x (i.e. where individuals smaller than or equal to $(-x)$ and all their descendants are deleted), satisfies the modified functional equation (2.1.1) with $c = 1$ and μ as a certain measure that is proportional to a probability measure. Moreover, the function f here happens to be such that it has not only all the properties of H that we have referred to above, but also it is increasing and left-continuous. (For the details, see the cited paper.) Hence, in view of (8.4.5), we have in our notation that if μ is nonarithmetic, then

$$f(x)e^{-\delta x} \to \left(1 + \sum_{n=1}^{\infty} \mu_{1n}(\mathbb{R}_+)\right) \bigg/ \left(-\delta \int_{(-\infty,0)} ye^{\delta y}\rho(dy)\right) \qquad (8.4.6)$$

as $x \to \infty$. (For μ arithmetic, we have an analogous result.) It was shown in Biggins et al (1991) that $f(x)e^{-\delta x}$ tends to a positive constant as $x \to \infty$ (without mentioning explicitly what the constant was in terms of μ and δ). If μ is concentrated on $\{-1, 0, 1, 2, \ldots\}$, then ρ is concentrated on $\{-1\}$ (as in the example in Remark 2.3.3 (iv)) and we get from the version of (8.4.4) in the present case that

$$f(x)e^{-\delta x} = \begin{cases} f(x-1)e^{-\delta(x-1)} + e^{-\delta x}\left(1 + \sum_{n=1}^{\infty} \mu_{1n}(\mathbb{R}_+)\right) & \text{if } x > 0 \\ 0 & \text{otherwise} \end{cases} \qquad (8.4.7)$$

with $f(x) = f(n+1)$ for each $x \in (n, n+1]$ and $n = 0, 1, \ldots$. In this example, in view of (8.4.7), we can obtain an exact expression for $f(x)$. The equation in question gives for each $n = 1, 2, \ldots$

$$f(n) = K(e^{n\delta} - 1), \qquad (8.4.8)$$

where $K = (1 + \Sigma_{m=1}^{\infty} \mu_{1m}(\mathbb{R}_+))/(e^{\delta} - 1)$. Noting that in Biggins *et al* (1991), $\mu(\mathrm{d}x) = b P_F(\mathrm{d}x)$, where P_F is the probability measure determined by the d.f. F, and substituting (8.4.8) in the functional equation in the present case, we can immediately see that K in (8.4.8) equals $(b - 1)^{-1}$. In that case, (8.4.8) determines the function f completely. Biggins *et al* (1991) used a somewhat different approach to arrive at the solution in the present case.

Finally, it is worth noting at this point that the functional equation studied in Biggins *et al* (1991) is somewhat restrictive and hence it is not surprising that the arguments given in the cited reference to arrive at the results in question turn out to be simpler than those used in our Chapter 2 to obtain Theorem 2.3.1.

8.4(c)   *Some problems in queueing theory*

In view of Theorem 2.3.4, it follows that if the function H is as in the theorem satisfying an additional condition that $H(x) = 0$ for $x < 0$, then we have for $x \in \mathbb{R}_+$

$$H(x) = \xi(x) \int_{[-x,0]} e^{\eta(x+y)} \left(\sum_{n=0}^{\infty} \rho^{*n} \right) (\mathrm{d}y) \tag{8.4.9}$$

with η, ρ and ξ as in the statement of Theorem 2.3.4. The solution (8.4.9) under some further restrictions was identified in the context of a random walk by Spitzer (1960). It now easily follows that if the weak ascending ladder height measure is a proper probability measure and the descending ladder height measure is not, then there exists a unique solution to (2.1.2) such that it is a probability distribution function concentrated on $\mathbb{R}_+$. This happens to be a compound geometric d.f. Also, it is clear that if μ is a probability measure, then the existence of a solution to (2.1.2) with the desired property implies that the weak ascending ladder height measure and the descending ladder height measure are as stated above. (If μ is a probability measure and $\int_{\mathbb{R}} x\mu(\mathrm{d}x)$ exists, then the ladder height measure requirements are equivalent to the condition that $\int_{\mathbb{R}} x\mu(\mathrm{d}x) > 0$.) The results contained in these latter observations have appeared essentially in Lindley (1952), Feller (1966) and elsewhere, and show, in particular, that in a $GI/G/1$ queueing system (with interarrival time distribution mean or service time distribution mean finite and P {an interarrival time = a service time} < 1) has a stationary waiting time distribution if and only if the relative traffic intensity of the system is less than 1, and that this distribution is uniquely given by a compound geometric distribution.

Theorem 2.3.1 or (more easily) Corollary 2.3.10 also provides one with some interesting applications in queueing theory. This could be applied to obtain certain conclusions of Kendall (1953) and Smith (1953), and Wishart (1956) concerning $GI/M/s$ and $GI/E_k/1$ queueing systems, respectively. The result gives, in each case, that the stationary queue length distribution exists if and only if the corresponding relative traffic intensity is less than 1, and that it is of a certain form. Also it yields the known results that in a $GI/M/1$ queueing system the stationary queue length distribution is geometric and in a $GI/M/s$ system the stationary waiting time distribution is exponential but for a discontinuity at zero. We shall illustrate the theme of the approach here by restricting ourselves to the queue length processes of $GI/M/s$ and $GI/E_k/1$ respectively.

Example 1 ($GI/M/s$) The process $\{Q_n\}$, where Q_n is the queue length at the nth arrival epoch (excluding the arriving customer) is, in this case, a temporally homogeneous irreducible aperiodic Markov chain. Kendall (1953) has given the one-step transition probabilities P_{ij}'s for this chain explicitly. These satisfy the conditions that for each i, $P_{ij} > 0$ if $j \leqslant i + 1$ and $P_{ij} = 0$ otherwise, and that for each $i \geqslant s - 1$, $j \geqslant s$, we have $P_{ij} = v(\{i + 1 - j\})$ with v as a certain probability measure such that $\mathrm{supp}[v] = \{0, 1, \ldots\}$ and $\Sigma_{x=0}^{\infty} x v(\{x\}) = \gamma^{-1}$, where γ is the relative traffic intensity. In view of Theorem 2.3.1 or Corollary 2.3.10, it easily follows that $\{Q_n\}$ has a stationary distribution $\{\Pi_j\}$ if and only if there exists a number $\beta \in (0, 1)$ such that $\beta = \Sigma_{x=0}^{\infty} \beta^x v(\{x\})$, i.e. if and only if $\gamma < 1$, and, if it exists, $\{\Pi_j\}$ is given by

$$\Pi_j = \begin{cases} K\beta^{j-s+1}, & j = s - 1, s, \ldots \\ K \sum_{i=s-1}^{\infty} \sum_{n=1}^{\infty} {}^{(s-1)}P_{ij}^{(n)} \beta^{i-s+1}, & \text{otherwise,} \end{cases} \qquad (8.4.10)$$

where K is the normalizing constant and ${}^{(s-1)}P_{ij}^{(n)}$ is the n-step transition probability corresponding to the transition $i \to j$ avoiding the states greater than or equal to $s - 1$ in between. (The second expression in (8.4.10) is to be read only for $s > 1$). From (8.4.10), we not only get the stationary distribution corresponding to $\{Q_n\}$, but also after a minor calculation that the stationary waiting time distribution corresponding to a $GI/M/s$ queueing system (with 'first come, first served' queue discipline) is exponential but for a discontinuity at the origin.

Example 2 ($GI/E_s/1$) As the results in question for $GI/E_s/1$ follow easily from those for a $GI/M/1$ system with batch arrivals where each of the batches is of deterministic size s, we shall consider the latter system here. As in the previous case, let $\{Q_n\}$ be the queue length process at arrival epochs (excluding the customers just arrived); once again, the process is seen to be a temporally homogeneous irreducible aperiodic Markov chain. The one-step transition probabilities P_{ij} in this case are such that for each i, we have $P_{ij} = v(\{i + s - j\})$ if $j \geqslant 1$ and $P_{i0} = 1 - \Sigma_{j=1}^{\infty} P_{ij}$, where v is a certain probability measure with $\mathrm{supp}[v] = \{0, 1, \ldots\}$ and $\Sigma_{x=0}^{\infty} x v(\{x\}) = s\gamma^{-1}$, γ being the relative traffic intensity. We have here the condition that $\gamma < 1$ to be equivalent to that there exists a number $\beta \in (0, 1)$ such that $\beta^s = \Sigma_{x=0}^{\infty} \beta^x v(\{x\})$. From Theorem 2.3.1 or Corollary 2.3.10, we then see that there is no stationary distribution for $\{Q_n\}$ when $\gamma \geqslant 1$, and that there exists a unique stationary distribution $\{\Pi_j\}$ for the chain when $\gamma < 1$ and it is such that for all $j \geqslant 1$

$$\Pi_j = \Pi_0 m_j \qquad (8.4.11)$$

(and $\Pi_0 = 1 - \sum_{j=1}^{\infty} \Pi_j$), where m_j is the measure relative to the first passage from j to 0 for the nonnegative matrix (in the sense of Senata (1973), as described in Section 1.3.10), i.e., the transpose of the stochastic matrix of $\{Q_n\}$. Clearly the Π_0 in (8.4.11) is such that

$$(\Pi_0 = (1 + \sum_{j=1}^{\infty} m_j)^{-1}, \tag{8.4.12}$$

because of the relation $\sum_{j=0}^{\infty} \Pi_j = 1$. In view of (8.4.11) and (8.4.12), the distribution $\{\Pi_j\}$ is determined.

It is worth noting in this place that m_j's appearing in (8.4.11) could be expressed in terms of a descending ladder height measure. If we denote by ρ the descending ladder height measure corresponding to a random walk with independent identically distributed jumps X_i, where the distribution of $X_i + s$ is given by $\{v(\{j\})\}$, then we can express these m_j's as

$$m_j = \sum_{k=1}^{\infty} \rho^{*k}(\{-j\}). \tag{8.4.13}$$

From what we have seen here, we easily get that for a $GI/E_s/1$ queueing system the stationary queue length distribution (at an arrival epoch) exists if and only if $\rho < 1$. Moreover, we could now get a certain expression for this distribution as well as that for the stationary waiting time distribution. Incidentally, in view of (8.4.11) and (8.4.13), it follows that $\{\Pi_j\}$ is a compound geometric distribution; this compares well with the property of the stationary waiting time distribution of a $GI/G/1$ system that it is compound geometric and could also be seen from the actual form of this latter distribution.

Wishart (1956) has also obtained the aforementioned results, but his expressions are in terms of complex variables with the approach based mainly on Rouché's theorem.

The topic of queueing departure processes provides us with further scope of applying the tools discussed in this monograph to problems in queueing networks. The study in this respect is of importance especially if one is dealing with networks that have queues in series; if the departure process of the nth queue is the arrival process of the $(n + 1)$th queue, then, unless this process is renewal or Markovian or in particular Poisson, in general, the analysis of the $(n + 1)$th queue turns out to be intractable. Daley (1976) and Kelly (1979) among others have reviewed the literature on departure processes of queues; Alamatsaz (1983) has also reported some new results in the area (which are expected to be published elsewhere in a joint paper with one of us). We now give a theorem on queueing departure processes, linked with a result of Daley (see Daley (1976) for the reference); we use Theorem 2.2.2 to get this theorem.

8.4.1   Theorem

Let $\{D_n : n = 1, 2, \ldots\}$ be the stationary departure process of a $GI/G/1/0$ system with interarrival time d.f. A such that $A(0) = 0$ and the service time d.f. B such that $B(0) = 0$ and $B(x) > 0$ for all $x > 0$. (Here D_n's denote interdeparture times.) Then D_1 and D_2 (or equivalently any two successive D_n's) are independent if and only if the system is $M/G/1/0$.

Proof   The 'if' part is known and easy to prove. We shall now deal with the 'only if' part. Assume hence that D_1 and D_2 are independent. Let I be a steady state idle period (i.e. an idle period other than the initial one.) Then, we have, under the assumption

$$((I^*B)(x))((I^*B)(y)) = \int_{[0,x]} I(x - z)(\int_{[z,z+y]}$$

$$B(y + z - z')A(\mathrm{d}z'))B(\mathrm{d}z) + o((I^*B)(x)), \quad x, y > 0, \quad (8.4.14)$$

where $o(I^*B(x))$ denotes the smaller order function of the quantity in question as $x \to l$ from above, where l is the left extremity of I. Dividing both sides of (8.4.14) by $(I^*B)(x)$ for $x > l$ and letting $x \to l$, we get that

$$(I^*B)(y) = (B^*A)(y), \quad y > 0,$$

which, in turn, implies that

$$I(y) = A(y), \quad y > 0 \qquad (8.4.15)$$

(8.4.15) is indeed a special case of the integral equation in Theorem 2.2.2 with the measure in question to be nonarithmetic and hence, by the theorem, we get that A is exponential. Hence, we have the assertion. ∎

8.4.2   Remarks

(i) It is clear, in view of Theorem 2.2.2, that for a member of a fairly general class of queueing systems, a steady state idle period is equal in distribution to an interarrival time if and only if its arrival process is Poisson; this property has played a key role in the characterization of $M/G/1/0$ given by Theorem 8.4.1.

(ii) Shanbhag (1973a) has characterized the queueing system $M/G/\infty$ in the class of $M/G/s/N$ (with initial queue length as zero and service time distribution not concentrated on $\{0\}$) on the basis of the property that the joint distribution of the number of arrivals during $(0, t]$ and the number of departures during the same interval is infinitely divisible for a certain fixed t. Corollary 8.3.7 implies that if we ask the property to be valid for all t, then the characterization in question holds in the class of $GI/G/s/N$ systems, provided we have the interarrival time distribution meeting the regularity condition stated in the corollary. In view of what is revealed in Remark 8.3.11(iii), one could also now give an analogous extended version of the characterization of $M/G/\infty$, appearing in Shanbhag (1973a), based on the property that the steady state joint distribution of the number of arrivals and the number of departures during a fixed nondegenerate interval be infinitely divisible.

8.4(d)   Williams's examples

On pp. 173 and 174 in Williams (1979) two examples identifying regular functions of certain Markov chains are dealt with using, amongst other things, the Martin-boundary arguments. In what follows, we shall show that the conclusions of these examples could also be reached via the results of Chapter 3.

Example 1   Let $\{X_n : n = 0, 1, \ldots\}$ be a simple random walk on $\mathbb{Z}^d$ such that the corresponding one-step transition probabilities are given by

$$\Pi(i, j) = \begin{cases} (2d)^{-1} & \text{if } \|j - i\| = 1 \\ 0 & \text{otherwise.} \end{cases}$$

In view of Dény's theorem or Corollary 3.4.4, it easily follows that every regular function (i.e., in obvious notation, every solution of $\Pi f = f \geqslant 0$) is constant; note that, in the present case, an exponential function is μ-harmonic if and only if it is identically equal to 1, μ being the measure on $\mathbb{Z}^d$ such that

$$\mu(\{i\}) = \begin{cases} (2d)^{-1} & \text{if } \|i\| = 1 \\ 0 & \text{otherwise.} \end{cases}$$

Example 2   Consider 'space-time coin-tossing' and let $X_n = (H_n, n), n = 0, 1, \ldots,$ where H_n represents the number of heads in n tosses. Obviously $\{X_n : n = 0, 1, \ldots\}$ is a temporally homogeneous Markov chain with state space $\{(m, n) \in \mathbb{Z}^2 : 0 \leqslant m \leqslant n\}$ and transition probabilities such that for each (m, n) in the state space,

$$\Pi((m, n), (m + 1, n + 1)) = \Pi((m, n), (m, n + 1)) = \tfrac{1}{2}.$$

From Ressel's specialized version of Corollary 3.4.4 or from Corollary 3.4.5, it easily follows that f is a regular function relative to $\{X_n\}$ only if there exists a probability measure on $[0,1]$ (i.e., on the Borel σ-field $[0,1]$) such that

$$f(m, n) \propto \int_{[0,1]} 2^n t^m (1 - t)^{n-m} \nu(dt), \quad m = 0, 1, \ldots, n; n = 0, 1, \ldots.$$

(The converse of the assertion is trivial.) This is seen on noting that the H defined by

$$H(m, m') = f(m, m + m'), \quad m, m' \in \mathbb{N}_0$$

satisfies (3.1.1) with $S = \mathbb{N}_0^2$ and μ concentrated on $\{(0, 1), (1, 0)\}$ such that $\mu(\{(0, 1)\}) = \mu(\{(1, 0)\}) = \tfrac{1}{2}$; in this case, as $\{\int_{[0,1]} 2^m t^m \nu(dt)\}$ is the moment sequence of a bounded real random variable, it follows that ν is determined uniquelly by f (see, also Remark 3.4.10, v)).

There are several other places in applied probability where one comes across variants and special cases of the functional equations that we have covered. These include studies of fractal measures, particle system and time series.

CHAPTER 9

Characterizations Based on Regression and Related Statistical Properties

9.1 INTRODUCTION

In Chapter 6, while dealing with properties of elliptical and stable distributions, we briefly came across characterizations based on constant regression, i.e. on the condition that the regression of one variable (or vector) on another variable (or vector) is constant (a.s.). These involved variables that were themselves linear functions of independent identically distributed (i.i.d.) random variables. There also exist in the literature characterizations of the type mentioned with variables possibly as some nonlinear Borel measurable functions of independent or exchangeable random variables that are not necessarily i.i.d. Amongst the distributions that have been characterized in this latter case are gamma, normal, Meixner hypergeometric, inverse gaussian, Poisson, binomial and negative binomial. Some of these characterizations lead us to results with applications in statistical inference; these include those related to Bhattacharyya inequality and Bayesian inference.

Kagan *et al* (1973) have dealt with most of the earlier literature on the subject. One of the most important contributions to the literature is that due to Laha and Lukacs (1960). Shanbhag (1972, 1978, 1979a, 1979b), Blight and Rao (1974), Morris (1982, 1983), Bondesson (1973, 1974), Heller (1983), Kingman (1972), Smith (1981), Kagan and Rao (1990), Kagan (1989), Alzaid *et al* (1990), Rao and Shanbhag (1989c) and several others have since obtained variants and extensions of the Laha–Lukacs result. These results answer natural questions arising in distribution theory and statistics, and are well-motivated. Some of these results extend other well known results such as the Kagan–Linnik–Rao theorem as well, and lead us to characterizations of nonidentically distributed or nonindependent random variables with interesting distributions.

We deal in this chapter with various characterizations based on constant regression including mainly those that are recent additions to the literature. An attempt is made to provide simplified techniques to arrive at these results wherever possible. Although approaches based on Fourier and Laplace transforms are still most popular and obvious to use for arriving at these results, we see that in many places much simpler approaches based on moments enable us to complete the required task.

The notation to be used in this chapter is, unless stated otherwise, as follows. Let $(\Omega, \mathcal{E}, P)$ be a probability space. If X is a random variable defined on this space and $\mathcal{F}$ is a sub-σ-field of $\mathcal{E}$, then, by the conditional expectation of X given $\mathcal{F}$, we mean (in the usual notation) an extended real-valued $\mathcal{F}$-measurable function f, whenever it exists, satisfying the condition

$$\int_A X \, dP = \int_A f \, dP \quad \text{for each } A \in \mathcal{F}$$

(where the identity is taken in the sense that if one side exists so does the other and the two sides are equal). We follow the usual notation to denote the conditional expectation and understand by the conditional expectation of X given $\{Y_\gamma : \gamma \in \Gamma\}$, where $\{Y_\gamma : \gamma \in \Gamma\}$ is a family of random variables defined on the probability space, $E(X|\sigma(\{Y_\gamma : \gamma \in \Gamma\}))$ whenever it exists; this latter expectation is also denoted by $E(X|Y_\gamma, \gamma \in \Gamma)$ (or by $E(X|Y_{\gamma_1}, \ldots, Y_{\gamma_n})$ if $\Gamma = \{\gamma_1, \ldots, \gamma_n\}$, a finite set). Further, we take here the equation $E(X|\mathcal{F}) = c$ a.s. to mean that $E(X|\mathcal{F})$ exists and equals a constant almost surely, which obviously implies that $E(|X|) < \infty$. (Shanbhag (1979a) and Alzaid *et al* (1990) have used the present formulation for the equations corresponding to the conditional expectations in stating their results; with this formulation, it is not necessary to mention separately the integrability conditions of the type in Kagan and Rao (1988) and Kagan (1989).)

9.2   THE LAHA–LUKACS RESULT AND ITS VARIANTS

The Laha–Lukacs result (1960) is the end product of several of the earlier investigations by Laha, Lukacs and others leading to partial results in this connection. Most of the literature in question is cited either in Lukacs and Laha (1964) or Kagan *et al* (1973). We revisit below the Laha–Lukacs result and some of its more recent variants that are of relevance to the theme of this monograph; the techniques that are given here to get these results are simple ones involving mainly moment arguments.

The main thrust of the Laha–Lukacs result is contained in the following theorem, which is a specialized version of it. As we shall see later, this simple result, in turn, gives the conclusion of the Laha–Lukacs result as an obvious corollary.

9.2.1   Theorem

Let n be an integer $\geqslant 2$ and $X_1, \ldots, X_n$ be i.i.d. nondegenerate random variables. Then, with a, b, c as real numbers, we have

$$E\left\{ X_1^2 - aX_1X_2 - bX_1 \,\middle|\, \sum_{j=1}^{n} X_j \right\} = c \quad \text{a.s.} \tag{9.2.1}$$

if and only if one of the following is valid:

(i) $a = 1$, $b = 0$ and X_1 is normal with variance c (implying that $c > 0$).

(ii) $a = 1$, $b \neq 0$ and $1/b(X_1 + (c/b))$ is Poisson.

(iii) $a > 1$, $4c(a - 1) = b^2$, and

$$\left(X_1 + \frac{b}{2(a - 1)} \right) \quad \text{or} \quad -\left(X_1 + \frac{b}{2(a - 1)} \right)$$

is gamma with index $(a - 1)^{-1}$.

(iv) $a < 1$ and there exists a number $\delta > 0$ such that $4c(a - 1) = b^2 - \delta^2$, and $\delta^{-1}(X_1 + ((b - \delta)/2(a - 1)))$ has a binomial $((1 - a)^{-1}, \bullet)$ distribution (implying that $(1 - a)^{-1}$ is a positive integer).

(v) $a > 1$ and there exists a number $\delta \neq 0$ such that $4c(a - 1) = b^2 - \delta^2$, and $\delta^{-1}(X_1 + ((b - \delta)/2(a - 1)))$ has a negative binomial $((a - 1)^{-1}, \bullet)$ distribution.

(vi) $a > 1$ and there exists a number $\delta \neq 0$ such that $4c(a - 1) = b^2 + \delta^2$, and $2\delta^{-1}(X_1 + (b/2(a - 1)))$ has a Meixner hypergeometric distribution with index $(a - 1)^{-1}$, i.e. has a distribution that is absolutely continuous with respect to the Lebesgue measure with density of the form

$$f(x) = (\cos \alpha)^\rho \frac{2^{\rho - 2}}{\pi \Gamma(\rho)} e^{\alpha x} \Gamma\left(\frac{\rho}{2} + \frac{ix}{2} \right) \Gamma\left(\frac{\rho}{2} - \frac{ix}{2} \right), \quad -\infty < x < \infty$$

with $\rho = (a - 1)^{-1}$ and α real. (The moment generating function corresponding to the distribution in question is defined for

$$t \in \left(-\frac{\pi}{2} - \alpha, \frac{\pi}{2} - \alpha \right), \quad \text{by } (\cos \alpha)^\rho (\cos(\alpha + t))^{-\rho}.)$$

Proof   The 'if' part of the theorem follows, in view of the 'if' part of Corollary 1.1.3, on noting that for each of the distributions in the list, the corresponding characteristic function ϕ satisfies

$$\{(\phi''(t) - bi\phi'(t))\phi(t) - a(\phi'(t))^2 + c(\phi(t))^2\}(\phi(t))^{n-2} = 0, \quad -\infty < t < \infty,$$
$$(9.2.2)$$

where ϕ' and ϕ'' are the first two derivatives of ϕ. (The ϕ in question has derivatives of all order). One could also prove the result via an analogous moment generating function argument (with obvious alternations).

To prove the 'only if' part, we note first that (9.2.1) implies by induction that X_1, has moments of all order, because it implies that $E(X_1^2) < \infty$ and that, for each integer $k \geqslant 0$,

$$E\left(X_1^2 \left| \sum_{j=1}^{n} X_j \right|^k \right) \leqslant E\left\{ (|a||X_1||X_2| + |b||X_1| + |c|) \left(\sum_{j=1}^{n} |X_j| \right)^k \right\},$$

where the right-hand side of the inequality is finite (implying that $E(|X_1|^{2+k})$ is finite) if $E(|X_1|^{k+1}) < \infty$. (The arguments of this type to show that certain moments are defined and finite have been used earlier by Shanbhag (1978, 1979a) and others.) Given any $\mu \in \mathbb{R}$ and (a, b, c) such that $(a - 1)\mu^2 + b\mu + c > 0$ with $a \geqslant 1$ we have, amongst the distributions for X_1 that are listed, a distribution with mean μ for which X_j's satisfy (9.2.1). In view of Remark 1.1.14(v), we have that this distribution is determined by its moments. If (9.2.1) holds with $E(X_1) = \mu$, then we have that $(a - 1)\mu^2 + b\mu + c > 0$ and that the moment sequence of X_1 is determined, given a, b, c and μ. Consequently, the 'only if' part of the theorem follows, provided it is shown that if (9.2.1) is valid with $a < 1$, then (iv) of the theorem holds. (9.2.1) gives that for any integer $k \geqslant 1$,

$$E\left\{\left(\frac{1 + a_k}{2}\right)(X_{1,k}^2 + X_{2,k}^2) - 2a_k\left(\frac{X_{1k} + X_{2,k}}{2}\right)^2 - b_k X_{1,k} - c_k\left|\sum_{j=1}^{n} X_{j,k}\right.\right\} = 0 \text{ a.s.,}$$
(9.2.3)

where $a_k = (a + k - 1)/k$, $b_k = b$ and $c_k = ck$, and $X_{1,k,...}X_{n,k}$ are i.i.d. random variables such that their common distribution is the k-fold convolution of the distribution of X_i. If $a < 1$, then, for a sufficiently large k, we get in (9.2.3), $0 < a_k < 1$, and, hence, it is easily seen that in this case the equation cannot be valid unless $X_{i,k}$ is bounded. (Note that the variable whose conditional expectation appears in (9.2.3) is greater than or equal to $((1 - a_k)/2)(X_{1,k}^2 + X_{2,k}^2) - b_k X_{1,k} - c_k$.)

This, in turn, implies that if $a < 1$, then (9.2.1) cannot be valid unless X_j's are bounded. Hence, it follows that if (9.2.1) is valid with $a < 1$, then the moment generating function, M, of X_1 is defined with domain of definition $\mathbb{R}$, and satisfies

$$\frac{d^2}{dt^2}\log M(t) = (a - 1)\left(\frac{d}{dt}\log M(t)\right)^2 + b\frac{d}{dt}\log M(t) + c, \quad t \in \mathbb{R}. \quad (9.2.4)$$

(9.2.4) implies that

$$\frac{d^2}{dt^2}\log M^*(t) = (a - 1)\left(\frac{d}{dt}\log M^*(t)\right)^2 + c^*, \quad t \in \mathbb{R}, \quad (9.2.5)$$

where

$$M^*(t) = \left(\exp\left\{\frac{bt}{2(a - 1)}\right\}\right)M(t), \quad t \in \mathbb{R} \quad \text{and} \quad c^* = c - \frac{b^2}{4(a - 1)}.$$

Noting that the left-hand side and (hence) the right-hand side of the equation are positive for each t (as a by-product of which we have $c^* > 0$), we can easily solve (9.2.5) to see that the moment generating function M^* is of the form

$$M^*(t) = e^{-\{(c^*/(1-a))^{1/2}t\}}q^{(1-a)^{-1}}\left(1 + \frac{p}{q}e^{2(c^*(1-a))^{1/2}t}\right)^{(1-a)^{-1}}, \quad t \in \mathbb{R} \quad (9.2.6)$$

with $q = 1 - p$ and $p \in (0, 1)$. (The solution to (9.2.5) is obtained by first determining the form of $(d/dt \log M^*(t)$, $t \in \mathbb{R}$, and then from it that of M^*). Considering $t < 0$ such that $|t|$ is sufficiently large, we can express, using the binomial theorem, the right-hand side of (9.2.6) as a series with terms of the form $\lambda_n \exp\{(2n - (1 - a)^{-1})(c^*(1 - a))^{1/2}t\}$ with λ_n real; it is seen that in the series, we have some of the λ_n's to be negative, unless $(1 - a)^{-1}$ is an integer. As M^* is a moment generating function, we cannot have corresponding to it the series expansion with some $\lambda_n < 0$. Recalling the relation between M and M^* and also the relation $c^* = c - b^2/4(a - 1)$, we hence get that (iv) is implied by (9.2.6). This completes the proof of the theorem. ∎

9.2.2   Corollary

Under the assumptions in the theorem, we have, with a, b, c and d real and at least one of them nonzero,

$$E \left\{ dX_1^2 - aX_1X_2 - bX_1 \,\middle|\, \sum_{j=1}^{n} X_j \right\} = c \text{ a.s.} \tag{9.2.7}$$

if and only if $d \neq 0$ and one of (i), (ii), ..., (vi) in the theorem holds with a, b and c replaced respectively by a/d, b/d, and c/d.

Proof   It is sufficient if we show that the validity of (9.2.7) with $d = 0$ implies that X_j's are degenerate, contradicting one of the assumptions. Assume then that (9.2.7) holds with $d = 0$ and at least one of a, b and c as nonzero. It is then immediate that at least one of a and b is nonzero and that X_j's are integrable. There is no loss of generality in assuming that $E(X_j) = 0$; this, in turn, implies that $c = 0$ and we get for $t \in (-\delta, \delta)$ for a sufficiently small δ, the characteristic function, ϕ, of X_1, to be such that

$$a(\phi'(t))^2 + bi\phi(t)\phi'(t) = 0 \tag{9.2.8}$$

from which it follows that the restriction to some interval $(-\delta', \delta')$, with $0 < \delta' \leqslant \delta$, of ϕ' vanishes identically. (Note that the claim made after (9.2.8) follows trivially when $b = 0$, and follows on noting that for some δ', $a\phi'(t) + bi\phi(t) \neq 0$ if $t \in (-\delta', \delta')$, when $b \neq 0$.) As this implies that $\phi \equiv 1$ and hence that $P\{X_1 = 0\} = 1$, we have the aforementioned contradiction. Thus, we have the corollary. ∎

9.2.3   Corollary

Suppose the assumptions in the theorem are met with an addition that $E(X_1^2) < \infty$. Define

$$\overline{X} = \frac{1}{n} \sum_{i=1}^{n} X_i \quad \text{and} \quad S^2 = \frac{1}{n - 1} \sum_{i=1}^{n} (X_i - \overline{X})^2.$$

Then

$$E\{S^2|\overline{X}\} = a_0 + a_1\overline{X} + a_2\overline{X}^2 \quad \text{a.s.,} \tag{9.2.9}$$

with a_0, a_1 and a_2 as real numbers, if and only if $a_2 \neq n$ and one of (i), (ii), ..., (vi) in the theorem holds with

$$a = \frac{n + (n-1)a_2}{n - a_2}, \qquad b = \frac{na_1}{n - a_2} \quad \text{and} \quad c = \frac{na_0}{n - a_2}.$$

Proof   It is obvious that (9.2.9) is equivalent to

$$E\left\{ d'X_1^2 - a'X_1X_2 - b'X_1 \,\bigg|\, \sum_{j=1}^{n} X_j \right\} = c' \text{ a.s.}$$

with $d' = 1 - a_2/n$, $a' = 1 + a_2(n-1)/n$, $b' = a_1$ and $c' = a_0$. Note that, in this equation, at least one of a' and d' is nonzero. Hence, the corollary follows from Corollary 9.2.2. ∎

9.2.4   Corollary

Let $\{F_\theta : \theta \in \Theta\}$ be a family of nondegenerate probability distribution functions on $\mathbb{R}$ that are absolutely continuous with respect to some σ-finite measure ν such that for each θ

$$F_\theta(x) \propto \int_{(-\infty,x]} \exp\{g(\theta)y\}\nu(\mathrm{d}y), \quad x \in \mathbb{R},$$

where $g : \Theta \to \mathbb{R}$ satisfying the condition that the set of values of g is dense in an open interval. (The condition on g obviously places a restriction on Θ implicitly, and the family that we have here is a version of an exponential family.) For each $\theta \in \Theta$ denote by μ_θ and σ_θ^2 the mean and the variance corresponding to F_θ respectively. Then

$$\sigma_\theta^2 = a_0 + a_1\mu_\theta + a_2\mu_\theta^2 \quad \text{for all } \theta \tag{9.2.10}$$

with a_0, a_1 and a_2 as real and independent of θ if and only if given a point $\theta_0 \in \Theta$, we have one of (i), (ii), ..., (vi) of the theorem to be valid with $a = a_2 + 1$, $b = a_1$, $c = a_0$ and X_1 as a random variable with d.f. F_{θ_0}.

Proof   Suppose for each $\theta \in \Theta$, we denote by $X_\theta^{(1)}$ and $X_\theta^{(2)}$ two independent random variables distributed with d.f. F_θ. Then, it easily follows that (9.2.10) holds if and only if for any fixed $\theta_0 \in \Theta$ we have (in obvious notation)

$$E_{\theta_0}\left\{ \left(X_{\theta_0}^{(1)}\right)^2 - (a_2 + 1)X_{\theta_0}^{(1)}X_{\theta_0}^{(2)} - a_1X_{\theta_0}^{(1)} \,\bigg|\, X_{\theta_0}^{(1)} + X_{\theta_0}^{(2)} \right\} = a_0 \quad \text{a.s.} \tag{9.2.11}$$

The result of the corollary now follows trivially from the theorem (i.e. from Theorem 9.2.1). ∎

9.2.5   Remarks

(i) The family $\{F_\theta\}$ of Corollary 9.2.4 is obviously such that given any $\theta_0 \in \Theta$, for each $\theta \in \Theta$, the moment generating function M_θ of F_θ is given by

$$M_\theta(\bullet) = \frac{M_{\theta_0}(g(\theta) - g(\theta_0) + \bullet)}{M_{\theta_0}(g(\theta) - g(\theta_0))}$$

with an appropriate domain of definition and M_{θ_0} as the moment generating function of F_{θ_0}. Hence, from the corollary, it follows trivially that (9.2.10) is valid if and only if either:

(a) $a_1 = a_2 = 0$ and $X_\theta^{(1)}$ (for each θ) is normal with variance a_0, or

(b) $a_1 \neq 0$, $a_2 = 0$ and $a_1^{-1}(X_\theta^{(1)} + a_0 a_1^{-1})$ is Poisson, or

(c) $a_2 > 0$, $4a_0 a_2 = a_1^2$ and $\delta(X_\theta^{(1)} + (2a_2)^{-1} a_1)$ for $\delta = 1$ or -1 with δ independent of θ is gamma with index a_2^{-1}, or

(d) $a_2 < 0$ and there exists a number $\delta > 0$ such that $4a_0 a_2 = a_1^2 - \delta^2$ and $\delta^{-1}\left(X_\theta^{(1)} + (2a_2)^{-1}(a_1 - \delta)\right)$ has a binomial $((-a_2)^{-1}, \bullet)$ distribution, or

(e) $a_2 > 0$ and there exists a number $\delta \neq 0$ independently of θ such that $4a_0 a_2 = a_1^2 - \delta^2$ and $\delta^{-1}(X_\theta^{(1)} + (2a_2)^{-1}(a_1 - \delta))$ has a negative binomial $(a_2^{-1},.)$ distribution, or

(f) $a_2 > 0$ and there exists a number $\delta \neq 0$ independently of θ such that $2\delta^{-1}(X_\theta^{(1)} + (2a_2)^{-1} a_1)$ has a Meixner hypergeometric distribution with index a_2^{-1}, where $X_\theta^{(1)}$ is as defined in the proof of Corollary 9.2.4.

(ii) Corollary 9.2.4 or a somewhat different version of it appearing in (i) above was implicitly or explicitly given in Shanbhag (1979b) and Morris (1982), and was partially given by many others including Gokhale (1980). (Letac(1992) gives a good account of the literature on such results and their generalizations.) The argument given here to arrive at the characterization as a corollary to the Laha-Lukacs result is essentially due to Shanbhag (1979b). Recently, Letac and Mora (1990) have extended this result by considering a modified version of (9.2.10) with a cubic expression of μ_θ in place of the quadratic one; they have shown that in this latter case a characterization result holds with six additional distributions. We are currently exploring in our joint work with E. Fosam the possibility of obtaining the Letac-Mora result via simpler arguments linking it with a regression property of the type in Theorem 9.2.1 (but with $n \geqslant 3$ and an additional term of the form $dX_1 X_2 X_3$ in the equation).

Essentially the argument used to prove Corollary 9.2.3 produces the following modified version of the corollary. The result in question is indeed a version of the general result of Laha and Lukacs and could be viewed, but for notational differences, as a restatement of Corollary 9.2.3.

9.2.6   Corollary

Let the assumptions in Corollary 9.2.3 be met and let $\overline{X}$ be as defined in the corollary. Define

$$Q(X_1, \ldots, X_n) = \sum_{i,j=1}^{n} a_{ij} X_i X_j + \sum_{i=1}^{n} b_i X_i + c,$$

and

$$A_0 = \sum_{i=1}^{n} a_{ii}, \qquad A_1 = \sum_{i \neq j=1}^{n} a_{ij} \quad \text{and} \quad B = \sum_{i=1}^{n} b_i,$$

assuming a_{ij}, b_i and c as given real numbers. Let a_0, a_1 and a_2 be real numbers such that at least one of $A_0 - a_2/n$, $A_1 - (n-1)a_2/n$, $B - a_1$ and $c - a_0$ is nonzero. Then

$$E\{Q(X_1, \ldots, X_n)|\overline{X}\} = a_0 + a_1\overline{X} + a_2\overline{X}^2 \quad \text{a.s.} \qquad (9.2.12)$$

if and only if $A_0 - a_2/n \neq 0$ and one of (i), (ii), ..., (vi) of Theorem 9.2.1 is valid with a, b and c replaced respectively by

$$\left(\frac{(n-1)a_2}{n} - A_1 \right) \bigg/ \left(A_0 - \frac{a_2}{n} \right), \ (a_1 - B) \bigg/ \left(A_0 - \frac{a_2}{n} \right) \text{ and } (a_0 - c) \bigg/ \left(A_0 - \frac{a_2}{n} \right).$$

9.2.7   Remarks

(i) If we extend the terminology used for conditional expectations and equations involving them further and understand by (9.2.12) that the conditional expectation on its left-hand side exists and equals the expression on its right-hand side almost surely (as in the case of an equation of the type (9.2.1) having a constant right-hand side), then Corollary 9.2.6 holds with the restriction that $E(X_1^2) < \infty$ replaced by that X_1 has a moment generating function (which by definition should have its domain of definition to be a nondegenerate interval). Note that nonpositive or nonnegative or extreme stable random variables in addition to many standard ones have moment generating functions. The version of Corollary 9.2.6 that we have mentioned in this remark subsumes the result appearing in Shanbhag (1970). (Observe that, under the stated assumption, (9.2.12) holds with $c = a_0$ and $B \neq a_1$, then we cannot have X_1 to be continuous.)

(ii) One may now raise the question as to whether Corollary 9.2.6 holds with $E(X_1^2) < \infty$ dropped (with no replacement and with the interpretation for (9.2.12) as in (i) above). That the modified assertion does not hold is shown by the following example.

Example   Let X_1 and X_2 be independent Cauchy random variables and let $\overline{X} = \frac{1}{2}(X_1 + X_2)$. Then

$$E(1 + X_1^2|\overline{X}) = 1 + \overline{X}^2 \quad \text{a.s.}$$

(We have involved the number 1 in the two expressions above just to make the verification slightly easier).

(iii) From Theorem 9.2.1 (and Remark 1.1.14 (v)), it is clear that under the given assumptions (9.2.1) is valid if and only if $a \geqslant 1$, or $a < 1$ and $(1 - a)^{-1}$ is an integer, and X_1 has moments μ_m of all order satisfying

$$\sum_{r=0}^{m} \binom{m}{r} \{\mu_{r+2}\mu_{m-r} - a\mu_{r+1}\mu_{m-r+1} - b\mu_{r+1}\mu_{m-r} - c\mu_r\mu_{m-r}\} = 0, \quad m = 0, 1, \ldots .$$

$$(9.2.13)$$

Moreover, when $a < 1$, if it is assumed that a moment sequence $\{\mu_m\}$ of a nondegenerate probability distribution, satisfying (9.2.13), exists, then we get that $(a - 1)\mu_1^2 + b\mu_1 + c > 0$ and hence that

$$c - \frac{b^2}{4(a - 1)} > (1 - a)\left(\mu_1 + \frac{b}{2(a - 1)}\right)^2 \geqslant 0,$$

and that for each m, μ_m is the coefficient of $t^m/m!$ in the Taylor series expansion in a neighborhood of the origin of the right-hand side of (9.2.6) multiplied by $\exp\{-bt/2(a - 1)\}$ for an appropriate p. As the latter function has a Taylor series expansion in some open interval about each point of the real line, it follows that this function is indeed the moment generating function of the distribution in question. In view of this (recalling the relevant observation in the proof of Theorem 9.2.1), we have now that if $a < 1$ and $(1 - a)^{-1}$ is a noninteger, then there is no moment sequence corresponding to a nondegenerate distribution, satisfying (9.2.13). It is of interest to note here that if we assume what is revealed in this last observation to be a valid assertion a priori, then the moment argument given in the proof of Theorem 9.2.1 to prove the 'only if' part for $a \geqslant 1$ also holds for $a < 1$, subject to obvious modifications (inclusive of an observation that if there is no $\{\mu_m\}$ as stated, (9.2.1) cannot be valid under the given assumptions).

(iv) In a private communication to us sometime in 1980, Professor B. Gyires revealed that there was an amendment required for the Laha–Lukacs proof for Corollary 9.2.6 and that he took care of it. In the context of Theorem 9.2.1, his revelations imply, amongst other things, that the argument provided by Laha and Lukacs for establishing that (9.2.1) is met, when $a < 1$, only if a linear function of X_1 is binomial is somewhat inadequate.

While we are on the topic of the Laha–Lukacs result, we may make yet another observation that is of relevance to it: through a note sent to us privately, Heller (1991) shows that a result of Lukacs (1963) in the spirit of the Laha–Lukacs result is erroneous and gives a corrected version of it. In essence, the following is the corrected version of Lukacs's theorem.

9.2.8   Theorem

Let $n \geqslant 3$ and $X_1, \ldots, X_n$ be i.i.d. nondegenerate random variables. Define, for $k = 1, 2, 3, 4$, $L_k = \sum_{j=1}^{n} X_j^k$. Then

$$E\{nL_4 + (n-4)L_3L_1 + (3-2n)L_2^2 + L_2L_1^2 - nL_3 + (n+1)L_2L_1 - L_1^3 | L_1\} = 0 \text{ a.s.}$$
$$(9.2.14)$$

if and only if either X_1 is Poisson or binomial or negative binomial, or, for some $\gamma > 0$, $-(X_1 + \gamma)$ is a negative binomial $(\gamma, \bullet)$ random variable. (Lukacs does not insist that X_i's be nondegenerate, but misses the case when X_i's are Poisson in his result; this is also the case with Kagan *et al* (1973; Theorem 6.4.1, p. 217).)

Proof   It is easily seen that (9.2.14) implies X_1^3 to be integrable, and that it is equivalent to

$$E\{X_1^3X_2 - 2X_1^2X_2^2 + X_1^2X_2X_3 + X_1^2X_2 - X_1X_2X_3 | L_1\} = 0 \quad \text{a.s.} \quad (9.2.15)$$

Clearly, in view of Corollary 1.1.3, we have (9.2.15) to be equivalent to

$$(\phi(t))^{n-3}\{\phi'''(t)\phi'(t) - 2(\phi''(t))^2 + \phi''(t)(\phi'(t))^2 + i\phi''(t)\phi'(t) - i(\phi'(t))^3\} = 0, \ t \in \mathbb{R},$$
$$(9.2.16)$$

with ϕ as the characteristic function of X_1 and ϕ', ϕ'' and ϕ''' as its first three derivatives. For a sufficiently small δ, (9.2.16) gives that

$$\psi'''(t)\psi'(t) - 2(\psi''(t))^2 + i\psi'(t)\psi''(t) = 0, \quad |t| < \delta, \quad (9.2.17)$$

with $\psi(t) = \log\phi(t)$ and ψ', ψ'' and ψ''' as the first three derivatives of ψ on $(-\delta, \delta)$. One could obviously establish the 'only if' part of the assertion from (9.2.17) directly. However, to show the link between the present result and the Laha–Lukacs result we shall use a deroute. As X_r's are nondegenerate, we can see that if (9.2.17) is valid, then there exists a $0 < \delta' \leqslant \delta$ such that on $(-\delta', \delta')$, ψ' is nonvanishing and the equation in (9.2.17) is valid, implying that ψ is differentiable any number of times and hence that X_1 has moments of all order. We have also in that case

$$\frac{\mathrm{d}}{\mathrm{d}t}(\psi''(t)(\psi'(t))^{-2}) - i\frac{\mathrm{d}}{\mathrm{d}t}((\psi'(t))^{-1}) = 0, \quad t \in (-\delta', \delta'),$$

yielding that

$$\psi''(t) - i\psi'(t) - a'(\psi'(t))^2 = 0, \quad t \in (-\delta', \delta'), \quad (9.2.18)$$

for some real a'. Clearly (9.2.18) implies that (9.2.13) is met with $a = a' + 1$, $b = 1$ and $c = 0$ by the moments μ_m of X_1. Hence, the 'only if' part of the theorem follows from Remark 9.2.7(iii). The 'if' part of the theorem follows on verifying that for the distributions involved in the assertion, (9.2.16) is valid. Hence, we have the theorem. ∎

9.2.9   Remark

Gyires's point in remark 9.2.7(iv) on the Laha–Lukacs proof of their result also applies to Lukacs's proof of his latter result. In view of this, one could find our indirect proof appearing above more informative than its predecessor.

We have already come across moment arguments in this section. The proof of the next theorem provides us with further evidence of the role played by these in studies dealing with identifiability and characterization problems in probability and statistics.

9.2.10 Theorem

Let $X_1, \ldots, X_n$ be a sequence of i.i.d. nondegenerate random variables. Define $\overline{X} = (1/n)\Sigma_{i=1}^n X_i$, and $S_2 = \Sigma_{i=1}^n (X_i - \overline{X})^2$ and $S_3 = \Sigma_{i=1}^n (X_i - \overline{X})^3$. If $n \geqslant 3$ then the following are equivalent:

(i) There exists a real number c such that

$$E((\overline{X} - c)S_3^l | S_2) = 0, l = 0, 1 \quad \text{a.s.}$$

(ii) There exists a real number c such that

$$E(\overline{X} | S_2, S_3) = c \quad \text{a.s.}$$

(iii)
$$E(S_3 | \overline{X}) = 0 \quad \text{a.s.}$$

(iv) X_1 is normal.

Moreover, if $n \geqslant 2$, then the following is equivalent to (iv).

(v) There exist real numbers c_1 and c_2 satisfying

$$E(\overline{X} | S_2) = c_1 \quad \text{a.s.}$$

and

$$E(\overline{X}^2 | S_2) = c_2 \quad \text{a.s.}$$

(From Corollary 9.2.3, it further follows that, in the latter case, (iv) is also equivalent to the condition that $E(S_2 | \overline{X}) = c$ a.s. for some real number c which obviously has to be positive; the result in question is essentially due to Lukacs (1942).)

Proof If X_1 is normal, $(X_1 - \overline{X}, \ldots, X_n - \overline{X})$ is independent of $\overline{X}$ and the distribution of X_i for each i has moments of all order. Consequently, it follows that (iv) implies each of (i), (ii), (iii) and (v) (obviously when n meets the respective condition). To prove the reverse implications, we may break the argument into the following steps.

(v) $\Rightarrow$ (iv) As for any $m \geqslant 0$,

$$E(|X_1|^{2m}) < \infty \Leftrightarrow E((S_2)^m) < \infty$$

and

$$E(\overline{X}^2 (S_2)^m) < \infty \Leftrightarrow E(|X_1|^{2m+2}) < \infty$$

and the second equation in (v) implies $E(\overline{X}^2(S_2)^m) = c_2 E((S_2)^m)$, we can see by induction that $E(|X_1|^{2m}) < \infty$ for $m = 1, 2, \ldots$ and hence that the distribution of X_1 has moments of all order, assuming that (v) is valid. Note that in view of the assumption on X_j's we have $\overline{X}$ to be nondegenerate and hence (v) implies that $c_2 > c_1^2$. If we are given that (v) is met, then it is clear that $E(X_1) = c_1$ and $\mathrm{var}(X_1) = n(c_2 - c_1^2)$ for some c_1 and c_2 and, by induction, all the moments of the distribution of X_1 are determined given c_1 and c_2. As $N(c_1, n(c_2 - c_1^2))$ for each c_1, c_2 (obviously satisfying $c_2 > c_1^2$) has the equations in (v) met and it is determined by its moments, we can then immediately claim that X_1 is normal. Thus, we have that (v) $\Rightarrow$ (iv).

(i) $\Rightarrow$ (iv)   Observe that from (i) we have that $E(X_1^4) < \infty$ and that for some real c

$$E(X_1^4|S_2) = E\left\{X_1^4 - \frac{n^2(X_1 - c)}{(n-1)(n-2)}S_3 \Big| S_2\right\} \quad \text{a.s.} \qquad (9.2.19)$$

It is a simple exercise to verify that for any positive integer r

$$E(X_1^4(S_2)^{r/2}) < \infty \Leftrightarrow E(|X_1|^{r+4}) < \infty, \qquad (9.2.20)$$

and if, for any fixed integer $k \geqslant 4$, we have $E(|X_1|^k) < \infty$, then

$$\left\{X_1^4 - \frac{n^2(X_1 - c)}{(n-1)(n-2)}S_3\right\}(S_2)^{(k-3)/2}$$

is an integrable random variable for each c and, hence by (9.2.19), $E(X_1^4(S_2)^{(k-3)/2}) < \infty$. If $E(X_1^4(S_2)^{(k-3)/2}) < \infty$, then (9.2.20) implies that $E(|X_1|^{k+1}) < \infty$, and hence if (i) is valid, we have by induction that $E(|X_1|^r) < \infty$, for all r ($\geqslant 4$ and consequently) $\geqslant 0$. Given any real c and $\sigma^2 > 0$, there is a normal random variable X_1 relative to which (i) is valid, and any normal distribution is determined by its moments. The moments of the distribution of any arbitrary random variable X_1 with mean c and variance σ^2 for which (i) holds, should obviously agree with those of $N(c, \sigma^2)$; this can be verified inductively. Consequently we have that if (i) is valid, then the random variable X_1 is normal. Hence, we have the stated implication.

(ii) $\Rightarrow$ (iv)   In view of the implication dealt with immediately above, it is sufficient if we show that $E(X_1^4) < \infty$ if (ii) is valid. (It would then follow that (ii) $\Rightarrow$ (i).) Assume then that (ii) holds. We have then $E(|X_1|) < \infty$ and, taking without loss of generality $c = 0$,

$$E(X_1|S_2, S_3) = 0 \quad \text{a.s.} \qquad (9.2.21)$$

Define a random variable Y such that

$$Y = \begin{cases} 1 & \text{if } S_3 \geqslant 0 \\ -1 & \text{if } S_3 < 0. \end{cases}$$

Take $A_\alpha = \{|X_1| > \alpha \Sigma_{i=2}^n |X_i|\}$ for each $\alpha > 0$. We can choose a sufficiently large $\alpha_0 (> 0)$ such that on A_{α_0}, $YX_1 = |X_1|$. From (9.2.21), we have

$$E\{X_1 Y I_{A_{\alpha_0}} ||S_3|\} = -E\{X_1 Y I_{A_{\alpha_0}^c} ||S_3|\} \quad \text{a.s.,} \tag{9.2.22}$$

where the notation I_B is for the indicator function of B. (9.2.22) implies that

$$E\left\{|X_1| I_{A_{\alpha_0}} ||S_3|\right\} \leqslant E\left\{|X_1| I_{A_{\alpha_0}^c} ||S_3|\right\} \quad \text{a.s..} \tag{9.2.23}$$

Observe that for every $k \geqslant 1$, there exists a $c_k > 0$ such that

$$|X_1| I_{A_{\alpha_0}^c} |S_3|^{k/3} \leqslant c_k |X_1| \sum_{i=2}^n |X_i|^k.$$

Hence, if, for some integer $k \geqslant 1$, we have $E(|X_1|^k) < \infty$, then

$$E\{|X_1| I_{A_{\alpha_0}^c} |S_3|^{k/3}\} < \infty,$$

which implies because of (9.2.23) that

$$E\{|X_1| I_{A_{\alpha_0}} |S_3|^{k/3}\} < \infty. \tag{9.2.24}$$

From (9.2.24), it is easily seen that $E(|X_1|^{k+1}) < \infty$. Hence, by induction, we get that $E(X_1^4) < \infty$ (or indeed that $E(|X_1|^k) < \infty$ for all $k \geqslant 1$). This completes the proof of the result sought.

(iii) $\Rightarrow$ (iv)   Assume that (iii) is valid. It then follows that $E(|X_1|^3) < \infty$ and

$$E((X_1 - \overline{X})^3 | \overline{X}) = 0 \quad \text{a.s.,}$$

from which it follows that

$$E\{X_1^2 \overline{X} + Z | \overline{X}\} = 0 \quad \text{a.s.,} \tag{9.2.25}$$

where

$$Z = \frac{n}{(n-1)(n-2)} \left\{ (X_1 - \overline{X})^3 - \left(\frac{n-1}{n}\right)^3 X_1^3 + \frac{1}{n^3} \sum_{j=2}^n X_j^3 \right\} - \left(\frac{n-1}{n}\right) X_1^2 X_2.$$

(9.2.25) implies that if for some $k \geqslant 3$, $E(|X_1|^k) < \infty$, then for each $\alpha > 0$

$$E\{X_1^2 \overline{X}^2 |\overline{X}|^{k-3} I_{\{|\overline{X}| \leqslant \alpha\}}\} \leqslant E(|Z||\overline{X}|^{k-2})$$

$$< \infty. \tag{9.2.26}$$

As the middle term in (9.2.26) is independent of α, the monotone convergence theorem then implies that $E\{X_1^2 \overline{X}^2 |\overline{X}|^{k-3}\} < \infty$ and hence that $E\{|X_1|^{k+1}\} < \infty$.

Hence, by induction, we can conclude that the distribution of X_1 has moments of all order. (We could also see this via an argument essentially of the type in the proof of the implication above, but with obvious alterations including, in particular, that here $Y = 1$ if $\overline{X} \geqslant 0$ and $= -1$ otherwise.) Given $E(X_1) = c \in \mathbb{R}$ and $\text{var}(X_1) = \sigma^2 > 0$ and, for the distribution of X_1, (iii) holds, then, as in the previous case, we can see that the moments of the distribution of X_1 are the same as those of $N(c, \sigma^2)$ (because (iii) is valid for this latter distribution as well, and it determines the moment sequence of the distribution when its mean and variance are given). As any normal distribution is determined by its moments, we can then immediately claim that X_1 is normal, and that the stated implication is valid.

The proof of the theorem is now complete. ∎

9.2.11   Corollary (Kawata and Sakamoto, 1949)

Let n be an integer $\geqslant 2$ and $X_1, \ldots, X_n$, $\overline{X}$ and S_2 be as defined in the statement of the theorem. Then, $\overline{X}$ and S_2 are independent if and only if X_1 is normal.

Proof   The 'if' part follows on using the fact that if X_1 is normal, $(X_1 - \overline{X}, \ldots, X_n - \overline{X})$ is independent of $\overline{X}$; the result is standard. To prove the 'only if' part, we may proceed as follows. Assume that $\overline{X}$ and S_2 are independent. We have then for a fixed $\lambda > 0$, and $l = 0, 1, 2$,

$$E\left\{ e^{-\lambda \sum_{i=1}^{n} X_i^2} \overline{X}^l \Big| S_2 \right\} = E\left(e^{-n\lambda \overline{X}^2} \overline{X}^l \right) e^{-\lambda S_2} \quad \text{a.s.} \qquad (9.2.27)$$

From (9.2.27) for $l = 0$ and $l = 1$ and similarly for $l = 0$ and $l = 2$, we arrive at

$$E\left\{ e^{-\lambda \sum_{i=1}^{n} X_i^2} \overline{X}^l \Big| S_2 \right\} = c_l^* E\left\{ e^{-\lambda \sum_{i=1}^{n} X_i^2} \Big| S_2 \right\}, \quad l = 1, 2 \quad \text{a.s.} \qquad (9.2.28)$$

where

$$c_l^* = E\left(e^{-n\lambda \overline{X}^2} \overline{X}^l \right) \Big/ E\left(e^{-n\lambda \overline{X}^2} \right), \quad l = 1, 2.$$

Denote by F the d.f. of X_1 and let $X_1^*, \ldots, X_n^*$ be i.i.d. random variables, each with d.f. F^*, where

$$F^*(x) \propto \int_{(\infty, x]} e^{-\lambda y^2} F(dy), \quad x \in \mathbb{R}. \qquad (9.2.29)$$

Define $\overline{X}^*$ and S_2^* to be the counterparts of $\overline{X}$ and S_2 respectively for X_i^*'s. From (9.2.28), it then easily follows that

$$E\{(\overline{X}^*)^l | S_2^*\} = c_l^*, \quad l = 1, 2 \quad \text{a.s.}$$

From Theorem 9.2.10, it is now immediate that F^* is normal. As (9.2.29) implies that for some $c > 0$, $F(dx) = c(\exp\{\lambda x^2\})F^*(dx)$, $X \in \mathbb{R}$, we can then easily check that F is normal. Hence, we have the assertion. ∎

9.2.12 Remarks

(i) Theorem 9.2.10 is indeed a collection of the results established or implied by Shanbhag (1978, 1979a) and Kagan (1990). The methods that we have used here to prove the theorem are adapted from Shanbhag's two papers. A result stated in Kagan's paper, with an implicit proof, is that the assertion (iii) ⇔ (iv) holds; the proof in question is based on an entirely different approach.

(ii) It is worth pointing out now that, as pointed out by Shanbhag (1978), $E(\overline{X}|S_2) = c$ a.s. (with $c \in \mathbb{R}$)) is not a characteristic property of $N(c, \bullet)$; the equation obviously holds if the sample is from a distribution that is symmetric about c with X_1, (and hence $\overline{X}$) integrable.

(iii) The theme of the proof of Corollary 9.2.10 is not new, though it has not appeared in this form in the existing proofs. Applying this, one could also arrive at the result of the corollary via Lukacs's (1942) result mentioned in the statement of Theorem 9.2.10. (Note that, in obvious notation, if $\overline{X}$ and S_2 are independent, $E(S_2^*|\overline{X}^*) = c^*$ a.s. for some c^*.)

One may now ask the question as to whether Corollary 9.2.6 holds with the assumption of i.i.d. replaced by that the random variables be independently distributed or with $\overline{X}$ replaced by $\overline{X}_2 = (X_1 + X_2)/2$. That the answers to the questions are in the negative is shown by the following two examples.

Example 1   Let X_1 and X_2 be square integrable independent nondegenerate random variables with moment generating functions given respectively by M_1 and M_1^λ, where $\lambda > 0$ and $\neq 1$ and obviously the domain of definition of M_1 is a nondegenerate interval. Then (from the obvious version of Corollary 1.1.3)

$$E\{\lambda X_1^2 - X_2^2 + (\lambda - 1)X_1 X_2|(X_1 + X_2)/2\} = 0 \quad \text{a.s.} \tag{9.2.30}$$

Moreover, taking in place of the present X_2 an appropriate shifted version of it (i.e. a variable obtained by adding an appropriate constant to it) as our X_2 we can even produce random variables for which (9.2.30) holds with 0 replaced by a nontrivial linear function of $(X_1 + X_2)/2$.

Example 2   Let $n \geqslant 3$ and X_i's and Q be as in Corollary 9.2.6 with an additional requirement that Q does not involve X_1 and X_2 (i.e. the coefficients a_{ij}'s and b_i's are such that Q can be expressed as a function of $X_3, \ldots, X_n$ only). If we define $\overline{X}_2$ as stated above, then, with a_0, a_1, a_2 real,

$$E(Q|\overline{X}_2) = a_0 + a_1\overline{X}_2 + a_2\overline{X}_2^2 \quad \text{a.s.}$$

if and only if $a_1 = a_2 = 0$ and $E(Q) = a_0$ (because the distribution of $\overline{X}_2$ has at least three distinct support points and $E(Q|\overline{X}_2) = E(Q)$ a.s.).

Example 2 above also tells us that Theorems 2.1 and 3.1 of Wang (1972) are false. Specialized versions of this example had appeared in Shanbhag (1973) (though it had remained to spell out in these explicitly that the coefficients of $X_1 X_j$ and $X_2 X_j$ in the quadratic expression were all zero).

In view of Examples 1 and 2 above, it is clear that if we have X_i's to be not necessarily identically distributed or we have in place of $\overline{X}$ a linear function of X_i's, then the conclusion of Corollary 9.2.6 does not remain valid and the problem of identifying the class of distributions for which (9.2.12) holds turns out to be more complex. However, with appropriate modifications, one can reduce the problem in this latter case to a more manageable form and arrive at a positive solution. The next theorem shows one such situation, and provides one with a procedure for obtaining further results that are similar in spirit.

9.2.13   Theorem

Let X_1 and X_2 be independent nondegenerate random variables. Then, with α, β, a, b, c, d real and at least one of a, b, c, d nonzero,

$$E\{X_2 - \alpha X_1 | X_1 + X_2\} = \beta \quad \text{a.s.} \tag{9.2.31}$$

and

$$E\{dX_1^2 - aX_1 X_2 - bX_1 | X_1 + X_2\} = c \quad \text{a.s.} \tag{9.2.32}$$

if and only if $\alpha > 0$, $d \neq 0$ one of (i), (ii), ..., (vi) of Theorem 9.2.1 is valid with a, b and c replaced by $a\alpha/d$, $(b + a\beta)/d$ and c/d respectively, and X_2 is such that, for each t lying in the domain of definition of the moment generating function of X_1,

$$E\left(e^{tX_2}\right) = (\exp\{\beta t\})(M_1(t))^{\alpha}, \tag{9.2.33}$$

where M_1 is the moment generating function of X_1, requiring additionally $\alpha(1 - a\alpha/d)^{-1}$ to be an integer if $(a\alpha)/d < 1$.

Proof   We shall first establish the 'only if' part. (9.2.31) implies that X_2 is integrable and if $\alpha \neq 0$, X_1 is also integrable. From Corollary 1.1.3, we hence get for t in a sufficiently small neighborhood of the origin (in which ϕ_1 and ϕ_2 do not vanish)

$$\frac{\phi_2'(t)}{\phi_2(t)} = \alpha \frac{\phi_1'(t)}{\phi_1(t)} + i\beta \tag{9.2.34}$$

and

$$d\frac{\phi_1''(t)}{\phi_1(t)} - a\frac{\phi_1'(t)}{\phi_1(t)}\frac{\phi_2'(t)}{\phi_2(t)} - bi\frac{\phi_1'(t)}{\phi_1(t)} = -c, \tag{9.2.35}$$

where ϕ_1 and ϕ_2 are the characteristic functions of X_1 and X_2 respectively, and the equations are to be read with the terms that have coefficients equal to zero deleted. As X_2 is nondegenerate, the validity of (9.2.33) for t lying in a small neighborhood of the origin implies that $\alpha \neq 0$. Substituting the right-hand side of (9.2.33) for $\phi_2'(t)/\phi_2(t)$ in (9.2.35), we get

$$d\frac{\phi_1''(t)}{\phi_1(t)} - a\alpha \left(\frac{\phi_1'(t)}{\phi_1(t)}\right)^2 - (b + a\beta)i\frac{\phi_1'(t)}{\phi_1(t)} = -c. \qquad (9.2.36)$$

Essentially, as in the proof of Corollary 9.2.2, we can then see that unless $d \neq 0$, we have a contradiction to the assumption that X_1 is nondegenerate. On the other hand, if (9.2.36) holds for t lying in a sufficiently small neighborhood of the origin, then it follows easily that the moment sequence $\{\mu_m\}$ for X_1 exists and satisfies (9.2.13) with a, b and c replaced respectively by $a\alpha/d$, $(b + a\beta)/d$ and c/d. Appealing to Theorem 9.2.1 and taking into account the information in Remark 9.2.7(iii), we hence get that X_1 is as stated in the 'only if' part of the theorem. To prove the remainder of the 'only if' part there are various arguments. We shall use one of these: from (9.2.34), it now follows that X_2 is square integrable and that $\text{var}(X_2) = \alpha\,\text{var}(X_1)$; as X_1 and X_2 are nondegenerate (or as X_1 is nondegenerate and $\alpha \neq 0$), we hence get that $\alpha > 0$. If M_1^α is a moment generating function then (9.2.34) implies inductively that its moments agree with the moments of the distribution of $X_2 - \beta$; clearly, the moment generating function has zero as an interior point of its domain of definition (in view of the form of the distribution of X_1) and hence, in view of Remark 1.1.14(v), we have that the moment generating function of $X_2 - \beta$ is M_1^α. Also, as (9.2.34) implies the moment sequence of $X_2 - \beta$ always exists and satisfies (9.2.13) with a replaced by $1 - \alpha^{-1} + d^{-1}a$, and b and c replaced respectively by appropriate constants, we have from what is revealed in Remark 9.2.7(iii) that the case with $\alpha(1 - d^{-1}a\alpha)^{-1}$ as a noninteger and $a\alpha/d < 1$ cannot occur. This completes the proof of the 'only if' part. The 'if' part can be easily verified via the moment generating functions or the characteristic functions of X_1 and X_2 (in view of Corollary 1.1.3). ∎

9.2.14   Remarks

(i) It is possible to give a somewhat different version of the statement of Theorem 9.2.13 spelling out the distribution of X_2 explicitly. However, we have chosen the present version of the statement for brevity.

(ii) Bolger and Harkness (1965), Shanbhag (1971), Alzaid (1983) and several others have studied some versions of Theorem 9.2.13. If $X_1 + X_2$ is assumed a priori to have a moment generating function or that $E(X_1^2) < \infty$ and $E(X_2^2) < \infty$, then Theorem 9.2.13 holds with (9.2.31) and (9.2.32) replaced respectively by

$$(1 + \alpha)E(X_1|X_1 + X_2) = (X_1 + X_2) - \beta \quad \text{a.s.}$$

and

$$(1+\alpha)(a+d)E(X_1^2|X_1+X_2) = (c(1+\alpha)-b\beta)+(b-a\beta)(X_1+X_2)+a(X_1+X_2)^2 \quad \text{a.s.}$$

with the identities interpreted as mentioned before.

(iii) Incidentally, in (9.2.32) of the statement of Theorem 9.2.13 we have not involved on the left-hand side terms with X_2^2 and X_2 only because in the presence of (9.2.31), we can reduce (assuming that it is nontrivial) the identity with such additional terms, under an appropriate moment condition on X_1, X_2, to an identity of the form (9.2.32). Conditions of the type referred to here appear naturally when one deals with processes with stationary independent increments.

(iv) Lai (1982b) has reviewed the literature on Theorem 9.2.13 prior to 1982, and more recently, Pusz and Wesolowski (1992) have studied some variants of the result. Wang (1991) has also given some interesting results in this connection.

(v) The following result essentially in the spirit of Theorem 9.2.13 is worth mentioning here. If X_1 and X_2 are positive independent random variables, then, for some $\alpha \in (0, 1)$ and $c > 0$,

$$E(X_1|X_1 + X_2) = \alpha(X_1 + X_2) \quad \text{a.s.}$$

and

$$E(X_1^{-1}|X_1 + X_2) = \alpha^{-1}(X_1 + X_2)^{-1} + c \quad \text{a.s.}$$

if and only if X_1 and X_2 are either inverse Gaussian, or stable with exponent $\frac{1}{2}$ and support $\mathbb{R}_+$, and the moment generating function of X_2 is some power of that of X_1. (For fixed α, c, the condition is equivalent to that the moment generating functions of X_1 and X_2 are given for some $k \geqslant 0$, respectively by

$$M_1(t) = \exp\left\{ - \left(\frac{2(1-\alpha)}{c} \right)^{1/2} (\sqrt{k-t} - \sqrt{k}) \right\}, \quad t \leqslant k$$

and

$$M_2(t) = \exp\left\{ - \left(\frac{1-\alpha}{\alpha} \right) \left(\frac{2(1-\alpha)}{c} \right)^{1/2} (\sqrt{k-t} - \sqrt{k}) \right\}, \quad t \leqslant k$$

We leave it as an exercise to the reader to verify the assertion. For a review on related characterizations of inverse Gaussian distributions, including that given by Khatri (1962), see Chikara and Folks (1989) and Seshadri (1993). The result we have mentioned was, however, implied in Shanbhag (1983).

9.3   EXTENSIONS OF THE KAGAN–LINNIK–RAO THEOREM

In this section, we deal with various extensions of the Kagan–Linnik–Rao theorem; these results could also be viewed as variants of the Laha–Lukacs result met in the previous section and are taken mostly from Rao and Shanbhag (1989c). To prove these, we need some lemmas.

9.3(a)   Some auxiliary lemmas

9.3.1   Lemma

Let k, r, $m \geq 1$ be integers and $P_k(x) = a_0 + \cdots a_k x^k$, $a_k \neq 0$, $x \in \mathbb{R}$ be a polynomial of degree k. Let $Z_1, \ldots, Z_m$ be random variables on the space $(\Omega, \mathcal{E}, P)$ and $\mathcal{E}_1, \ldots, \mathcal{E}_m$ be sub-σ-fields of $\mathcal{E}$ such that if $m > 1$, $\sigma(\sigma(Z_i) \cup \mathcal{E}_i)$, $i = 1, 2, \ldots, m$ are independent. Then, if $m \geq k$ and $k \leq r \leq m$, the condition that

$$E\left(P_k\left(\sum_{i=1}^{m} Z_i \right) \middle| \sigma\left(\bigcup_{i=1}^{r} \mathcal{E}_i \right) \right) = c \quad \text{a.s.} \tag{9.3.1}$$

holds only if either k is odd and for each $i = 1, 2, \ldots, r$

$$E(Z_i | \mathcal{E}_i) = c \quad \text{a.s.} \tag{9.3.2}$$

or for some $i \geq 1$ and $\leq r$

$$E(Z_i | \mathcal{E}_i) = c \quad \text{a.s.} \quad \text{and} \quad E(Z_i^2 | \mathcal{E}_i) = c \quad \text{a.s.} \tag{9.3.3}$$

Moreover, if $m \geq k$ and $[(k+1)/2] \leq r \leq m$, then (9.3.1) is valid only if there exists a real number a such that for some $1 \leq i \leq [(k+1)/2]$, we have

$$E(Z_i^2 + a Z_i | \mathcal{E}_i) = c \quad \text{a.s.}, \tag{9.3.4}$$

or k is odd and for each $i = [(k+1)/2], \ldots, r$, (9.3.2) is valid, where $[(k+1)/2]$ is the integral part of $(k+1)/2$. (Here as well as in what follows, the notation c appearing in different equations does not imply the constants corresponding to the equations to be identical.)

Proof   (Since (9.3.1) implies $E(|Z_i^k|) < \infty$, $i = 1, 2, \ldots, m$, the steps involved herein are justified; we shall not discuss this point again.) Since the assertions of the lemma are trivially valid for $k = 1$, it is sufficient if we take $k \geq 2$. We shall now establish the first part of the lemma. Assume that $m \geq k$, $k \leq r \leq m$, (9.3.1) is valid and that (9.3.3) is not valid for any $i \geq 1$ and $\leq r$. If $E \in \mathcal{E}_r$ with $P(E) > 0$, then appealing to the binomial theorem, we obtain that

$$\sum_{i=0}^{k} a_i \sum_{j=0}^{i} \binom{i}{j} E\left(\left(\sum_{l=1}^{r-1} Z_l \right)^{i-j} \middle| \sigma\left(\bigcup_{l=1}^{r-1} \mathcal{E}_l \right) \right) \left\{ E\left(\left(\sum_{l=r}^{m} Z_l \right)^{j} \middle| \mathcal{E}_r \right) \right.$$

$$\left. - E\left(\left(\sum_{l=r}^{m} Z_l \right)^{j} \middle| E \right) \right\} = 0 \quad \text{a.s.}, \tag{9.3.5}$$

which implies that either (9.3.2) is valid for $i = r$ or an equation similar to (9.3.1) is valid with $k - 1$ in place of k and $r - 1$ for r and m. This, in turn, implies inductively that (9.3.2) is valid for at least one $i(\geq 1$ and $\leq r)$. For $k = 2$, this leads to a contradiction since it implies readily that (9.3.3) is valid at least for one i. Consequently, we can assume $k \geqslant 3$. We have now to show that our assumptions imply that k is odd and (9.3.2) is valid for all i in question. There is no loss of generality in assuming that at least for $i = r$, (9.3.2) holds. In view of our assumption that (9.3.3) is not valid for any i and hence $i = r$, (9.3.5) implies that an equation analogous to (9.3.1) with $k - 2$ in place of k and $r - 1$ in place of r as well as of m is valid. We can then conclude inductively that k is odd and (9.3.2) is valid for all $i \geqslant 1$ and $\leqslant r$. This establishes the first part of the lemma. To prove the second part of the lemma, consider the case when $m \geqslant k$, $[(k + 1)/2] \leqslant r \leqslant m$, (9.3.1) is valid and there is no a yielding the validity of (9.3.4) for some $1 \leqslant i \leqslant [(k + 1)/2]$. Since we have already agreed to assume $k \geqslant 2$, our assumption concerning the nonexistence of a yields that $k \geqslant 3$. Since we do not have an a for which (9.3.4) holds for $i = 1$, from an equation essentially of the form of (9.3.5), it follows that there exists a polynomial P_{k-2} of degree $k - 2$ such that

$$E \left(P_{k-2} \left(\sum_{i=2}^{m} Z_i \right) \middle| \sigma \left(\bigcup_{i=2}^{r} \mathcal{E}_i \right) \right) = c \quad \text{a.s.} \tag{9.3.6}$$

Using the same argument (but for obvious notational changes) repeatedly, we see that (9.3.6) implies that either k is even and for some polynomial P_2 of degree 2

$$E \left(P_2 \left(\sum_{i=k/2}^{m} Z_i \right) \middle| \sigma \left(\bigcup_{i=k/2}^{r} \mathcal{E}_i \right) \right) = c \quad \text{a.s.,} \tag{9.3.7}$$

or k is odd and

$$E \left(\sum_{i=[(k+1)/2]}^{m} Z_i \middle| \sigma \left(\bigcup_{i=[(k+1)/2]}^{r} \mathcal{E}_i \right) \right) = c \quad \text{a.s.} \tag{9.3.8}$$

However, in view of the nonexistence of a satisfying (9.3.4) for $i = [(k + 1)/2]$, it follows that (9.3.7) cannot be valid. Consequently, we claim that k is odd with (9.3.8) to be valid. (9.3.8) implies the validity of (9.3.2) for $i = [(k + 1)/2], \ldots, r$. ∎

9.3.2 Remark

We have neither given the strongest possible version of Lemma 9.3.1 nor have used the full power of the present version in our arguments to follow. We have chosen the version given here for the lemma only because it appealed to us most. Incidentally, the second assertion of the lemma implies as its corollary that the result remains valid if $1 \leqslant i \leqslant [(k + 1)2]$ is replaced by $i = 1, 2, \ldots, r$ and simultaneously $i = [(k + 1)/2], \ldots, r$ is replaced by $i = 1, 2, \ldots, r$.

9.3.3   Lemma

Let k, m, r and P_k be as defined in Lemma 9.3.1. Let $(Y_1, Z_1), \ldots, (Y_m, Z_m)$ be independent random vectors. Then, if $k + 2 \leqslant r \leqslant m$, the condition that

$$E\left(P_k\left(\sum_{l=1}^{m} Z_l\right) \middle| Y_2 - Y_1, \ldots, Y_r - Y_1\right) = c \quad \text{a.s.} \tag{9.3.9}$$

is valid only if $(E(|Z_l|^k) < \infty, l = 1, 2, \ldots, m$ and) at least $r - k + 1$ of the vectors (Y_l, Z_l), $l = 1, 2, \ldots, r$ satisfy

$$E\left(e^{itY_l} Z_l\right) = (\alpha_l + \beta_l t) E\left(e^{itY_l}\right), \quad |t| < \varepsilon, \tag{9.3.10}$$

where $i = \sqrt{-1}$, for some real $\varepsilon > 0$ and real α_l and some β_l.

(Subtracting, as in the proof of Corollary 1.3.7, the relation (9.3.10) for $-t$ from that for t, we can see the β_l has to be of the form $i\gamma_l$ with γ_l real.)

Proof   Assume that $k + 2 \leqslant r \leqslant m$ and that (9.3.9) holds. (Obviously we have $E(|Z_l|^k) < \infty, l = 1, 2, \ldots, m$.) Define

$$\mathbf{Y}^{(j)} = (Y_1 - Y_2, Y_2 - Y_3, \ldots, Y_j - Y_{j+1}), \quad j = 1, 2, \ldots, r - 1.$$

We can now rewrite (9.3.9) as

$$E\left(P_k\left(\sum_{l=1}^{m} Z_l\right) \middle| \mathbf{Y}^{(r-1)}\right) = c \quad \text{a.s.} \tag{9.3.11}$$

and also note that there is no loss of generality if we take $m = r$. Let $(-\alpha, \alpha)$ be an open interval on which none of $E(e^{itY_l})$, $l = 1, 2, \ldots, r$ vanishes. Defining the conditional expectation of a complex-valued integrable Borel measurable function in the usual way, we can obtain the conditional expectation of $\exp\{it(Y_{r-1} - Y_r)\}$ times the left-hand side of (9.3.11) given $\mathbf{Y}^{(r-2)}$. In view of (9.3.11), there is no loss of generality in taking this to be equal to zero almost surely. After a minor manipulation, this implies that there exist continuous complex-valued functions $a_j(\bullet)$, $j = 0, 1, \ldots, k$ on $(-\alpha, \alpha)$ such that $a_k(\bullet) \equiv 1$ and for each $t \in (-\alpha, \alpha)$

$$E\left(e^{itY_{r-1}} \sum_{j=0}^{k} a_j(t) \left(\sum_{l=1}^{r-1} Z_l\right)^j \middle| \mathbf{Y}^{(r-2)}\right) = 0 \quad \text{a.s.} \tag{9.3.12}$$

Suppose the lemma does not hold. Then for at least k of the (Y_l, Z_l), $l = 1, 2, \ldots, r$, we cannot have α_l, β_l and ε_l such that (9.3.10) (with obviously ε replaced by ε_l) is valid. There is no loss of generality in that case to assume that (Y_l, Z_l), $l = r - k, \ldots, r - 1$ are k such vectors. Consider the case of $k \geqslant 2$. Using an inductive argument, we can now show that, for each $l^* = 1, 2, \ldots, k + 1$, we have an open interval

$$I_{l^*} \subset \left(-\frac{\alpha}{2^{k-l^*+1}}, \frac{\alpha}{2^{k-l^*+1}}\right)$$

and continuous functions $a_j^{(l^*)}(\bullet)$, $j = 0, 1, \ldots, k - l^* + 1$ on I_{l^*} such that $a_{k-l^*+1}^{(l^*)}(\bullet) \equiv 1$ and

$$
E\left(e^{itY_{r-l^*}} \sum_{j=0}^{k-l^*+1} a_j^{(l^*)}(t) \left(\sum_{l=1}^{r-l^*} Z_l\right)^j \middle| \mathbf{Y}^{(r-l^*-1)}\right) = 0 \quad \text{a.s.}, \quad t \in I_{l^*}, \quad (9.3.13)
$$

where $\mathbf{Y}^{(0)}$ is defined to be equal to a constant almost surely. This result is clearly valid for $l^* = 1$. Assume then that it is valid for a particular $l^* \leqslant k$. From (9.3.13), we get in view of the binomial theorem

$$
E\left(e^{i(t-s)Y_{r-l^*-1}} \sum_{j=0}^{k-l^*+1} a_j^{(l^*)}(t) \sum_{j'=0}^{j} \binom{j}{j'} \left(\sum_{l=1}^{r-l^*-1} Z_l\right)^{j'} \phi_{j-j'}^{(r-l^*)}(s) \middle| \mathbf{Y}^{(r-l^*-2)}\right) = 0
$$

$$
\text{a.s.}, \quad t \in I_{l^*}, \quad s \in \mathbb{R}, \quad (9.3.14)
$$

where $\phi_{j-j'}^{(r-l^*)}(s) = E(e^{isY_{r-l^*}} Z_{r-l^*}^{j-j'})$. Denoting $t - s$ by t' and expressing t in terms of t' and s (i.e. taking $t = t' + s$), we arrive at from (9.3.14) a new equation in terms of t' and s. Restricting to $s \in (-\alpha, \alpha)$, divide this resulting equation by $E(e^{isY_{r-l^*}})$ (i.e. by $\phi_0^{(r-l^*)}(s)$). Refer to the equation obtained as (9.3.14)*. Note that for any t'-interval, we cannot have an ε such that

$$
(k - l^* + 1)\frac{\phi_1^{(r-l^*)}(s)}{\phi_0^{(r-l^*)}(s)} + a_{k-l^*}^{(l^*)}(t' + s) \quad (9.3.15)
$$

is independent of s for $|s| < \varepsilon$; this follows because in view of Lemma 1.3.4 (or a substantially simpler special case of it) (9.3.15) leads us to a contradiction to our assumption that (Y_{r-l^*}, Z_{r-l^*}) is one of the vectors corresponding to which the constants for (9.3.10) to be met do not exist. We can then find an open interval

$$
I_{l^*+1} \subset \left(-\frac{\alpha}{2^{k-l^*}}, \frac{\alpha}{2^{k-l^*}}\right)
$$

and an $s_0 \in (-\alpha, \alpha)$ such that

$$
(k - l^* + 1)\left(\frac{\phi_1^{(r-l^*)}(0)}{\phi_0^{(r-l^*)}(0)} - \frac{\phi_1^{(r-l^*)}(s_0)}{\phi_0^{(r-l^*)}(s_0)}\right) + a_{k-l^*}^{(l^*)}(t') - a_{k-l^*}^{(l^*)}(t' + s_0) \neq 0
$$

$$
\text{for each } t' \in I_{l^*+1} \quad (9.3.16)
$$

and $t' + s_0 \in I_{l^*}$ for each $t' \in I_{l^*+1}$. For each $t' \in I_{l^*+1}$, subtracting the conditional expection in (9.3.14)* for $s = s_0$ from that for $s = 0$ and dividing it by the left-hand side of (9.3.16), we get a conditional expectation that is equal to 0 almost surely. The resulting equation implies that the result corresponding to (9.3.13) is valid with $l^* + 1$ in place of l^* and hence by induction for all $l^* = 1, 2, \ldots, k + 1$. This, in turn, implies that

$$E\left(e^{itY_{r-k-1}} \mid \mathbf{Y}^{(r-k-2)}\right) = 0 \quad \text{a.s.} \tag{9.3.17}$$

for each t contained in some open subinterval I of $(-\alpha, \alpha)$, contradicting our assumption that none of $E(e^{itY_l})$ vanishes on $(-\alpha, \alpha)$. From this we conclude that our assumption that the lemma does not hold cannot be valid. Hence we have the lemma. ∎

9.3.4 Lemma

Let $k \geqslant 1$ and P_k be as defined in Lemmas 9.3.1 and 9.3.3. Then, if Z_1 and Z_2 are independent gamma random variables satisfying

$$E(P_k(Z_1 + Z_2) \mid \log Z_1 - \log Z_2) = c \quad \text{a.s.,} \tag{9.3.18}$$

we have that Z_1 and Z_2 have the same scale parameter. Furthermore, if Z_1 and Z_2 are independent normal random variables satisfying

$$E(P_k(Z_1 + Z_2) \mid Z_1 - Z_2) = c \quad \text{a.s.,} \tag{9.3.19}$$

then Z_1 and Z_2 have the same variance.

Proof If Z_1 and Z_2 are as in the first assertion of the lemma, then there exists a neighborhood of the origin such that $E(e^{\theta Z_i}) < \infty$, $i = 1, 2$ for every θ in it. Furthermore, for θ in this neighborhood, we have in view of (9.3.18)

$$E(Z_1^\theta Z_2^{-\theta} P_k(Z_1 + Z_2))/E(Z_1^\theta Z_2^{-\theta}) = c. \tag{9.3.20}$$

Noting that $E(Z_1^\theta Z_2^{-\theta} P_k(Z_1 + Z_2)) = \Sigma_{j=0}^k a_j \Sigma_{r=0}^j \binom{j}{r} E(Z_1^{\theta+r} Z_2^{-\theta+j-r})$ in (9.3.20), we can easily check that the left-hand side of (9.3.20) is a polynomial in θ with the coefficient of θ^k as $a_k((1/\lambda_1) - (1/\lambda_2))^k$ assuming the density of Z_r to be $\exp(-\lambda_r z_r)z_r^{\alpha_r - 1}\lambda_r^{\alpha_r}/\Gamma(\alpha_r)$, $z_r > 0$ (with λ, $\alpha_r > 0$) for $r = 1, 2$. Furthermore, if Z_1 and Z_2 are as in the second assertion, i.e. to be independent normal satisfying (9.3.19), then we can see that for all $\theta \in \mathbb{R}$

$$E(e^{\theta Z_1 - \theta Z_2} P_k(Z_1 + Z_2))/E(e^{\theta Z_1 - \theta Z_2}) = c. \tag{9.3.21}$$

It is easily observed that the left-hand side of this latter identity is a polynomial in θ with the coefficient of θ^k to be equal to $a_k(\sigma_1^2 - \sigma_2^2)^k$ where σ_1^2 and σ_2^2 are variances of Z_1 and Z_2 respectively. Equating the coefficients of θ^k in (9.3.20) and similarly in (9.3.21), we then arrive at the results claimed. ∎

9.3(b) The Main Theorems

We shall now discuss the main results referred to earlier.

9.3.5   Theorem

Let k and P_k be as defined in the lemmas in Section 9.3(a). Let $X_1, \ldots, X_n, n \geqslant 2$ be independent random variables. If $n \geqslant 2k$, then for any particular $r \geqslant k$ and $\leqslant n/2$, we have

$$E\left(P_k\left(\sum_{l=1}^{n} X_l\right) \Big| X_{2l-1} - X_{2l}, \quad l = 1, 2, \ldots, r\right) = c \quad \text{a.s.}, \tag{9.3.22}$$

only if at least for $r - k + 1$ of the l's in $\{1, 2, \ldots, r\}$, X_{2l-1} and X_{2l} are normal satisfying $\mathrm{var}(X_{2l-1}) = \mathrm{var}(X_{2l})$, or k is odd and

$$E(X_{2l-1} + X_{2l}|X_{2l-1} - X_{2l}) = c \quad \text{a.s.}, \quad l = 1, 2, \ldots, r. \tag{9.3.23}$$

(Here as well as in what follows we view degenerate random variables as normal with zero variance.) If, additionally, the X_l's are identically distributed then the assertion also holds with $X_{2l-1} - X_{2l}$ in (9.3.22) and (9.3.23) replaced by $(X_{2l-1} - X_{2l})^2$.

Proof   There is no loss of generality in assuming that $n = 2r$. Define $m = r$, $Z_l = X_{2l-1} + X_{2l}$, $l = 1, 2, \ldots, m$ and $\mathcal{E}_l = \sigma(X_{2l-1} - X_{2l})$, $l = 1, 2, \ldots, m$. Using Shanbhag's (1978) argument based on moments, met in the last section, or a familiar argument based on characteristic functions, it is easily seen that the validity of (9.3.3) for an i yields X_{2i-1} and X_{2i} to be normally distributed such that $\mathrm{var}(X_{2i-1}) = \mathrm{var}(X_{2i})$. (Also this follows as a corollary to Theorem 5.7.1 of Kagan *et al* (1973).) Hence we have easily the first assertion in view of the first part of Lemma 9.3.1 (on noting that it is sufficient if it is proved for $r = k$). The second assertion follows via essentially the same argument on defining m and Z_l as earlier and $\mathcal{E}_l$ to be $\sigma((X_{2l-1} - X_{2l})^2)$ for $l = 1, 2, \ldots, m$ in view of a corollary of Shanbhag's (1978) result that if X_1 and X_2 are identically distributed random variables, then $E(X_1 + X_2|(X_1 - X_2)^2) = c$ a.s. and $E((X_1 + X_2)^2|(X_1 - X_2)^2) = c$ a.s. only if X_1 and X_2 are normal. ∎

9.3.6   Remarks

(i) On noting that if $X_1, \ldots, X_n$ are independent observations from a symmetric distribution with a finite kth order moment, then for any odd $k \geqslant 1$ and $n \geqslant 2k$,

$$E\left(\left(\sum_{l=1}^{n} X_l\right)^k \Big| (X_1 - X_2)^2, \ldots, (X_{2k-1} - X_{2k})^2\right) = 0 \quad \text{a.s.},$$

it follows that the result for even k in Theorem 9.3.5 does not hold for odd k. It is also worth pointing out at this stage that the result in the theorem for even k does not hold if we take in place of k in the condition $k \leqslant r \leqslant n/2$ some integer smaller than k; this follows on noting that if we let X_1, X_2, X_3, X_4 be i.i.d. random variables taking values 0 and 1 with probabilities $\sqrt{3}/(1 + \sqrt{3})$ and $1/(1 + \sqrt{3})$

respectively, then

$$E\left(\left(\sum_{l=1}^{4} X_l - (2/(1+\sqrt{3}))\right)^2 \Big| X_1 - X_2\right) = 1 + \frac{2}{1+\sqrt{3}}\left(1 - \frac{1}{1+\sqrt{3}}\right) \quad \text{a.s.}$$

(The situation here does not alter even when $2k$ in the inequality $n \geqslant 2k$ is replaced by something smaller.)

(ii) If we have the random variables $X_1, \ldots, X_n$ to be identically distributed, then (9.3.23) is equivalent to

$$E(X_1 + X_2 | X_1 - X_2) = c \quad \text{a.s.} \tag{9.3.24}$$

Under (9.3.24) or its weaker version when $X_1 - X_2$ is replaced by $(X_1 - X_2)^2$, if either $n \geqslant k$ and k is even or $n \geqslant k + 2$ and k is odd, the i.i.d. random variables $X_1, \ldots, X_n$ satisfy

$$E\left(P_k\left(\sum_{l=1}^{n} X_l\right) \Big| \Delta_k\right) = c \quad \text{a.s.,} \tag{9.3.25}$$

where $\Delta_k = ((X_1 - X_2)^2, \ldots, (X_{k-1} - X_k)^2)$ if k is even and equals

$$\left((X_1 - X_2)^2, \ldots, (X_{k-2} - X_{k-1})^2, X_k - \frac{X_{k+1} + X_{k+2}}{2}\right) \quad \text{otherwise,}$$

only if X_l's are normal. This latter result follows from the second assertion of Lemma 9.3.1 on taking m, Z_l and $\mathcal{E}_l$ appropriately in view of Shanbhag's (1978) result mentioned in the proof of Theorem 9.3.5 and the Ramachandran–Rao result appearing as Theorem 5.4.3 in Kagan et al (1973) yielding that the condition

$$E\left(X_k + X_{k+1} + X_{k+2} \Big| X_k - \frac{X_{k+1} + X_{k+2}}{2}\right) = c \quad \text{a.s.} \tag{9.3.26}$$

implies the random variables X_l to be normal. One could obviously have several other versions of Δ_k under which the result mentioned above is valid.

In view of Theorem 9.3.5 and Remark 9.3.6(ii) above it follows that if $X_1, \ldots, X_n$ are i.i.d. random variables, $\overline{X} = (1/n)\sum_{l=1}^{n} X_l$ and $\max(3, 2k) \leqslant r \leqslant n$, then

$$E(P_k(\overline{X}) | X_1 - \overline{X}, \ldots, X_r - \overline{X}) = c \quad \text{a.s.} \tag{9.3.27}$$

only if X_l are normal. The result is also clearly valid if $(X_1 - \overline{X}, \ldots, X_r - \overline{X})$ in (9.3.27) is replaced by $(X_2 - X_1, \ldots, X_r - X_1)$. These results are arrived at under more restrictive conditions by Wesolowski (1987) and Kagan and Rao (1988). (Incidentally, if $k \neq 1$, (9.3.25) implies that $E(X_l^2) < \infty$ and hence the proof of the assertion that (9.3.26) implies X_l to be normal simplifies considerably.) One may now ask the question as to whether these results are valid if $\max(3, 2k)$ is replaced by $k + 2$. The next theorem appearing in the present section answers this question. However, before dealing with it, let us give an example illustrating that (even when the supplementary condition $E(X_1 + X_2|(X_1 - X_2)^2) = c$ a.s. is used) the results do not hold if $\max(3, 2k)$ is replaced by anything smaller than $k + 2$.

Example   Let $k \geqslant 3$ be an odd integer and $n = k + 1$. Further, let $X_1, \ldots, X_n$ be i.i.d. Bernoulli random variables such that $P\{X_1 = 0\} = \frac{1}{2} = P\{X_1 = 1\}$, and let $P_k(x) = \Pi_{i=1}^{k}(x - i/(k + 1))$, $x \in \mathbb{R}$. We have

$$E(P_k(\overline{X})|X_1 - \overline{X}, \ldots, X_n - \overline{X}) = 0 \quad \text{a.s.,}$$

since the conditional expectation of $P_k(\overline{X})$ given $X_1 = \ldots = X_n$, equals zero and $P_k(\overline{X}) \equiv 0$ on $\{X_1 = \ldots = X_n\}^c$. (Also note that in the present example, $E(X_1 + X_2|(X_1 - X_2)^2) = 1$ a.s.)

We may also throw some further light on the theorem. Kagan (1989) gives, under some moment conditions, the analogue corresponding to the gamma distributions of the Kagan–Rao result mentioned earlier. This latter result has the restriction that $n \geqslant 2k$, $k \geqslant 2$. Once again, one can ask the question as to whether the Kagan result remains valid if $n \geqslant 2k$ is replaced by $n \geqslant k + 2$. That the result in this case is also not valid if we take in place of $k + 2$ anything smaller can easily be illustrated via an example analogous to that given above. Observe that if we take $Y_l = X_l + 1$, $l = 1, 2, \ldots, n$ and

$$P_k^*(x) = P_k(x - 1), \quad x \in \mathbb{R},$$

where n, k, X_l and P_k are as in the example given above, then

$$E(P_k^*(\overline{Y})|Y_1/\overline{Y}, \ldots, Y_n/\overline{Y}) = 0 \quad \text{a.s.}$$

The theorem answers both the questions raised above in the affirmative.

9.3.7   Theorem

Let k, r and n be positive integers and P_k be as defined earlier in Lemma 9.3.1. Let $X_1, \ldots, X_n$ be independent random variables and $Y_1, \ldots, Y_n$ be independent positive random variables. Then, if $k + 2 \leqslant r \leqslant n$, we have

$$E\left(P_k\left(\sum_{l=1}^{n} X_l\right)\Big| X_2 - X_1, \ldots, X_r - X_1\right) = c \quad \text{a.s.} \tag{9.3.28}$$

only if at least $r - k + 1$ of the X_l's are normal with the same variance. Also, in that case, we have

$$E\left(P_k\left(\sum_{l=1}^{n} Y_l\right)\Big| Y_2/Y_1, \ldots, Y_r/Y_1\right) = c \quad \text{a.s.} \tag{9.3.29}$$

only if at least $r - k + 1$ of the Y_l's are either all degenerate, or all gamma with the same scale parameter. (The second assertion clearly extends Theorem 6.2.1 in Kagan *et al* (1973).)

Proof   Assume that $k + 2 \leqslant r \leqslant n$. If (9.3.28) holds, then on taking $(Y_l, Z_l) = (X_l, X_l)$, Lemma 9.3.3 implies that at least $r - k + 1$ of the X_l's are normal. This, in turn, establishes the first part of the theorem in view of Lemma 9.3.4. If (9.3.29) holds, then viewing $(\log Y_l, Y_l)$ as (Y_l, Z_l) of Lemma 9.3.3, we see, in view of Corollaries 1.3.6 and 1.3.7, that at least $r - k + 1$ of the Y_l's are gamma or degenerate. Lemma 9.3.4 and a simple result that if $E(P_k(Y)|Y) = c$ a.s. for any random variable Y, then the support of the distribution of Y cannot have more than k points easily imply the second part of the theorem. ∎

9.3.8   Remark

(i) In the case of $k = 1$, $r = n$ (with obviously $n \geqslant 3$), (9.3.28) cannot be valid unless all the X_l, $l = 1, 2, \ldots, n$ are normal with the same variance; this is obviously a modified version of the celebrated Kagan–Linnik–Rao theorem. Similarly, in this special case, (9.3.29) cannot be satisfied unless the Y_l, $l = 1, 2, \ldots, n$ are either all degenerate, or all gamma with the same scale parameter. However, if $k \geqslant 2$, we come across situations wherein (9.3.28) or (9.3.29) is satisfied with some of the X_l or Y_l as nonnormal or nondegenerate nongamma. The following example provides us with an illustration.

Example   Let X_1, X_2, X_3 and X_4 be independent such that X_1, X_2 and X_3 are standard normal random variables and X_4 is a random variable with characteristic function $\cos(ct)e^{-t^2/2}$, $t \in \mathbb{R}$ for some $c \neq 0$. Also, let Y_1, Y_2, Y_3 and Y_4 be independent random variables such that Y_1, Y_2 and Y_3 are identically distributed gamma random variables with unit scale and index $\frac{1}{4}$ and Y_4 is a random variable distributed according to a mixture of unit scale gamma distributions with indices $\frac{1}{4}$ and $\frac{5}{4}$ respectively. Observe that we have here

$$E\left\{ \left(\sum_{l=1}^{4} X_l \right)^2 - (4 + c^2) \Big| X_2 - X_1, \ldots, X_4 - X_1 \right\} = 0 \quad \text{a.s.}$$

and

$$E\left\{ \left(\sum_{l=1}^{4} Y_l \right)^2 - 4 \sum_{l=1}^{4} Y_l^4 + 2 \Big| Y_2/Y_1, \ldots, Y_4/Y_1 \right\} = 0 \quad \text{a.s.}$$

(ii) In view of Theorem 9.3.7, Kagan's (1989) Theorem 2 follows even when his a priori moment condition $\int_0^\infty x^{-\varepsilon} dF(x) < \infty$ is dropped. It is also clear that an analogous result characterizing the normal distribution holds on considering the location parameter in place of the scale parameter.

One could also make some further interesting observations of relevance to the characterization problems being studied in the present section. To do this, we need a lemma.

9.3.9 Lemma

Let $m \geqslant 3$ and $(Y_1, Z_1), \ldots, (Y_m, Z_m)$ be independent two-component random vectors with all Z_l's integrable with nonzero expectations. Then, for $3 \leqslant r \leqslant m$,

$$E\left\{ \prod_1^m Z_l | Y_2 - Y_1, \ldots, Y_r - Y_1 \right\} = c \quad \text{a.s.} \tag{9.3.30}$$

only if the vectors (Y_l, Z_l), $l = 1, 2, \ldots, r$ satisfy

$$E\left(e^{itY_l} Z_l \right) = \alpha_l e^{i\gamma t} E\left(e^{itY_l} \right), \quad |t| < \varepsilon \tag{9.3.31}$$

for some real $\varepsilon > 0$ and $\alpha_l \neq 0$ and real γ.

Proof   There is no loss of generality in assuming that $r = m = 3$. Suppose $\alpha > 0$ is such that for each $t \in (-\alpha, \alpha)$ and $l = 1, 2, 3$, $E(e^{itY_l}) \neq 0$. Then, in view of Corollary 1.1.3, (9.3.30) with $r = m = 3$ implies that

$$\psi_1(-t_1 - t_2)\psi_2(t_1)\psi_3(t_2) = 1, \quad t_1, t_2, t_1 + t_2 \in (-\alpha, \alpha), \tag{9.3.32}$$

where

$$\psi_l(t) = E\left(Z_l e^{itY_l} \right) / (E(Z_l)E(e^{itY_l})), \quad t \in (-\alpha, \alpha).$$

As a corollary to Lemma 1.3.4, we then get that there exist $0 < \varepsilon \leqslant \alpha$ and some constant β such that for each l

$$\psi_l(t) = \exp\{\beta t\}, \quad |t| < \varepsilon. \tag{9.3.33}$$

(Indeed, it is an easy exercise to see without appealing to the lemma that (9.3.33) holds with $\varepsilon = \alpha$.) This, in turn, implies that (9.3.31) holds with β in place of $i\gamma$. Equating the real parts on both sides of this identity, we get that for $l = 1, 2, \ldots, r$

$$(E(Z_l))^{-1} E(Z_l \cos(tY_l)) = e^{\beta_0 t} E(\cos(t(Y_l + \gamma_l))), \quad |t| < \varepsilon \tag{9.3.34}$$

if β is of the form $\beta_0 + i\gamma$ with β_0 and γ as real. Subtracting from (9.3.34) the equation obtained from it by replacing t by $-t$, we arrive at a contradiction unless $\beta_0 = 0$ (because for any t such that $|t|$ is sufficiently small, we cannot have $E(\cos(t(Y_l + \gamma_l))) = 0$). Hence, we have the lemma. ∎

9.3.10 Corollary

If in Lemma 9.3.9, we have additionally

$$E(Z_l | Y_l) = c_0^{(l)} + c_1^{(l)} Y_l \quad \text{a.s.},$$

where $c_1^{(l)} \neq 0$, for $l = 1, 2, \ldots, r$, then (9.3.30) holds if and only if Y_l, $l = 1, 2, \ldots r$ are all degenerate random variables or are such that for some $\gamma \neq 0$, $(Y_l + c_0^{(l)}/c_1^{(l)})/\gamma$ are Poisson random variables.

Proof   There is no loss of generality in assuming that $c_0^{(l)} = 0$ and $c_1^{(l)} = 1$. Also the 'if' part follows trivially in view of Corollary 1.1.3. We need now a proof only for the 'only if' part under the stated assumption. Lemma 9.3.9 gives that if (9.3.30) is valid, then (9.3.31) is valid. In the present case (9.3.31) yields for $l = 1, 2, \ldots, r$

$$\phi_l'(t)/\phi_l(t) = i\alpha_l e^{i\gamma t}, \qquad |t| < \varepsilon, \tag{9.3.35}$$

where $\phi_l(t) = E\{\exp\{itY_l\}\}$. (9.3.35) implies that Y_l has moments of all order, with variance equal to $\gamma\alpha_l$ and other moments the same as those of a Poisson random variable with mean $\alpha_l\gamma^{-1}$, multiplied by γ. (We view here a degenerate random variable at 0 also as a Poisson random variable.) As the moments in question determine the distribution it is immediate that the assertion holds and we have the corollary. ∎

9.3.11   Remarks

(i) Corollary 9.3.10 is essentially due to Wesolowski (1989). In view of the results discussed in Section 9.2, it is easily seen that several variants and extended versions of this corollary follow from Lemma 9.3.9. For example, if we assume that (Y_l, Z_l), $l = 1, 2, \ldots, r$ satisfy additionally

$$(Y_l, Z_l) \stackrel{\mathrm{d}}{=} (\overline{X}_l, S_l^2), \qquad l = 1, 2, \ldots, r$$

where for each l, $\overline{X}_l$ and S_l^2 are the mean and the variance of a random sample of size $n_l \geqslant 2$ from a probability distribution (not necessarily the same for different l), then we get that (9.3.30) is met if and only if Y_l, $l = 1, 2, \ldots, r$ are normal or are of the form characterized in Corollary 9.3.10.

(ii) There is now the question as to whether Corollary 9.3.10 or each of the results implied in (i) above holds for sufficiently large m and r if $\Pi_{i=1}^m Z_i$ is replaced by a polynomial of $\Pi_{i=1}^m Z_i$ of degree greater than or equal to 2. That the answer to this question is in the negative is shown by the next example.

Example   To get a negative result, it is sufficient if we know that for independent Poisson random variables with mean λ, the condition that

$$E\left\{ \left(\prod_{l=1}^2 Y_l \right)^2 \Big| Y_1 - Y_2 \right\} = c \quad \text{a.s.} \tag{9.3.36}$$

does not hold. (This, in turn, would imply that for any $n \geqslant 2$, if $Y_1, \ldots, Y_n$ are independent Poisson random variables with mean λ, then the condition that

$$E\left\{ \left(\prod_{l=1}^2 Y_l \right)^2 \Big| Y_2 - Y_1, \ldots, Y_n - Y_1 \right\} = c \quad \text{a.s.}$$

does not hold.) (9.3.36) is seen, in view of Corollary 1.1.3, to be equivalent to

$$(\phi(t)\phi(-t))^{-1} \left(\frac{\mathrm{d}^2\phi(t)}{\mathrm{d}t^2} \right) \left(\frac{\mathrm{d}^2\phi(-t)}{\mathrm{d}t^2} \right) = (E(Y_1^2))^2, \quad t \in \mathbb{R}, \tag{9.3.37}$$

where

$$\phi(t) = \mathrm{e}^{-\lambda+\lambda e^{it}}, \quad t \in \mathbb{R}.$$

It is an easy exercise now to see that (9.3.37) does not hold and hence that the answer to the question is in the negative.

9.4 CHARACTERIZATIONS OF MIXTURES BASED ON EXCHANGEABILITY

Kingman (1972) and more recently Smith (1981) have characterized mixtures of normal distributions based on infinite sequences of exchangeable random variables via the specialized version of Theorem 1.2.7 for $S = \mathbb{R}$. Alzaid *et al* (1990) have shown, using the specialized version of the theorem for $S = \mathbb{R}$ (or that for $S = \mathbb{R}^p$) that characterizations of the type met in Sections 9.2 and 9.3 (as well as in Section 6.4) based on independent identically distributed random variables (or random vectors) can easily be translated into characterizations of mixtures of probability distributions based on sequences of exchangeable random variables (or random vectors). Results given by these latter authors subsume those given by Kingman, Smith and several others. In view of what is revealed in Smith (1981), it is clear that these results have a strong statistical motivation in Bayesian inference.

   We devote the present section to studying mainly the findings of Alzaid *et al* (1990), and their implications.

   Let m be a positive integer and B be a nonempty Borel subset of $\mathbb{R}^m$. Further, let $\{F(\bullet|\mathbf{b}) : \mathbf{b} \in B\}$ be a family of probability distributions such that $F(x|\bullet)$ is Borel measurable on B for each $x \in \mathbb{R}$ and $\left\{ \left(\psi_\gamma^{(1)}, \phi_\gamma^{(1)}, \psi_\gamma^{(2)}, \phi_\gamma^{(2)} \right) : \gamma \in \Gamma \right\}$ be a countable family of vectors of real-valued Borel measurable functions on $\mathbb{R}^\infty$ such that for every probability measure on the Borel σ-field of $\mathbb{R}^\infty$ for which the projection maps are i.i.d. and for every $\gamma \in \Gamma$, $(\psi_\gamma^{(1)}, \phi_\gamma^{(1)})$ and $(\psi_\gamma^{(2)}, \phi_\gamma^{(2)})$ are i.i.d. Let $\xi = (\xi_1, \ldots, \xi_m)$ be a vector of extended real-valued tail Borel measurable functions on $\mathbb{R}^\infty$.

   Using the notation introduced, we can now give the following.

9.4.1 Theorem

Suppose any infinite sequence $\{X_n : n = 1, 2, \ldots\}$ of i.i.d. random variables satisfies the condition that

$$E\{\phi_\gamma^{(1)}(\mathbf{X})|\psi_\gamma^{(1)}(\mathbf{X})\} = 0, \quad \gamma \in \Gamma \quad \text{a.s.}, \tag{9.4.1}$$

where $\mathbf{X} = (X_1, X_2, \ldots)$, only if there exists a point $\mathbf{b} \in B$ such that $P\{\xi(\mathbf{X}) = \mathbf{b}\} = 1$ and the d.f. of X_1 is $F(\bullet|\mathbf{b})$. Then if $\{Y_n : n = 1, 2, \ldots\}$ is a sequence of exchangeable random variables, the equation

$$E\left\{\phi_\gamma^{(1)}(\mathbf{Y})\phi_\gamma^{(2)}(\mathbf{Y})|\psi_\gamma^{(1)}(\mathbf{Y}) - \psi_\gamma^{(2)}(\mathbf{Y})\right\} = 0, \quad \gamma \in \Gamma \quad \text{a.s.}, \qquad (9.4.2)$$

where $\mathbf{Y} = (Y_1, Y_2, \ldots)$, is valid only if $\xi(\mathbf{Y}) \in B$ a.s. and conditional upon $\xi(\mathbf{Y})$, the random variables Y_i's are i.i.d. with d.f. $F(\bullet|\xi(\mathbf{Y}))$ a.s.

Proof   In view of Corollary 1.1.3, it follows that (9.4.2) is equivalent to

$$E\left\{e^{it(\psi_\gamma^{(1)}(\mathbf{Y}) - \psi_\gamma^{(2)}(\mathbf{Y}))}\phi_\gamma^{(1)}(\mathbf{Y})\phi_\gamma^{(2)}(\mathbf{Y})\right\} = 0, \quad \gamma \in \Gamma \quad -\infty < t < \infty, \quad (9.4.3)$$

where the left-hand side of the identity is to be understood as the one that is well defined and equal to the right-hand side of the identity. (Note that (9.4.3) assumes implicitly that $E\left\{|\phi_\gamma^{(1)}(\mathbf{Y})\phi_\gamma^{(2)}(\mathbf{Y})|\right\} < \infty$ for each $\gamma \in \Gamma$.) The specialized version of Theorem 1.2.7 for $S = \mathbb{R}$ implies that if (9.4.3) holds with Y obviously as mentioned in the statement of the theorem, then $\phi_\gamma^{(1)}(\mathbf{Y})$ are integrable for all $\gamma \in \Gamma$ and

$$\left|E\left\{e^{it\psi_\gamma^{(1)}(\mathbf{Y})}\phi_\gamma^{(1)}(\mathbf{Y})|\mathcal{I}\right\}\right|^2 = 0, \quad \gamma \in \Gamma \quad \text{a.s.} \quad -\infty < t < \infty, \qquad (9.4.4)$$

where $\mathcal{I}$ is the tail σ-field relative to $\{Y_n : n = 1, 2, \ldots\}$. Appealing to (9.4.4) we can see that there exists, in that case, a regular conditional probability corresponding to Y given $\mathcal{I}$, such that relative to it coordinate mappings are i.i.d., and

$$\left|E\left\{e^{it(\psi_\gamma^{(1)}(\mathbf{Y})}\phi_\gamma^{(1)}(\mathbf{Y})|\mathcal{I}\right\}\right|^2 = 0, \quad -\infty < t < \infty, \quad \gamma \in \Gamma \qquad (9.4.5)$$

with the expectation computed through the regular probability. In view of Corollary 1.1.3 and the assumption in the theorem, it then follows that this regular conditional probability is such that $P\{\xi(Y) \in B|\mathcal{I}\} = 1$ and the conditional d.f. of Y_1 given $\mathcal{I}$ is given by $E\{F(\bullet|\xi(Y))|\mathcal{I}\}$ and, hence except on a null set, by $F(\bullet|\xi(Y))$, where $F(\bullet|\xi(Y))$, for $\xi(Y) \notin B$ is to be understood as some fixed d.f. on $\mathbb{R}$ and the quantities involving $\mathcal{I}$ as those computed through the regular conditional probability. The assertion of the theorem is now obvious. ■

9.4.2   Remarks

(i) If $\psi_\gamma^{(1)}$, $\psi_\gamma^{(2)}$ for each $\gamma \in \Gamma$ in the above theorem are both nonnegative or both nonpositive, then the theorem is also valid with $\psi_\gamma^{(1)}(\mathbf{Y}) - \psi_\gamma^{(2)}(\mathbf{Y})$ replaced by $\psi_\gamma^{(1)}(\mathbf{Y}) + \psi_\gamma^{(2)}(\mathbf{Y})$ for some or all $\gamma \in \Gamma$.

(ii) If $\psi_\gamma^{(1)}$, $\psi_\gamma^{(2)}$ are independent of γ and Γ is finite, then the theorem remains valid if (9.4.2) is replaced by

$$\sum_{\gamma \in \Gamma} E\left\{\phi_\gamma^{(1)}(\mathbf{Y})\phi_\gamma^{(2)}(\mathbf{Y})|\psi^{(1)}(\mathbf{Y}) - \psi^{(2)}(\mathbf{Y})\right\} = 0, \quad \text{a.s.}, \qquad (9.4.6)$$

with $\phi_\gamma^{(1)}(\mathbf{Y})\phi_\gamma^{(2)}(\mathbf{Y})$ integrable for all γ, where $\psi^{(1)}$ and $\psi^{(2)}$ denote the functions $\psi_\gamma^{(1)}$ and $\psi_\gamma^{(2)}$ that are independent of γ. Additionally, if we have $\psi^{(1)}$ and $\psi^{(2)}$ both nonnegative or both nonpositive, then the result in question remains valid when $\psi^{(1)}(\mathbf{Y}) - \psi^{(2)}(\mathbf{Y})$ in (9.4.6) is replaced by $\psi^{(1)}(\mathbf{Y}) + \psi^{(2)}(\mathbf{Y})$.

(iii) If the sequence $\{Y_n : n = 1, 2, \ldots\}$ of exchangeable random variables is taken such that for two distinct real values of θ

$$E\left\{e^{\theta(\psi_\gamma^{(1)}(\mathbf{Y})+\psi_\gamma^{(2)}(\mathbf{Y}))}\,|\phi_\gamma^{(1)}(\mathbf{Y})\phi_\gamma^{(2)}(\mathbf{Y})|\right\} < \infty, \quad \gamma \in \Gamma,$$

then the above theorem remains valid with the left hand side of (9.4.1) replaced by the means of the conditional distributions of $\phi_\gamma^{(1)}(\mathbf{X})$ given $\psi_\gamma^{(1)}(\mathbf{X})$ for $\gamma \in \Gamma$ and simultaneously the left-hand side of (9.4.2) replaced by the means of the conditional distributions of $\phi_\gamma^{(1)}(\mathbf{Y})\phi_\gamma^{(2)}(\mathbf{Y})$ given $\psi_\gamma^{(1)}(\mathbf{Y}) + \psi_\gamma^{(2)}(\mathbf{Y})$ for $\gamma \in \Gamma$. (If additionally Γ is finite and $\psi_\gamma^{(1)}$ and $\psi_\gamma^{(2)}$ are independent of γ, then in place of the modified (9.4.2), we can even take the condition that the sum of the expected values involved in the condition be equal to zero a.s.)

(iv) If it is assumed that $E\{|\phi_\gamma^{(1)}(\mathbf{Y})\phi_\gamma^{(2)}(\mathbf{Y})|\} < \infty$ for all $\gamma \in \Gamma$, then the results of Theorem 9.4.1 and (iii) above hold with 'only if' in two places replaced by 'if and only if'.

(v) Versions of Theorem 9.4.1 and of the various results appearing in (i) to (iv) above when Y_i's are p-component vectors and/or when $\psi_\gamma^{(1)}$ and $\psi_\gamma^{(2)}$ are functions with values in $\mathbb{R}^{n_\gamma}$ for each $\gamma \in \Gamma$ are now easy to obtain.

(vi) Theorem 9.4.1 and Remarks 9.4.2, (i)–(iv) also clearly hold if we take ξ_r's as invariant Borel measurable functions in place of that they are tail Borel measurable functions.

Kingman (1972) showed that if $\{Y_n : n = 1, 2, \ldots\}$ is a sequence of exchangeable random variables such that $(Y_1, \ldots, Y_4)$ has a spherical distribution, then there exists a nonnegative random variable V, such that conditional on V the Y_n's are independent $N(0, V)$ random variables (a.s.). (A somewhat different version of Kingman's result appears in Kelker (1970).) Smith (1981) extended Kingman's ideas to show essentially that if $\{Y_n : n = 1, 2, \ldots\}$ is a sequence of exchangeable random variables such that $(Y_1, \ldots, Y_8)$ has a central spherically symmetric distribution, then there exist random variables M, V with $V \geqslant 0$ such that conditional on (M, V), the Y_n's are independent $N(M, V)$ random variables. (It may be noted that in Kingman's and Smith's results degenerate distributions are allowed to be called normal; also it is easily seen that the resulting sequence $\{Y_n\}$ in Kingman's case has for each n the distribution of $(Y_1, \ldots, Y_n)$ to be spherical and, in Smith's case, for each n the distribution of the vector to be centered spherically symmetric.)

From Box and Hunter (1957), in view of the representation (6.3.3) with $C = I$ relative to s.s.d's, it is clear that if (Y_1, Y_2, Y_3, Y_4) is spherically symmetric, then (for some version)

$$\mu_{1,1,0,0}^{(t)} + \mu_{4,4,0,0}^{(t)} - 6\mu_{4,2,2,0}^{(t)} + 9\mu_{2,2,2,2}^{(t)} = 0, \quad t \geqslant 0, \tag{9.4.7}$$

where $\mu_{l_1 l_2 l_3 l_4}^{(t)}$ are the product moments corresponding to the conditional distribution of (Y_1, Y_2, Y_3, Y_4) given $Y_1^2 + \cdots + Y_4^2 = t$. From the definition of the centered spherical symmetry in Smith (1981), it follows that if $Y_1, \ldots, Y_8$ is centered spherically symmetric, (9.4.7) is valid with $\mu_{l_1 l_2 l_3 l_4}^{(t)}$ as the product moments corresponding to the conditional distribution of

$$(Y_1 - Y_2, Y_3 - Y_4, Y_5 - Y_6, Y_7 - Y_8),$$

given

$$(Y_1 - Y_2)^2 + (Y_3 - Y_4)^2 + (Y_5 - Y_6)^2 + (Y_7 - Y_8)^2 = t.$$

The next two results may then be viewed as corollaries to Theorem 9.4.1, subject to the modification stated in Remark 9.4.2(iii). These results extend respectively those given by Kingman (1972) and Smith (1981).

9.4.3   Corollary

Let $\{Y_n : n = 1, 2, \ldots\}$ be a sequence of exchangeable random variables. Then, defining $\mu_{l_1 l_2 l_3 l_4}^{(t)}$ to be the product moments corresponding to the conditional distribution of (Y_1, Y_2, Y_3, Y_4) given $Y_1^2 + Y_2^2 + Y_3^2 + Y_4^2 = t$, (9.4.7) is valid if and only if there exists a nonnegative random variable V such that conditional on V, the random variables Y_n's are independently distributed $N(0, V)$ random variables (a.s.).

Proof   The 'if' part is trivial. We shall now establish the 'only if' part. Take $m = 1$, and that $\xi_1(\mathbf{x}) = \lim_{n \to \infty}(1/n)\sum_1^n x_i^2$ if the limit exists, and equals zero otherwise. Define $\{F(\bullet|b)\}$ to be the family of normal probability distribution functions with zero mean. Define

$$\Gamma = \{1, 2\},$$

$$\phi_1^{(1)}(\mathbf{x}) = x_1^4 - 3(x_1 x_2)^2, \qquad \phi_2^{(1)}(\mathbf{x}) = x_1,$$

$$\phi_1^{(2)}(\mathbf{x}) = x_3^4 - 3(x_3 x_4)^2, \qquad \phi_2^{(2)}(\mathbf{x}) = x_3,$$

$$\psi_\gamma^{(1)}(\mathbf{x}) = x_1^2 + x_2^2, \qquad \gamma = 1, 2,$$

$$\psi_\gamma^{(2)}(\mathbf{x}) = x_3^2 + x_4^2, \qquad \gamma = 1, 2.$$

We have here the situation of Remark 9.4.2(iii) (with $\psi_\gamma^{(1)}$ and $\psi_\gamma^{(2)}$ independent of γ and Γ finite) with a replacement for (9.4.2) as specified. For any sequence $\{X_n : n = 1, 2, \ldots\}$ of i.i.d. random variables, (9.4.1) with the modification of Remark 9.4.2(iii) implies inductively that X_1 has moments of each order with odd-order moments equal to zero and with even-order moments to be either all equal to zero or the same as those of the square root of a gamma random variable with index $\frac{1}{2}$. (Observe that

$$E(\phi_1^{(1)}(\mathbf{X})|\psi_1^{(1)}(\mathbf{X})) = 0 \quad \text{a.s.} \quad \Rightarrow \alpha''(\theta)\alpha(\theta) - 3(\alpha'(\theta))^2 = 0,$$

where $\alpha(\theta) = E(e^{-\theta X_1^2}) \Rightarrow X_1^2$ is a constant multiple of a χ^2 random variable with one degree of freedom, and $E(\phi_2^{(1)}(\mathbf{X})|\psi_2^{(1)}(\mathbf{X})) = 0$ a.s. means that $E(X_1|X_1^2 + X_2^2) = 0$ a.s. implying, when $E(X_1^{2m}) < \infty$ for all positive integer m, that all odd-order moments of X_1 equal zero.) Consequently, we have X_1 to be normal with mean 0 and variance $\xi_1(\mathbf{X})$ a.s. The result of the corollary is now obvious. ∎

9.4.4　Corollary

Let $\{Y_n : n = 1, 2, \ldots\}$ be a sequence of exchangeable random variables and let $\mu_{l_1 l_2 l_3 l_4}^{(t)}$ be the product moments corresponding to the conditional distribution of $(Y_1 - Y_2, Y_3 - Y_4, Y_5 - Y_6, Y_7 - Y_8)$, given

$$(Y_1 - Y_2)^2 + (Y_3 - Y_4)^2 + (Y_5 - Y_6)^2 + (Y_7 - Y_8)^2 = t.$$

Then (9.4.7) is valid if and only if there exists a random vector (M, V), with $V \geq 0$, such that conditional upon (M, V) the Y_i's are independently distributed as normal with mean M and variance V (a.s.).

Proof　(The 'if' part is trivial.) To prove the 'only if' part, take $m = 2$, $\xi_1(\mathbf{x}) = \lim_{n \to \infty}(1/n)\Sigma_1^n x_i$ if the limit exists, and equals zero otherwise, take $\xi_1(\mathbf{x}) = \lim_{n \to \infty}(1/n)\Sigma_1^n x_i^2 - (\xi_1(\mathbf{x}))^2$ if the limit exists, and equals zero otherwise, and take $\{F(\bullet|b_1, b_2)\}$ to be the family of normal probability distribution functions. Define

$$\Gamma = \{1, 2\},$$

$$\phi_1^{(1)}(\mathbf{x}) = (x_1 - x_2)^4 - 3(x_1 - x_2)^2(x_3 - x_4)^2,$$

$$\phi_1^{(2)}(\mathbf{x}) = (x_5 - x_6)^4 - 3(x_5 - x_6)^2(x_7 - x_8)^2,$$

$$\phi_1^{(2)}(\mathbf{x}) = (x_1 - x_2), \qquad \phi_2^{(2)}(\mathbf{x}) = (x_5 - x_6),$$

$$\psi_\gamma^{(1)}(\mathbf{x}) = \sum_{i=1}^{2}(x_{2i-1} - x_{2i})^2, \qquad \gamma = 1, 2,$$

$$\psi_\gamma^{(2)}(\mathbf{x}) = \sum_{i=3}^{4}(x_{2i-1} - x_{2i})^2, \qquad \gamma = 1, 2.$$

Essentially the argument in the proof of Corollary 9.4.3 implies that $X_1 - X_2$ is normal with zero mean and variance $2\xi_2(X)$ a.s. if (9.4.1) with the modification of Remark 9.4.2(iii) is valid. Consequently, it follows that if the modified version of (9.4.1) holds, then X_1 is normal with mean $\xi_1(X)$ and variance $\xi_2(X)$ a.s. The required result is then immediate. ∎

　　In view of the results given in Sections 9.2 and 9.3. of the present monograph, and the results in several other places, it is obvious that one can arrive at various other characterizations on mixtures of probability distributions on the basis of exchangeable random variables. In particular, the resulting sequence of Corollary 9.4.4 is characterized, under a mild restriction that $Y_1 Y_2$ is square integrable, by the condition that $\{Y_n\}_1^\infty$ is an exchangeable sequence of random variables satisfying for some $k \geqslant 2$,

$$E\{(S_1^2 - \xi_2(\mathbf{Y}))(S_2^2 - \xi_2(\mathbf{Y}))|\overline{Y}_1 - \overline{Y}_2\} = 0 \quad \text{a.s.},$$

where ξ_2 is as defined in the proof of Corollary 9.4.4,

$$\overline{Y}_1 = \frac{1}{k}\sum_{n=1}^k Y_n, \qquad \overline{Y}_2 = \frac{1}{k}\sum_{n=1}^k Y_{n+k},$$

$$S_1^2 = \frac{1}{k-1}\sum_{n=1}^k (Y_n - \overline{Y}_1)^2, \qquad S_2^2 = \frac{1}{k-1}\sum_{n=1}^k (Y_{n+k} - \overline{Y}_2)^2.$$

In the same notation and under the same restriction, we have $\{Y_n : n = 1, 2, \ldots\}$ to be a sequence of exchangeable random variables satisfying

$$E\{(S_1^2 - \overline{Y}_1)(S_2^2 - \overline{Y}_2)|\overline{Y}_1 - \overline{Y}_2\} = 0 \quad \text{a.s.}$$

(or, alternatively, satisfying

$$E\{(S_1^2 - \overline{Y}_1)(S_2^2 - \overline{Y}_2)|\overline{Y}_1 + \overline{Y}_2\} = 0 \quad \text{a.s.} \tag{9.4.8}$$

when the restriction of square integrability of $Y_1 Y_2$ is replaced by the restriction that Y_n's are nonnegative and the conditional expectations are taken as in Remark 9.4.2(iii)) if and only if there exists a nonnegative random variable W such that conditional upon W, Y_n's are independent Poisson random variables with mean W a.s. (We allow here the degenerate distribution at zero to be called Poisson; it is also worth pointing out that (9.4.8) is implied in particular by the condition that Y_n's are such that the conditional distribution of $(Y_1, \ldots, Y_{2k})$ given $\Sigma_1^{2k} Y_i$ is multinomial a.s.).

9.5   FURTHER STATISTICAL RESULTS

Restricting himself to a certain exponential family, Shanbhag (1972, 1979) has essentially obtained a variant of the result in Corollary 9.2.5 (expressed in the form in Remark 9.2.6(i)). He showed that the diagonality of the 3×3 Bhattacharyya matrix characterizes the class of the normal, Poisson, gamma, binomial, negative binomial and Meixner hypergeometric distributions and the distributions obtained from them or their conjugates by making scale and location changes. From what is seen in Seth (1949), it follows easily that for each distribution in the class just mentioned, the Bhattacharyya matrix of any order is (defined and) diagonal. As the Bhattacharyya bounds involve the inverse of the Bhattacharyya matrix, and

they improve the Cramér–Rao lower bound for the variance of an estimator of a
real parameter and become sharper and sharper as the order of the Bhattacharyya
matrix increases, the Seth–Shanbhag findings could be seen to be addressing a
natural and important question. Several authors including, among others, Whittaker
(1973), Blight and Rao (1974), Alzaid (1983), Bartoszewicz (1980), Lai (1982 a),
(1982 b), Morris (1982, 1983), Ghosh and Sathe (1987), Targhetta (1990) and
Ghosh and Mukerjee (1991) have since given variants, extensions and applications
of the Seth–Shanbhag results. (See also Alharbi (1994) for some observations in
this connection.)

In order to state the Seth–Shanbhag results explicitly, which we shall do through
a theorem, we need some notation: let $(\mathbb{R}^n, \mathfrak{B}_n, \mu)$ be a measure space and $\{F_\theta :
\theta \in (\theta_0, \theta_1)\}$ (where $\theta_0 = -\infty$ and/or $\theta_1 = +\infty$ are allowed) be a family of
(distinct) nondegenerate probability distribution functions such that

$$F_\theta(x) = \int_{\{y \in \mathbb{R}^n : y \leqslant x\}} f_\theta(y)\mu(\mathrm{d}y), \quad x \in \mathbb{R}^n, \qquad (9.5.1)$$

where $\mathfrak{B}_n$ is the Borel σ-field of $\mathbb{R}^n$ and f_θ is a nonnegative real Borel measurable
function. (We may view F_θ as the d.f. of a sample in which the observations are
not necessarily i.i.d., and f_θ as a probability density function of F_θ with respect to
the measure μ, although we have not assumed μ to be σ-finite here as is usually
done.) Consider henceforth that F_θ is given by (9.5.1) with

$$f_\theta(x) = \exp\{\phi(x)g(\theta)\}\psi(x)/B(g(\theta)), \quad x \in \mathbb{R}^n, \qquad (9.5.2)$$

where ϕ is a real Borel measurable function, ψ is a nonnegative real Borel measurable function and $g(\theta)$ is a thrice differentiable function of θ. In that case, the
vector $(f_\theta^{(1)}/f_\theta, f_\theta^{(2)}/f_\theta, f_\theta^{(3)}/f_\theta)$, where $f_\theta^{(r)} = \partial^r f_\theta/\partial\theta^r$, $r = 1, 2, 3$, defines
almost surely a vector of random variables on the probability space $(\mathbb{R}^n, \mathfrak{B}_n, P_\theta)$,
where P_θ is the probability measure determined by F_θ on $\mathfrak{B}_n$. It is easily seen
that the covariance matrix of this vector, which indeed is our 3×3 Bhattacharyya
matrix, is defined. It also follows that if $g(\theta)$ is differentiable s times, then the
$s \times s$ Bhattacharyya matrix, i.e. the covariance matrix of $(f_\theta^{(1)}/f_\theta, \ldots, f_\theta^{(s)}/f_\theta)$ (in
obvious notation), is defined. (Here, it is implicit that $(f_\theta^{(1)}/f_\theta, \ldots, f_\theta^{(s)}/f_\theta)$ defines
almost surely a vector of random variables on $(\mathbb{R}^n, \mathfrak{B}_n, P_\theta)$, with P_θ as defined
before.)

We are now in a position to give the theorem in question.

9.5.1   Theorem

Define for real c, d,

$$F_\theta^*(x|c, d) = P_\theta(\{c\phi + d \leqslant x\}), \quad x \in \mathbb{R},$$

with P_θ as stated above. If the 3×3 Bhattacharyya matrix is diagonal (i.e. on
(θ_0, θ_1)), then, for some $c \neq 0$ and some d, we have $F_\theta^*(.|c, d)$ to be either

$N(\theta, \sigma^2)$ for some σ^2 independent of θ or Poisson with mean θ^* or, with mean θ^* and index independent of θ, either gamma or binomial or negative binomial or Meixner hypergeometric, where the notation θ^* is taken to mean a linear function of θ (that could be different in different contexts) with a nonzero coefficient for θ. Moreover, in each of these six cases, the Bhattacharyya matrix of any order is defined and is diagonal.

Proof There is no loss of generality in assuming $n = 1$ and $\phi(x) = x$, $x \in \mathbb{R}$ with $\psi \equiv 1$, (in which case we have trivially μ to be σ-finite). To prove the first assertion, denote by $J_{rs}(= J_{rs}(\theta))$ the (r, s)th element of the 3×3 Bhattacharyya matrix, for $r, s = 1, 2, 3$. We have

$$J_{rs} = \text{cov}_\theta \left(\frac{f_\theta^{(r)}}{f_\theta}, \frac{f_\theta^{(s)}}{f_\theta} \right)$$

$$= E_\theta \left(\frac{f_\theta^{(r)}}{f_\theta} \frac{f_\theta^{(s)}}{f_\theta} \right)$$

because

$$E_\theta \left(\frac{f_\theta^{(k)}}{f_\theta} \right) = 0, \quad k = 1, 2, 3. \tag{9.5.3}$$

Clearly, if the 3×3 Bhattacharyya matrix is diagonal, then $J_{21} = J_{32} = 0$. We have

$$J_{21} = E_\theta \left\{ \frac{f_\theta^{(2)}}{f_\theta} \left(g'(\theta)X - \frac{B'(g(\theta))g'(\theta)}{B(g(\theta))} \right) \right\}$$

$$= g'(\theta) E_\theta \left\{ \frac{f_\theta^{(2)}}{f_\theta} X \right\} - \frac{B'(g(\theta))g'(\theta)}{B(g(\theta))} E_\theta \left\{ \frac{f_\theta^{(2)}}{f_\theta} \right\},$$

where B' and g' are the derivatives of B and g respectively and $X(= X_\theta)$ is a random variable with d.f. F_θ. Because of (9.5.3), we see that

$$J_{21} = g'(\theta) E_\theta \left\{ \frac{f_\theta^{(2)}}{f_\theta} X \right\}$$

$$= g'(\theta) \frac{d^2}{d\theta^2} E_\theta(X). \tag{9.5.4}$$

As, under the given assumptions, we cannot have a nondegenerate subinterval of (θ_0, θ_1) on which g' vanishes identically, and we have $(d^2/d\theta^2)E_\theta(X)$ to be continuous, it follows from (9.5.4) that $J_{21} = 0$ if and only if $(d^2/d\theta^2)E_\theta(X) = 0$, i.e. if and only if

$$E_\theta(X) = c_{11} + c_{21}\theta, \tag{9.5.5}$$

where c_{11} and c_{21} are constants (with $c_{21} \neq 0$) not depending on θ, implying that $g'(\theta) = c_{21}/\mathrm{var}_\theta(X)$ and hence that g' is nonvanishing. Writing $f_\theta^{(2)}(x)/f_\theta(x)$ as $\Sigma_{l=0}^2 d_l(\theta)x^l$, where $d_l(\theta)$ is the coefficient of x^l we see that

$$J_{32} = E_\theta \left(\frac{f_\theta^{(3)}}{f_\theta} \sum_{l=0}^2 d_l(\theta)X^l \right)$$

$$= \sum_{l=0}^2 d_l(\theta) \left(\frac{f_\theta^{(3)}}{f_\theta} X^l \right)$$

$$= \sum_{l=1}^2 d_l(\theta) \frac{d^3}{d\theta^3} E_\theta(X^l)$$

because of (9.5.3). Since $d_2(\theta) = (g'(\theta))^2$, it follows that if $(d^2/d\theta^2)E_\theta(X) = 0$, or equivalently if (9.5.5) is valid, then

$$J_{32} = (g'(\theta))^2 \frac{d^3}{d\theta^3} E_\theta(X^2)$$

and hence $J_{32} = 0$ if and only if

$$\frac{d^3}{d\theta^3} E_\theta(X^2) = 0.$$

This last equation is equivalent to

$$E_\theta(X^2) = c_{12} + c_{22}\theta + c_{32}\theta^2, \qquad (9.5.6)$$

where $c_{j2}(j = 1, 2, 3)$ do not depend on θ. Following the procedure used to establish (9.5.4), we see that

$$J_{31} = g'(\theta) \frac{d^3}{d\theta^3} E_\theta(X)$$

and hence that (9.5.5) implies that $J_{31} = 0$. Thus, it follows, from what we have seen, that the 3×3 Bhattacharyya matrix is diagonal if and only if (9.5.5) and (9.5.6) are met. The first assertion is then obvious, in view of the result of Corollary 9.2.5 (taken in the form in Remark 9.2.6(i)).

To prove the second assertion, note that in the six cases involved,

$$E_\theta(X^r) = \sum_{l=0}^r c_l^{(r)}\theta^l, \qquad (9.5.7)$$

where $c_l^{(r)}$'s are constants not depending on θ. Furthermore, in these cases

$$\frac{f_\theta^{(r)}(x)}{f_\theta(x)} = \sum_{l=0}^r d_l^{(r)}(\theta)x^l, \qquad (9.5.8)$$

where $d_l^{(r)}(\theta)$ is a function of θ only, with $d_r^{(r)}(\theta) = (g'(\theta))^r$. (9.5.7) and (9.5.8) imply that

$$E_\theta \left(\frac{f_\theta^{(r')}}{f_\theta} \frac{f_\theta^{(r)}}{f_\theta} \right) = \sum_{l=0}^{r} d_l^{(r)}(\theta) E_\theta \left(\frac{f_\theta^{(r')}}{f_\theta} X^l \right)$$

$$= \sum_{l=0}^{r} d_l^{(r)}(\theta) \frac{d^{r'}}{d\theta^{r'}} E_\theta(X^l)$$

$$= 0 \quad \text{if } r' = r+1, r+2, \ldots .$$

Because of symmetry, it then follows that

$$E_\theta \left(\frac{f_\theta^{(r)}}{f_\theta} \frac{f_\theta^{(r')}}{f_\theta} \right) = 0 \quad \text{if } r \neq r';$$

this implies that in all the cases that we have in the statement of the theorem, the Bhattacharyya matrix of any order is defined and is diagonal. Hence, we have the theorem. ∎

One may now ask the question as to whether the first assertion of Theorem 9.5.1 holds if 3×3 is replaced 2×2. That the result with the alteration does not hold is shown by the following example of Shanbhag and Kapoor (1993).

Example Let M be the moment generating function of a nondegenerate distribution with an open interval containing the origin as its domain of definition. Define g with appropriate domain of definition, such that, for some real c and k and each t,

$$kg^{-1}(t+c) = \frac{M'(t)}{M(t)};$$

note that g is well defined because M is strictly logconvex on its domain of definition. If we consider the exponential family

$$F_\theta(dx) = \frac{\exp\{g(\theta)x\}}{B(g(\theta))} \mu(dx), \quad x \in \mathbb{R},$$

where μ is such that $(exp\{cx\})\mu(dx)$ is the distribution corresponding to M and $B(g(\theta)) = M(g(\theta) - c)$ (with the domain of definition of g to be the parameter space), we get the corresponding mean function to be $k\theta$. In view of what we have seen in the proof of the first assertion of Theorem 9.5.1 above, it is then clear that the 2×2 Bhattacharyya matrix in this case is diagonal. The exponential family arrived at here is by no means as restrictive as that appearing in the statement of Theorem 9.5.1, relative to which the 3×3 Bhattacharyya matrix is defined and is diagonal.

9.5.2   Remarks

(i) Amongst the distributions that appear in Theorems 9.2.1 and 9.5.1, we have an absolutely continuous stable distribution (normal), a discrete stable distribution (Poisson), absolutely continuous nonstable self-decomposable distributions (gamma and Meixner hypergeometric), a discrete nonstable self-decomposable distribution (negative binomial) and a noninfinitely divisible distribution (binomial). Each of these distributions is either a lattice or an absolutely continuous unimodal distribution; some of these are even strongly unimodal distributions (either in the class of absolutely continuous distributions or in the class of lattice distributions). For the details of some of the definitions that we have assumed in this remark, especially those concerning discrete or lattice distributions, see van Harn (1978).

(ii) From the proof of the second assertion in Theorem 9.5.1, it is clear that for any positive integer r, the (r, r)th element of the $s \times s$ Bhattacharyya matrix, where $s \geqslant r$, in each of the six cases arrived at in the first assertion of the theorem, is given by

$$(g'(\theta))^r \frac{\mathrm{d}^r}{\mathrm{d}\theta^r} E_\theta((\phi(X))^r) \quad \text{or} \quad (g'(\theta))^r r! c_r^{(r)},$$

where $c_r^{(r)}$ is as in (9.5.7) (but with $\phi(X)$ in place of X). In Seth (1949) or Blight and Rao (1974), one could find more definite expressions in this connection; see also Alzaid (1987), who derives an expression in the case of the Meixner hypergeometric distribution. Blight and Rao (1974) and so also Morris (1982, 1983) contain some other interesting results related to Theorem 9.5.1.

(iii) Amongst the distributions that we have encountered in Theorem 9.5.1, the binomial distribution possesses the property that for some s, the corresponding $s \times s$ Bhattacharyya matrix is singular.

In addition to the statistical characterization result that we have met here, there exist several others that are of interest. In particular, the results of Section 9.4 allow us to characterize distributions of samples in Bayesian inference, that are nice in some sense. Chernoff's (1981) inequality and various extensions of it provide us with characterizations that have statistical implications; in many of their papers, Cacoullos, Prakasa Rao, Sreehari and coauthors have dealt with these. There are also some characterizations based on the Fisher information (see, for example, Johnstone and MacGibbon (1984) and Targhetta (1990)).

BIBLIOGRAPHY

The list of references given below covers the papers and books cited in the text, and also some not cited but related to the subject matter of discussion in the monograph.

Abdul-Razak, R.S. (1983) Power series distributions in mathematical statistics and applied probability, Ph.D. thesis, Penn State University.

Aczel, J. (1966) *Lectures on Functional Equations and their Applications.* Academic Press, New York.

Adatia, A., Law, A.G. and Wang, Q. (1991) Characterization of a mixture of gamma distributions via conditional finite moments. *Commun. Statist. Theory Method*, **20**, No. 5 and 6, 1937–1949.

Ahmed, A.N. (1991) Characterization of beta, binomial and Poisson distributions. *IEEE Trans. Reliability*, **3**, 290–295.

Ahsanullah, M. (1975) A characterization of the exponential distribution. *Statistical Distributions in Scientific Work*, Vol. 3, pp. 131–135, eds. Patil, G.P., Kotz, S. and Ord, J.K. Dordrecht, Reidel.

Ahsanullah, M. (1978) A characterization of the exponential distribution by spacing of order statistics. *J. Appl. Prob.*, **15**, 650–653.

Ahsanullah, M. (1984) A characterization of the exponential distribution by higher order gap. *Metrika*, **31**, 323–326.

Ahsanullah, M. (1987) Record statistics and the exponential distribution. *Pak. J. Statist.*, **3**, 17–40.

Ahsanullah, M. (1989) On characterizations of the uniform distribution based on functions of order statistics. *Aligarh J. Statist.*, **9**, 1–6.

Ahsanullah, M. and Holland, B. (1984) Record values and the geometric distribution. *Statistiche Hefte*, **25**, 319–327.

Ahsanullah, M. and Rahman, M. (1972) A characterization of the exponential distribution. *J. Appl. Prob.*, **9**, 457–461.

Akhiezer, N.I. (1965) *The Classical Moment Problem.* Hafner, New York.

Alharbi, A.A.G. (1994) On the convergence of the Bhattacharyya bounds in multiparametric case. To appear in *Applicationes mathematicae.*

Alamatasaz, M.H. (1983) On structural aspects of probability distributions and their mixtures with applications in queueing theory. Ph.D. thesis, Sheffield University.

Alamatasaz, M.H. (1985) A note on an article by Artikis. *Acta. Math. Acad. Sci. Hungar.*, **45**, 159–162.

Alamatsaz, M.H. (1993) On characterizations of exponential and gamma distributions. *Statistics and Probability Letters*, **17**, 315–320.

Aldous, D.J. (1985) Exchangeability and related topics. *Lecture Notes in Math.*, 1117, Springer, Berlin, pp. 1–198.

Ali, M.M. (1980) Characterization of the normal distribution among continuous symmetric spherical class. *J. Roy. Statist. Soc.*, Ser. B, **42**, 162–164.

Aly, M.A.H. (1988) Some contributions to characterization theory with applications in stochastic processes. Ph.D. Thesis, University of Sheffield.

Alzaid, A.A. (1983) Some contributions to characterization theory. Ph.D. Thesis, University of Sheffield.

Alzaid, A.A. (1986) Some results connected with the Rao–Rubin condition. *Sankhyā*, Ser. A, **48**, 104–108.

Alzaid, A.A. (1987) A note on the Meixner class. *Pak. J. Statist.*, **3**, 79–82.

Alzaid, A.A. and Al-Osh, M.A. (1991) Characterization of probability distributions based on the relation $X \overset{d}{=} U(X_1 + X_2)$. Sankhyā, **53**, Ser. B, 188–190.

Alzaid, A.A., Lau, K., Rao, C.R. and Shanbhag, D.N. (1988) Solution of Deny's convolution equation restricted to a halfline via a random walk approach. *J. Multivariate Analysis*, **24**, 309–329.

Alzaid, A.A., Rao, C.R. and Shanbhag, D.N. (1986a) An application of the Perron–Frobenius theorem to a damage model problem. *Sankhyā*, Ser. A, **48**, 43–50.

Alzaid, A.A., Rao, C.R. and Shanbhag, D.N. (1986b) Characterization of discrete probability distributions by partial independence. *Commun. Statist. Theory Meth.*, **15**, 643–656.

Alzaid, A.A., Rao, C.R. and Shanbhag, D.N. (1987a) An extension of Spitzer's integral representation theorem with an application. *Ann. Prob.*, 15, 1210–1216.

Alzaid, A.A., Rao, C.R. and Shanbhag, D.N. (1987b) Solution of the integrated Cauchy equation using exchangeability. *Shankhyā*, Ser. A, **49**, 189–194.

Alzaid, A.A., Rao, C.R. and Shanbhag, D.N. (1990) Elliptical symmetry and exchangeability with characterizations. *J. Mult. Anal.*, **33**, 1–16.

Andrews, C. (1988) Recent advances in damage models. M.Sc. dissertation, University of Sheffield.

Arnold, B.C. (1980) Two characterizations of the geometric distribution. *J. Appl. Prob.*, **17**, 570–573.

Arnold, B.C. and Ghosh, M. (1976) A characterization of the geometric distribution by properties of order statistics. *Scand. Actuar. J.*, **58**, 232–234.

Arnold, S.F. and Lynch, J. (1982) On Ali's characterization of spherical normal distribution. *J. Roy. Statist. Soc.*, Ser. B, **44**, 49–51.

Ash, R.B. (1972) *Real Analysis and Probability*. Academic Press, New York.

Athreya, K.B. and Ney, P.E. (1972) *Branching Processes*. Springer, Berlin.

Azencott, R. (1970) *Espaces de Poisson des Groups Localement Compacts*, Springer-Verlag Lecture Notes, No. 148, New York.

Azlarov, T.A. and Volodin, N.A. (1986) *Characterization Problems Associated with the Exponential Distribution*. Springer-Verlag.

Balasubramanian, K. and Beg, M.I. (1990) On expectations of functions of order statistics. *Sankhyā*, **52**, Ser. A, 103–114.

Barlow, R.E. and Proschan, F. (1965) *Mathematical Theory of Reliability*. McGraw-Hill, New York.

Barlow, R.E and Proschan, F. (1975) *Statistical Theory of Reliability and Life Testing*. Holt, Rinehart and Winston, New York.

Bartlett, M.S. (1934) The vector representation of a sample. *Proc. Cambridge Philos. Soc.*, **30**, 327–340.

Bartoszewicz, J. (1980) On the convergence of the Bhattacharyya bounds in the multiparameter case. *Zastosowania Mathematicae*, **16**, 601–608.

Basu, A.P. (1965) On the characterization of the exponential distribution by order statistics. *Ann. Inst. Stat. Math.*, **17**, 93–96.

Basu, A.P. (1971) Bivariate failure rate. *J. Amer. Statist. Assoc.*, **66**, 103–104.

Beg, M.I. and Balasubramanian, K. (1990) Distributions determined by conditioning on a single order statistics. *Metrika*, **37**, 37–43.

Beg, M.I. and Kirmani, S.N.U.A. (1974) On a characterization of exponential and related distributions. *Austral. J. Statist.*, **16**, 163–166.

Beg, M.I. and Kirmani, S.N.U.A. (1978) Characterization of the exponential distribution by a weak homoscedasticity. *Commun. Statist. Theory Meth.*, **7A**, 307–310.

Beg, M.I. and Kirmani, S.N.U.A. (1979) On characterizing the exponential distribution by a property of truncated spacing. *Sankhyā*, **41**, Ser. A, 278–284.

Behboodian, J. (1989) A note on skewness and symmetry. *The Statistician*, **38**, 21–24.

Benkherouf, L. and Bather, J.A. (1988) Oil exploration: sequential decisions in the face of uncertainty. *J. Appl. Prob.*, **25**, 529–543.

Berg, C., Christensen, J. and Ressel, P. (1984) *Harmonic Analysis on Semigroups, Theory of Positive Definite and Related Functions.* Springer-Verlag, New York.

Bhattacharyya, A. (1946) On some analogues of the amount of information and their use in statistical estimation. *Sankhyā*, **8**, 1–14.

Bhattacharyya, A. (1947) On some analogues of the amount of information and their use in statistical estimation. *Sankhyā*, **8**, Ser. A, 201–218.

Biggns, J.D., Lubachevsky, B.D., Schwartz, A. and Weiss, A. (1991) A branching random walk with a barrier. *Ann. Appl. Prob.*, **4**, 573–581

Billingsley, P. (1968) *Convergence of Probability Measures.* Wiley, New York.

Blackwell, D. (1948) A renewal theorem. *Duke Math J.*, **15**, 145–150.

Blackwell, D. (1953) Extension of a renewal theorem. *Pacific J. Math.*, **3**, 315–320.

Blackwell, D. (1955) On transient Markov processes with a countable number of states and stationary transition probabilities. *Ann. Math. Statist.*, **26**, 654–658.

Blight, B.J.N. and Rao, P.V. (1974) The convergence of the Bhattacharyya bounds. *Biometrika*, **61**, 137–142.

Bolger, E.M. and Harkness, W.L. (1965) Characterizations of some distributions by conditional moments. *Ann. Math. Statist.*, **36**, 703–705.

Bondesson, L. (1973) Characterizations of the normal and the gamma distributions. *Z. Wahrsch. Verw. Gebiete*, **26**, 335–344.

Bondesson, L. (1974) Characterizations of probability laws through constant regression. *Z. Wahrsch. Verw. Gebiete.*, **30**, 93–115.

Bondesson, L. (1975) Normal distribution, gamma distribution and quadratic polynomial statistics. *Scand. J. Statist.*, **2**, 138–144.

Borovkov, A.A. and Utev, S.A. (1983) On an inequality and a related characterization of the normal distribution. *Theory Prob. App.*, **28**, 219–228.

Box, G.E.P. and Hunter, J.S. (1957) Multivariate experimental designs for exploring response surfaces. *Ann. Math. Statist.*, **28**, 195–241.

Breiman, L. (1968) *Probability Theory.* Addison-Wesley, Reading, MA.

Brown, M. (1983) Approximating IMRL distributions by exponential distributions, with applications to first passage times. *Ann. Prob.*, **11**, 419–427.

Burrill, C.W. (1972) *Measures, Integration and Probability.* McGraw-Hill, New York.

Cacoullous, T. and Khatri, C.G. (1991) Correcting remarks on 'Characterization of normality within the class of elliptical contoured distributions'. *Stat. Prob. Letters*, **11**, 551–552.

Cacoullos, T. and Papageorgiou, H. (1983) Characterizations of discrete distributions by a conditional distribution and a regression function. *Ann. Inst. Statist. Math.*, **35**, 95–103.

Cacoullos, T. and Papathanasiou, V. (1982) On upper and lower bounds for the variance of a function of a random variable. *Ann. Prob.*, **10**, 799–809.

Cacoullos, T. and Papathanasiou, V. (1985) On upper bounds for the variance of functions of random variables. *Statist. Prob. Letters*, **3**, 175–184.

Cacoullos, T. and Papathanasiou, V. (1986) Bounds for the variance of functions of random variables by orthogonal polynomials and Bhattacharyya bounds. *Statist. Prob. Letters*, **4**, 21–23.

Cacoullos, T. and Papathanasiou, V. (1989) Characterizations of distributions by variance bounds. *Statist. Prob. Letters*, **7**, 351–356.

Chan, L.K. (1967) On a characterization of distributions by expected values of extreme order statistics. *Amer. Math. Monthly*, **74**, 950–951.

Chatterji, S.D. (1963) Some elementary characterizations of the Poisson distribution. *Amer. Math. Monthly*, **70**, 958–964.

Chernoff, H. (1981) A note on an inequality involving the normal distribution. *Ann. Prob.*, **9**, 533–535.

Chikkara, R.S. and Folks, J.L. (1989) *The Inverse Gaussian Distribution, Theory, Methodology and Applications*. Marcel Dekker, New York.

Chmielewski, M.A. (1981) Elliptically symmetric distribution: A review and bibliography. *Internat. Statist. Rev.*, **49**, 67–74.

Choquet, G. (1969) *Lectures on Analysis*, Vols I and II. Benjamin, New York.

Choquet, G. and Deny, J. (1960) Sur l'equation de convolution $\mu = \mu^* \sigma'$. *C. R. Acad. Sci. Paris*, **250**, 799–801.

Chow, Y.S. and Teicher, H. (1979) *Probability Theory, Independence, Interchangeability, Martingales*. Springer Verlag, New York.

Chung, K.L. (1972) The Poisson process as renewal process. Collection of articles dedicated to the memory of A. Rényi. *I. Period. Math. Hungar.*, 2, 41–48.

Çinlar, E. (1972) Superposition of point processes. *Stochastic Point Process*, pp. 549–606, ed. P.A.W. Lewis. Wiley, New York.

Çinlar, E. (1974/1975) Markov renewal theory: A survey. *Management Science*, 21, 727–752.

Çinlar, E. and Jagers, P. (1973) Two mean values which characterize a Poisson process. *J. Appl. Prob.*, 10, 678–681.

Consul, P.C. (1990) New class of location-parameter discrete probability distributions and their characterizations. *Commun. Statist. Theory Meth.*, **19**(12), 4653–4666.

Cox, D.R. (1961) *Renewal Theory*. Methuen, London.

Cox, D.R. (1972) Regression models and life tables (with discussion). *J.R. Statist. Soc.*, Ser. B, **34**, 187–220.

Cox, D.R. and Isham, V. (1980) *Point Processes*. Chapman and Hall.

Crawford, G.B. (1966) Characterization of geometric and exponential distributions. *Ann. Math. Statist.*, **37**, 1790–1795.

Daley, D.J. (1976) Queueing output processes. *Adv. Appl. Prob.*, **8**, 395–415.

Daley, D.J. and Shanbhag, D.N. (1975) Independent interdeparture times in $M/G/1/N$ queues. *J. Roy. Statist. Soc.*, Ser. B, **37**, 259–263.

Dallas (1981) Record values and the exponential distribution. *J. Appl. Prob.*, **18**, 949–951.

Daniels, H.E. (1975) The deterministic spread of a simple epidemic. In honour of M.S. Bartlett ed. Gani, J. *Perspectives in Probability and Statistics*, Academic Press, pp. 373–386.

David, H.A. (1981) *Order Statistics*. Wiley, New York, 2nd edn.

Davies, P.L. and Shanbhag, D.N. (1987) A generalization of a theorem of Deny with applications in characterization theory. *J. Math. Oxford* (2), **38**, 13–34.

Davies, L. and Shimizu, R. (1976) On identically distributed linear statistics. *Ann. Inst. Statist. Math.*, **28**, 469–489.

Davies, L. and Shimizu, R. (1980) On a functional equation in the theory of linear statistics. *Ann. Inst. Statist. Math.*, **32**, 17–23.

Deny, J. (1961) Sur l'equation de convolution $\mu = \mu^* \sigma$. Semin. Theory Potent.M. Brelot, Fac. Sci. Paris, 1959–1960. 4 ann.

Desu, M.M. (1971) A characterization of the exponential distribution by order statistics, *Ann. Math. Statist.*, **42** 437–438.

Doob, J.L. (1953) *Stochastic Processes*. Wiley, New York.

Doob, J.L., Snell, J.L. and Williamson, R.E. (1960) Application of boundary theory to sums of independent random variables. *Contributions to Probability and Statistics*. Stanford University Press, pp. 182–197.

Downton, F. (1969) An integral equation approach to equipment failure. *J.R. Statist. Soc.*, Ser. B, **31**, 335–349.

Dubins, L.E. and Freedman, D.A. (1979) Exchangeable processes need not be mixtures of independent, identically distributed random variables. *Z. Wahrsch. verw. Gebiete*, **48**, 115–132.

Dudley, R.M. (1989) *Real Analysis and Probability*. The Wadsworth and Brooks/Cole Mathematics Series.

Dudley, R.M. and Kanter, M. (1974) Zero–one laws for stable measures. *Proc. Amer. Math. Soc.*, **45**, 245–252.

Duncan, G. and Kadane, J.B. (1980) Advanced problem No. 6314, *Amer. Math. Monthly*, **87**, 676.

Eaton, M.L., Morris, C. and Rubin, H. (1971) On extreme stable laws and some applications. *J. Appl. Prob.*, **8**, 794–801.

Eberl, W. (1985) A characterization of the gamma distribution by optimal estimates. *Theory of Probability and its Applications*, **30**, 855–860.

El-Neweihi, E. and Govindarajulu, Z. (1979) Characterization of the geometric distribution and discrete IFR(DER) distributions using order statistics. *J. Statist. Plann. Inf.*, **3**, 85–90.

Feller, W. (1966) (Revised 1971) *An Introduction to Probability Theory and its Applications*, Vol. 2. Wiley, New York.

Feller, W. (1968) *An Introduction to Probability Theory and its Applications*, Vol. 1, 3rd ed. Wiley, New York.

Ferguson, T.S. (1964) A characterization of the exponential distribution. *Ann. Math. Statist.*, **35**, 1199–1207.

Ferguson, T.S. (1965) A characterization of the geometric distribution. *Amer. Math. Monthly*, **72**, 256–260.

Ferguson, T.S. (1967) On characterizing distributions by properties of order statistics. *Sankhyā*, Series A, **29**, 265–277.

Fisz, M. (1958) Characterization of some probability distributions. *Skand. Aktiarict.*, 65–67.

Fosam, E.B. (1993) Characterizations and structural aspects of probability distributions. Ph.D. thesis, Sheffield University.

Fosam, E.B., Rao, C.R. and Shanbhag, D.N. (1993) Comments on some papers involving the integrated Cauchy functional equation *Statistics and Probability Letters*, **17**, 299–302.

Fosam, E.B. and Shanbhag, D.N. (1994) Certain characterizations of exponential and geometric distributions. *J.R.S.S.*, Ser. B, **56**, 157–160.

Galambos, J. (1975a) Characterizations in terms of properties of the smaller of two observations. *Commun. Statist.*, **4**(3), 239–244.

Galambos, J. (1975b) Characterizations of probability distributions by properties of order statistics I and II. *Statistical Distributions in Scientific Work*, Vol. 3, pp. 71–101, eds Patil, G.P., Kotz, S. and Ord, J.K. Dordrecht, Reidel.

Galambos, J. and Kotz, S. (1978) Characterizations of probability distributions. *Lecture Notes in Mathematics*, 675, Springer-Verlag, Berlin.

Gather, U. (1989) On a characterization of the exponential distribution by properties of order statistics. *Statistics and Probability Letters*, **7**, 93–96.

Gather, U. and Szekely, G. (1989) Characterizations of distributions by linear forms of order statistics. Unpublished.

Geisser, S. (1973) Normal characterizations via the squares of random variables. *Sankhyā*, Ser. A, **35**, 492–494.

Ghosh, J.K. and Mukerjee, R. (1991) Characterization of priors under which Bayesian and frequentist Bartlett corrections are equivalent in the multivariate case. *J. Mult. Anal.*, **38**, 385–393.

Ghosh, J.K. and Sathe, Y.S. (1987) Convergence of the Bhattacharya bounds — revisited. *Sankhyā*, Ser. A, **49**, 37–42.

Gine, G.E. and Hahn, M.G. (1983) On stability of probability laws with univariate stable marginals. *Z. Wahrsch. verw. Gebiete.*, **64**, 157–163.

Glänzel, W., Telcs, A. and Schubert, A. (1984) Characterization by truncated moments and its applications to Pearson-type distributions. *Z. Wahrsch. Verw. Gebiete*, **66**, 173–183.

Gordon, F.S. (1973) Characterizations of populations using regression properties. *Ann. Statist.*, **1**, 114–126.

Goria, M.N. (1987) Some characterizations of the uniform distribution. *Commun. Statist. Theory Meth.*, **16**, 813–819.

Govindarajulu, Z. (1975) Characterization of the exponential distribution using lower moments of order statistics. *Statistical Distributions in Scientific Work*, Vol. 3, pp. 117–129, eds Patil, G.P., Kotz, S. and Ord, J.K. Dordrecht, Reidel.

Govindarajulu, Z. (1980) Characterization of the geometric distribution using properties of order statistics. *J. Statist. Inf.*, **4**, 237–247.

Grossward, E. and Kotz, S. (1981) An integrated lack of memory characterization of the exponential distribution. *Ann. Inst. Statist. Math.*, **33**, 205–214.

Grossward, E., Kotz, S. and Johnson, N.L. (1980) Characterizations of the exponential distribution by relevation-type equations. *J. Appl. Prob.*, **17**, 874–877.

Gu, H.M. and Lau, K.S. (1984) Integrated Cauchy functional equation with an error term and the exponential law. *Sankhyā*, Ser. A, **46**, 339–354.

Gupta, P.L. and Gupta, R.C. (1983) On the moments of residual life in reliability and some characterization results. *Commun. Statist. Theory Meth.*, **12**, 449–461.

Gupta, P.L. and Gupta, R.C. (1986) A characterization of the Poisson process. *J. Appl. Prob.*, **23**, 233–235.

Gupta, R.C. (1974) Characterizations of distributions by property of discrete order statistics. *Commun. Statist. Theory Meth.*, **3**, 287–289.

Gupta, R.C. (1975) On the characterization of distributions by conditional expectations. *Commun. Statist. Theory Meth.*, **4**, 99–103.

Gupta, R.C. (1979) On the characterization of survival distributions in reliability by properties of their renewal densities. *Commun. Statist. Theory Meth.*, **8**, 685–697.

Gupta, R.C. (1984) Relationships between order statistics and record values and some characterization results. *J. Appl. Prob.*, **21**, 425–430.

Gupta, R.D. and Richards, D. St. P. (1987) Multivariate Liouville distributions. *J. Mult. Anal.*, **23**, 233–256.

Gupta, R.D. and Richards, D. St. P. (1990) The Dirichlet's distributions and polynomial regression. *J. Mult. Anal.*, **32**, 95–102.

Gyires, B. (1980) Private communication.

Haberland, E. (1975) Infinitely divisible recurrent point processes. *Math Nachr.*, **70**, 259–264.

Hall, W.J. and Simons, G. (1969) On characterizations of the gamma distribution. *Sankhyā*, **31**, Ser. A, 385–390.

Hall, W.J. and Wellner, J.A. (1981) Mean residual life. *Statistics and Related Topics*, pp. 169-184, eds Csorgo, M., Dawson, S.A., Rao, J.N.K. and Saleh, M.K. Md. E.. Amsterdam, North-Holland.

Hamdan, M.A. (1972) On a characterization by conditional expectations. *Technometrics*, **14**, 497-499.

Hardy, G.H. (1967) *A Course of Pure Mathematics*. Cambridge University Press.

Hardy, G.H., Littlewood, J.E. and Polya, G. (1964) *Inequalities*. Cambridge University Press, 2nd edn.

Harkness, W.L. and Harkness, M.L. (1968) Generalized hyperbolic secant distributions. *J. Amer. Stat. Assoc.*, **63**, 329-337.

Hartman, P. and Wintner, A. (1940) On the spherical approach to the normal distribution law. *Amer. J. Math.*, **62**, 759-779.

Heller, B. (1983) Special functions and characterizations of probability distributions by zero regression properties. *J. Mult. Analysis*, **13**, 473-487.

Heller, B. (1991) Private Communication.

Hewitt, E. and Savage, L.J. (1955) Symmetric measures on cartesian products. *Trans. Amer. Math. Soc.*, **80**, 470-501.

Heyde, C.C. (1963) On a property of the lognormal distributions. *J. Roy. Statist. Soc.*, Ser. B, **25**, 392-393.

Heyde, C.C. (1970) Characterization of the normal law by a symmetry of a certain conditional distribution. *Sankhyā*, Ser A, **32**, 115-118.

Hitha, N. and Nair, U.N. (1989) Characterization of some discrete models by properties of residual life function. *Calcutta Statist. Assoc. Bull.*, **38**, 219-223.

Hoeffding, W. (1953) On the distribution of the expected values of the order statistics. *Ann. Math. Statist.*, **24**, 93-100.

Hollander, M. and Proschan, F. (1984) Nonparametric concepts and methods in reliability. *Handbook of Statistics*, Vol. 4, pp. 613-655, eds Krishnaiah, P.R. and Sen, P.K., North Holland, Amsterdam.

Holmes, P. (1974) Another characterization of the Poisson process. *Sankhyā*, Ser. A, **36**, 449-450.

Huang, J.S. (1978) On a "lack of memory" property, *Stat. Tech. Rept.*, Univ. of Guelph, Canada.

Huang, W.J. and Chen, L.S. (1989) Note on a characterization of gamma distributions. *Statist. Prob. Lett.*, **8**, 485-489.

Huang, W.J. and Chen, L.S. (1991) On a study of certain power mixtures. *Chinese J. Math.*, **19**, 95-104.

Huang, W.L. and Li, S.H. (1993) Characterization results based on record values. *Statistica Sinica*, **3**, 583-599.

Huang, W.J., Li, S. and Su, J. (1993) Some characterizations of the Poisson process and geometric renewal process. *J. Appl. Prob.*, **30**, 121-130.

Isham, V., Shanbhag, D.N. and Westcott, M. (1975) A characterization of the Poisson process using forward recurrence times. *Math. Proc. Cambridge Philos. Soc.*, **78**, 513-515.

Jacod, J. (1974) Multivariate point processes: predictable projection, Random-Nikodym derivatives, representation of martingales. *Z. Wahrsch. verw. Gebiete*, **31**, 235-253.

Janaradan, K.G. and Rao, B.R. (1982) Characterization of generalized Markov-Polya and generalized Polya-Eggenberger distributions. *Commun. Statist. Theory Meth.*, **11**, 2113-2124.

Johnson, N.L. and Kotz, S. (1975) A vector multivariate hazard rate. *J. Multivariate Anal.*, **5**, 53-66.

Johnson, N.L., Kotz, S. and Kemp, A.W. (1992) *Univariate Discrete Distributions (Revised Edition)*, Wiley, New York.

Johnstone, I. and MacGibbon, B. (1984) An information measure for Poisson characterization. *Mathematical Science Research Institute*, Nos 74–83, Berkeley, California.

Jupp, P.E. and Mardia, K.V. (1982) A characterization of the multivariate Pareto distribution. *Ann. Statist.*, **10**, 1021–1024.

Kac, M. (1939) On a characterization of the normal distribution. *Amer. J. Math.*, **61**, 726–728.

Kagan, A.M. (1989) Constancy of regression of a polynomial of sample average of positive random variables on their ratios characterizes gamma distribution. *Statist. Data Analy. Inf.* pp. 309–316. ed. Dodge, Y. Elsevier Science Publishers (North-Holland).

Kagan, A. (1990/1991) Linearity of regression of the third sample moment on the sample average. *Metron*, **48**, 33–37.

Kagan, A.M., Linnik, Yu V. and Rao, C.R. (1965) On a characterization of the normal law based on a property of the sample average. *Sankhyā*, **27A**, 405–406.

Kagan, A.M., Linnik, Yu V. and Rao, C.R. (1973) *Characterization Problems in Mathematical Statistics*. Wiley, New York.

Kagan, A.M. and Rao, C.R. (1991) Constancy of regression of sample average on residuals characterizes the normal distribution. In *Probability, Statistics and Design of Experiments* (Bose Symposium Volume, Ed. R.R. Bahadur), pp. 419–424, Wiley Eastern.

Kakosyan, A.V., Klebanov, L.B. and Melamed, J.A. (1984) *Characterization of Distributions by Method of Intensively Monotonic Operators*. Lecture Notes Math. 1088, Springer-Verlag, New York.

Kapoor, S. (1988) Some characterizations in statistical inference and related results. Ph.D. Thesis, University of Sheffield.

Karlin, S. (1966) *A First Course in Stochastic Processes*. Academic Press, New York.

Kawata, T. and Sakamoto, N. (1949) On the characterization of the normal population by the independence of the sample mean and the sample variance. *J. Math. Soc. Japan*, **1**, 111–115.

Kelker, D. (1970) Distribution theory of spherical distributions and location-scale parameter generalization. *Sankhyā*, Ser. A, **32**, 419–430.

Kelly, F.P. (1979) Reversibility and stochastic networks. *Wiley Series in Probability and Statistics*. Wiley.

Kendall, D.G. (1951) Some problems in the theory of queues. *J. Royal Statist. Soc.*, Ser. B, **13**, 151–185.

Kendall, D.G. (1953) Stochastic processes occurring in the theory of queues and their analysis by the method of imbedded Markoff chain. *Ann. Math. Statist.*, **24**, 338–354.

Kendall, D.G. (1963) Extreme points methods in stochastic analysis. *Z. Wahrsch. verw. Gebiete.*, **1**, 295–300.

Kendall, D.G. (1964) Functional equations in information theory. *Z. Wahrsch. verw. Gebiet.*, **2**, 225–229.

Kendall, D.G. (1967) On finite and infinite sequences of exchangeable events. *Studia. Sci. Math. Hungar.*, **2**, 319–327.

Khatri, C.G. (1962) A characterization of the inverse Gaussian distribution. *Ann. Math. Statist.*, **33**, 800–803.

Khatri, C.G. (1978a) Characterizations of some discrete distributions by linear regression. *J. Ind. Statist. Assoc.*, **16**, 49–58.

Khatri, C.G. (1978b) Characterizations of some multivariate distributions by conditional distributions and linear regression. *J. Ind. Statist. Assoc.*, **16**, 59–70.

Khatri, C.G. and Cocoullos, T. (1993) Characterizations of distributions within the elliptical class by a gamma distributed quadratic form. *Khatri Memorial Volume* (to appear).

Khatri, C.G. and Mukerjee, R. (1987) Characterization of normality within the class of elliptical contoured distributions. *Statist. Prob. Letters*, **5**, 187–190.

Kimeldorf, G. Plachky, D. and Sampson, A.R. (1981) A simultaneous characterization of the Poisson and Bernoulli distributions. *J. Appl. Prob.*, **18**, 316–320.

Kingman, J.F.C. (1965) Stationary measures for branching processes. *Proc. Amer. Math. Soc.*, **16**, 245–247.

Kingman, J.F.C. (1966) On the algebra of queues. *J. Appl. Prob.*, **3**, 285–326.

Kingman, J.F.C. (1972) On random sequences with spherical symmetry. *Biometrika*, **59**, 492–494.

Kingman, J.F.C. and Taylor, S.J. (1966) *Introduction to Measure Probability*. Cambridge University Press, London.

Kirmani, S.N.U.A. and Alam, S.N. (1980) Characterization of the geometric distribution by the form of a predictor. *Comm. Statist.*, A, **9**, 541–548.

Kirmani, S.N.U.A. and Beg, M.L. (1984) On characterization of distributions by expected records. *Sankhyā*, Ser. A, **46**, 463–465.

Klebanov, L.B. (1973a) The admissibility of the sample mean as an estimate of the location parameter for nonquadratic risk functions. *Theory of Prob. and its Applications*, **18**, 329–339.

Klebanov, L.B. (1973b) Inadmissibility of the polynomial estimators of a location parameter (Russian). *Mat. Zametki*, **14**, 885–893.

Klebanov, L.B. (1980) some results connected with characterizations of the exponential distribution. *Theor. Veoj. i Primenen.*, **25**, 628–633.

Klebanov, L.B. and Melamed, J.A. (1983) A method associated with characterizations of the exponential distribution. *Ann. Inst. Statist. Math.*, **35**, 105–114.

Kochar, S.C. (1992) A note on a characterization of symmetry about a point. *The Statistician*, **41**, 161–163.

Kotlarski, I.I. (1972) On a characterization of some probability distributions by conditional expectations. *Sankhyā*, Ser. A, **34**, 461–467.

Kotz, S. (1974) Characterizations of statistical distributions: A supplement to recent survey. *Int. Stat. Rev.*, **42**, No. 1, 39–65.

Kotz, S. and Johnson, N.L. (1974) A Characterization of exponential distributions by waiting time property. *Comm. Statist.*, **3**, 257–258.

Kotz, S. and Shanbhag, D.N. (1980) Some new approaches to probability distributions. *Adv. Appl. Prob.*, **12**, 903–921.

Kotz, S. and Steutel, F.W. (1988) Note on a characterization of the exponential distributions. *Statist. Prob. Letter*, **6**, 201–203.

Kourouklis, S. (1986) On a simultaneous characterization of the Poisson and Bernoulli distributions. *Austral. J. Statist.*, **28**, 414–417.

Krein, M.G. (1962) Integral equation on a half-line with kernel depending upon the difference of the arguments. *Amer. Math. Soc., transl.* **22**(2), 163–288. (In Russian, *Uspehi Mat. Nauk* (N.S.), **13**, (1958), No. 5(83), 3–120.)

Krishnaji, N. (1970) Characterization of the Pareto distribution through a model of under-reported incomes. *Econometrica*, **38**, 251–255.

Krishnaji, N. (1970b) A characteristic property of the Yule distribution. *Sankhyā*, Ser. A, **32**, 243–346.

Krishnaji, N. (1971) Note on a characterizing property of the exponential distribution. *Ann. Math. Statist.*, **42**, 361–362.

Krishnaji, N. (1974) Characterization of some discrete distributions based on damage models. *Sankhyā*, Ser. A, **36**, 204–213.

Kupka, J. and Loo, S. (1989) The hazard and vitality measures of ageing. *J. Appl. Prob.*, **26**, 532–542.

Laha, R.G. and Lukas, E. (1960) On a problem connected with quadratic regression. *Biometrika*, **47**, 335–343.

Lai, C.D. (1982a) A characterization of gamma, Meixner hypergeometric and negative binomial distributions based on canonical measures. *Ann. Inst. Statist. Math.*, **34**, 359–363.

Lai, C.D. (1982b) Meixner classes and Meixner hypergeometric distributions. *Austral. J. Statist.*, **24**, 221–233.

Lai, C.D. and Vere-Jones, D. (1979) Odd man out — the Meixner hypergeometric distribution. *Austral. J. Statist.*, **21**, 256–265.

Lau, K.S. (1992) Fractal measures and mean *p*-variations. *J. Funct. Anal.*, **108**, 421–457.

Lau, K.S. and Prakasa Rao, B.L.S. (1990) Characterization of the exponential distribution by the relevation transform. *J. Appl. Prob.*, **27**, 726–729.

Lau, K.S. and Rao, C.R. (1982) Integrated Cauchy functional equation and characterizations of the exponential law. *Sankhyā*, Ser. A, **44**, 72–90.

Lau, K. and Rao, C.R. (1984) Integrated Cauchy functional equation on the whole line. *Sankhyā*, Ser. A, **46**, 311–319.

Lau, K.S. and Zeng, W.B. (1990) The convolution equation of Choquet and Deny on semigroups. *Studia Math.*, **97**, 115–135.

Laurent, A.G. (1974) On characterization of some distributions by truncated moment properties. *J. Amer. Stat. Assoc.*, **69**, 823–827.

Lee, L. and Thompson, W.A. Jr. (1976) Failure rate — a unified approach. *J. Appl. Prob.*, **13**, 176–182.

Letac, G. (1981) Isotropy and sphericity: some characterizations of the normal distribution. *Ann. Statist.*, **9**, 408–417.

Letac, G. (1992) Lectures on natural exponential families and their variance functions. *Monografias de Mathematica*, No. 50, Instituto de Matematica Pura e Aplicada, Rio de Janeiro.

Letac, G. and Mora, M. (1990) Natural real exponential families with cubic variance functions. *Ann. Statist.*, **18**, 1–37.

Lévy, P. (1937) L'arithmétique des lois de probabilité et les produits finis de lois de Poisson. *C.R. Acad. Sci. Paris*, **204**, 944–946.

Liang, T.C. and Balakrishnan, N. (1992) A characterization of the exponential distributions through conditional independence. *J.R. Statist. Soc.*, Ser. B, **54**, 269–271.

Liberman, U. (1985) An order statistic characterization of the Poisson renewal process. *J. Appl. Prob.*, **22**, 717–722.

Lin, G.D. (1988) Characterizations of uniform distributions and exponential distributions. *Sankhyā*, Ser. A, **50**, 64–69.

Lin, G.D. (1990) Characterizations of continuous distributions via expected values of two functions of order statistics. *Sankhyā*, Ser. A, **52**, 84–90.

Lin, G.D. (1992) Characterizations of distributions via moments. *Sankhyā*, Ser. A, **54**, 128–132.

Lindley, D.V. (1952) Theory of queues with a single server. *Proc. Cambridge Philos. Soc.*, **48**, 277–279.

Linnik, Y.V. (1964) *Decomposition of Probability Distributions*. Oliver and Boyd, Edinburgh and London.

Loéve, M. (1963) *Probability Theory*, 3rd edn. Van Nostrand, New York.

Lukacs, E. (1942) A characterization of the normal distribution. *Ann. Math. Statist.*, **13**, 91–93.

Lukacs, E. (1955) A characterization of the gamma distribution. *Ann. Math. Statist.*, **26**, 319–324.

Lukacs, E. (1970) *Characteristic Functions*, 2nd edn. Griffin, London.

Lukacs, E. and Laha, R.G. (1964) *Applications of Characteristic Functions*, Hafner, New York.

Markus, D.J. (1983) Nonstable laws with all projections stable. *Z. Wahrsch. verw. Gebiete.*, **64**, 139–156.

Marsaglia, G.M. (1989) The $X + Y$ and X/Y characterization of the gamma distribution. *Contributions to Probability and Statistics: Essays in Honor of I. Olkin*, pp. 91–98, eds Gleser, L.J., Perlaman, M.D., Press, S.J. and Sampson, A.R., Springer Verlag, New York.

Marsaglia, G. and Tubilla, A. (1975) A note on the lack of memory property of the exponential distribution. *Ann. Prob.*, **3**, 352–354.

Marshall, A.W. (1989) A bivariate uniform distribution. *Contributions to Probability and Statistics: Essays in Honor of I. Olkin*, pp. 99–106, eds Gleser, L.J., Perlaman, M.D., Press, S.J. and Sampson, A.R. Springer Verlag, New York.

Marshall, A.W. and Olkin, I. (1967) A generalised bivariate exponential distribution. *J. Appl. Prob.*, **4**, 291–302.

Marshall, A.W. and Olkin, I. (1991) Functional equations for multivariate exponential distributions. *J. Mult. Anal.*, **39**, 209–215.

Maxwell, J.C. (1860) Illustration of the dynamical theory of gases — Part I. On the motions and collisions of perfectly elastic bodies. *Taylor's Philos. Mag.*, **19**, 19–32.

Meilijson, I. (1972) Limiting properties of the mean residual life function. *Ann. Math. Statist.*, **43**, 354–357.

Meyer, P.A. (1966) *Probability and Potentials*. Blaisdel, Waltham, Massachusetts.

Mohan, N.R. and Nayak, S.S. (1982) A characterization based on the equidistribution of the first two spacings of record values. *Z. Wahrsh. Verw. Gerb.*, **60**, 219–221.

Moran, P.A.P. (1952) A characteristic property of the Poisson distribution. *Prob. Cam. Phil. Soc.*, **48**, 206–207.

Morris, C.N. (1982) Natural exponential families with quadratic variance functions. *Ann. Statist.*, **10**, 65–80.

Morris, C.N. (1983) Natural exponential families with quadratic variance functions: statistical theory. *Ann. Statist.*, **11**, 515–529.

Morrison, D.G. (1978) On linear increasing mean residual life times. *J. Appl. Prob.*, **15**, 617–620.

Mukherjee, S.P. and Roy, D. (1986) Some characterizations of the exponential and related life distributions. *Calcutta Statist. Assoc. Bull.*, **35**, 189–197.

Mukherjee, S.P. and Roy, D. (1989) Characterizations of some bivariate life distributions. *J. Multivariate Anal.*, **19**, 1–8.

Nagaraja, H.N. (1975) Characterization of some distributions by conditional moments. *J. Indian Statist. Assoc.*, **13**, 57–61.

Nagaraja, H.N. (1977) On a characterization based on record values. *Austral. J. Statist.*, **19**, 70–73.

Nagaraja, H.N. and Srivastava, R.C. (1987) Some characterizations of geometric type distributions based on order statistics. *J. Stat. Plann. Inf.*, **17**, 181–191.

Nair, U.N. and Hitha, N. (1989) Characterizations of discrete distributions by models based on their partial sums. *Statist. Prob. Letters*, **8**, 335–337.

Nair, K.R.M. and Nair, N.U. (1988) On characterizing the bivariate exponential and geometric distributions. *Ann. Inst. Statist. Math.*, **40**, 267–271.

Nair, N.U. and Nair, K.R.M. (1990) Characterizations of a bivariate geometric distribution. *Statistica*, anno, L, No. 2, 247–253.

Nair, N.U. and Sankaran, P.G. (1991) Characterization of Pearson family of distributions. *IEEE Trans. Reliability*, **40**, 75–77.

Nash, D. and Klamkin, M.S. (1976) A spherical characterization of the normal distribution. *J. Math. Anal. Appl.*, **55**, 156–158.

Nassar, M.M. (1988) Two properties of the exponential distributions. *IEEE Trans.*, **37**, 383–385.

Nassar, M. and Mahmoud, M. (1985) On characterizations of a mixture of exponential distributions. *IEEE Trans. Reliability*, **R-37**, 379-382.

Neuts, M.F. and Resnick, S.I. (1971) On the times of births in a linear birth process. *J. Austral. Math. Soc.*, **12**, 473-475.

Norton, R.M. (1975) On properties of the arc-sine law. *Sankhyā*, Ser. A, **37**, 306-308.

Norton, R.M. (1978) Moment properties and the arc-sine law. *Sankhyā*. Ser. A, **40**, 192-198.

Obretenov, A. (1985) Characterization of the multivariate Marshall-Olkin exponential distribution. *Prob. Math. Statist.*, **16**(1), 51-56.

Olshen, R.A. (1973) A note on exchangeable sequences. *A. Wahrsch. Gebiete*, **28**, 317-321.

Osaki, S. and Li, X. (1988) Characterization of gamma and negative binomial distributions. *IEEE Trans. Reliability*, **37**, 379-382.

Ouyang, L.Y. (1983) On characterizations of distributions by conditional expectations. *Tamkang. J. Management Sci.*, **4**(1), 13-21.

Ouyang, L.Y. (1987) On characterizations of probability distributions based on conditional expected values. *Tamkang. J. Math.*, **18**(1), 113-122.

Pakes, A.G. (1992) A characterization of gamma mixtures of stable laws motivated by limit theorems. *Statistica Neerlandica*, **46**, 209-218.

Pakes, A.G. and Khatree, R. (1992) Length-biasing, characterization of laws and the moment problem. *Austral. J. Statist.*, **34**, 307-322.

Panaretos, J. (1977) Characterization of discrete distributions based on conditionality and damage models. Ph.D. thesis, University of Bradford.

Panaretos, J. (1982a) On a structural property of finite distributions. *J.R.S.S.*, Ser. B, **44**, 209-211.

Panaretos, J. (1982b) On characterizing some discrete distributions using an extension of the Rao-Rubin theorem. *Sankhyā*, Ser. A., **44**, 415-422.

Papageorgiou, H. (1983) On characterizing some bivariate discrete distributions. *Austral. J. Statist.*, **25**, 136-144.

Papageorgiou, H. (1985) On characterizing some discrete distributions by a conditional distribution and a regression function. *Biom. J.*, **27**, 473-479.

Parthasarathy, K.R. (1967) *Probability Measures on Metric Spaces*. Academic Press, New York.

Parzen, E. (1962) *Stochastic Processes*. Holden-Day, San Francisco.

Patil, G.P. and Rao, C.R. (1977) *The Weighted Distributions: A Survey and Their Applications, Applications of Statistics*, pp. 383-405, ed. Krishnaiah, P.R., Amsterdam, North-Holland.

Patil, G.P. and Rao, C.R. (1978) Weighted distributions and size biased sampling with applications to wildlife populations and human families. *Biometrics*, **34**, 179-189.

Patil, G.P. and Ratnaparkhi, M.V. (1975) Problems of damaged random variables and related characterizations. *Statistical Distributions in Scientific Work*, Vol. 3, pp. 255-270, eds Patil, G.P., Kotz, S., and Ord, J.K. Dordrecht, Reidel.

Patil, G.P. and Ratnaparkhi, M.V. (1977) Characterizations of certain statistical distributions based on additive damage models involving Rao-Rubin condition and some of its variants. *Sankhyā*, Ser. B, **39**, 65-75.

Patil, G.P. and Seshadri, V. (1964) Characterization theorems for some univariate discrete distributions. *J.R.S.S.*, Ser. B, **26**, 262-292.

Patil, G.P. and Taillie, C. (1979) On a variation of the Rao-Rubin condition. *Sankhyā*, Ser. A, **41**, 129-132.

Paulauskas, V.J. (1976) Some remarks on multivariate stable distributions. *J. Mult. Analysis*, **6**, 356-368.

Pfeifer, D. (1982) Characterizations of the exponential distributions by independence nonstationary record increments. *J. Appl. Prob.*, **19**, 127-135.

Phelps, R.R. (1966) *Lecture Notes on Choquet's Theorem.* Van Nostrand, Princeton.

Prakasa Rao, B.L.S. (1974) On a property of bivariate distributions. *J. Mult. Anal.,* **4**, 106–113.

Prakasa Rao, B.L.S. and Sreehari, M. (1986) Another characterization of multivariate distribution. *Stat. Prob. Letters*, **4**, 209–210.

Prakasa Rao, B.L.S. and Sreehari, M. (1987) On a characterization of Poisson distribution through an inequality of Chernoff type. *Aus. J. of Stat.*, **29**, 38–41.

Puri, P.S. and Rubin, H. (1970) A characterization based on absolute difference of two independent identically distributed random variables. *Ann. Math. Statist.*, **41**, No. 6, 2113–2122.

Puri, P.S. and Rubin, H. (1974) On a characterization of the family of distributions with constant multivariate failure rates. *Ann. Prob.*, **2**, 738–740.

Pusz, J. (1993) A characterization of the binomial and negative binomial distributions by regression of products. *Metrika*, **40**, 237–242.

Pusz, J. and Wesolowski, J. (1992) A non-Lukacsian regressional characterization of the gamma distribution. *Appl. Math. Lett.*, **5**, 81–84.

Raghavan, T.E.S. (1993) Private communication.

Raikov, D. (1937) A characteristic property of the Poisson laws. *C.R. Acad. Sci. U.S.S.R.*, **14**, 8–11.

Ramachandran, B. (1980) An integral equation in probability theory and its implications. *Tech. Report*, Indian Stat. Inst., New Delhi.

Ramachandran, B. (1982) On the integral equation $f(x) = \int_{[0,\infty)} f(x+y)\,d\mu(y)$. *Sankhyā*, Ser. A, **44**, 364–371.

Ramachandran, B. and Lau, K.S. (1990) Integrated Cauchy functional equation with an error term on $[0, \infty)$. *Proceedings of the R.C. Bose Memorial Symposium on Analysis and Probability*, New Delhi.

Ramachandran, B. and Lau, K.S. (1991) *Functional Equations in Probability Theory.* Academic Press, New York.

Ramachandran, B., Lau, K.S. and Gu, H.M. (1988) On characteristic functions satisfying a functional equation and related classes of simultaneous integral equations. *Sankhyā*, Ser. A, **50**, 190–198.

Ramachandran, B. and Prakasa Rao, B.L.S. (1984) On the equation $f(x) = \int_{-\infty}^{\infty} f(x+y)\,d\mu(y)$. *Sankhyā*, Ser. A, **46**, 326–338.

Rao, C.R. (1947) A note on the problem of Ragnar Frisch. *Econometrika*, **15**, 245–249.

Rao, C.R. (1963) On discrete distributions arising out of methods of ascertainment. Paper presented at the Montreal Conference on Discrete Distributions. (Printed in *Sankhyā* A, **27**, 311–324, 1965).

Rao, C.R. (1965) On Discrete Distributions Arising out of Methods of Ascertainment. Classical and Contagious Discrete Distributions, pp. 320–332 ed. Patil, G.P. Calcutta, Statistical Publishing Society.

Rao, C.R. (1983) An extension of Deny's theorem and its applications to characterization of probability distributions. *A Festchrift for Eric Lehmann*, pp. 348–366, eds Bickel, P.J., Doksum, K. and Hodges Jr, J.L. Wadsworth Statistics/Probability Series.

Rao, C.R. and Rubin, H. (1964) On a characterization of the Poisson distribution. *Sankhyā*, Ser. A, **26**, 295–289.

Rao, C.R., Sapatinas, T. and Shanbhag, D.N. (1994) The integrated cauchy functional equation: some comments on recent papers. *Adv. Appl. Prob.*, **26**, (to appear).

Rao, C.R. and Shanbhag, D.N. (1986) Recent results on characterization of probability distributions: a unified approach through extensions of Deny's theorem. *Adv. Appl. Prob.*, **18**, 660–678.

Rao, C.R. and Shanbhag, D.N. (1989a) Recent advances on the integrated Cauchy functional equation and related results in applied probability. *Papers in Probability, Statistics and Mathematics in Honor of Samuel Karlin*, pp. 239-253, eds Anderson, T.W., Athreya, K.B. and Iglehart, D.L. New York, Academic Press.

Rao, C.R. and Shanbhag, D.N. (1989b) Further extensions of the Choquet–Deny and Deny theorems with applications in characterization theory. *Quart. J. Math.*, Oxford, **40**, Series 2, 333-350.

Rao, C.R. and Shanbhag, D.N. (1989c) Characterizations based on regression properties: improved versions of recent results (submitted for publication).

Rao, C.R. and Shanbhag, D.N. (1991) An elementary proof for an extended version of the Choquet-Deny theorem. *J. Mult. Anal.*, **38**, 141-148.

Rao, C.R. and Shanbhag, D.N. (1992a) A stability theorem for the integrated Cauchy functional equation. *Khatri Memorial Volume* (to appear).

Rao, C.R. and Shanbhag, D.N. (1992b) Some observations on the integrated Cauchy functional equation. *Math. Nachr.*, **157**, 185-195.

Rao, C.R., Shanbhag, D.N. and Tang, S.Y. (1992) Extended stability theorems for the integrated Cauchy functional equation (unpublished).

Rao, C.R., Srivastava, R.C., Talwalker, S. and Edgar, G.A. (1980) Characterization of probability distributions based on generalized Rao–Rubin condition. *Sankhyā*, Ser. A, **42**, 161-169.

Rao, M.B. and Shanbhag, D.N. (1982) Damage models. *Encyclopedia of Statistical Science*, Vol 2, pp. 262-265, eds Johnson, N.L., and Kotz, S. New York, Wiley.

Rényi, A. (1953) On the theory of order statistics. *Acta. Math. Acad. Sci. Hung.*, **4**, 191-231.

Rényi, A. (1957) A characterization of Poisson processes. *Magyar Tud. Kutotó Int. Kozl.*, **1**, 519-527.

Ressel, P. (1985) De Finetti-type theorems: an analytical approach. *Ann. Prob.*, **13**, 898-922.

Revuz, D. (1975) *Markov chains*, North Holland, Amsterdam.

Riedel, M. (1985) On a characterization of the normal distribution by means of identically distributed linear forms. *J. Mult. Anal.*, 241-252.

Rogers, G.S. (1963) An alternative proof of the characterization of the density Ax^B. *Amer. Math. Monthly*, **70**, 857-858.

Rossberg, H.J. (1972) Characterization of the exponential and the Pareto distributions by means of some properties of the distributions which the difference and quotients of the order statistics are subject to. *Math. Operatonsforch Statist.*, **3**, 207-216.

Rossberg, H.J., Jesiak, B. and Siegel, G. (1985) *Analytic Methods of Probability Theory*. Akademie-Verlag, Berlin.

Roy, D. (1989) A characterization of Gumbel's bivariate exponential and Lindley and Singpurwalla's bivariate Lomax distributions. *J. Appl. Prob.*, **26**, 886-891.

Roy, L.K. and Wasan, M.T. (1969) A characterization of the inverse Gaussian distribution . *Sankhyā*, Ser. A., **31**, 217-218.

Ruben, H. (1978) On quadratic forms and normality. *Sankhyā*, Ser. A, **40**, 156-173.

Rudin, W. (1974) *Real and Complex Analysis*, 2nd edn. McGraw-Hill.

Sackrowitz, H. and Samuel-Cahn, E. (1984) Estimation of the mean of a selected negative exponential population. *J.R. Statist. Soc.*, Ser. B, **46**, 242-249.

Sahobov, O. and Geshev, A. (1974) Characteristic property of the exponential distribution. *Natura. Univ. Plovdiv.*, **7**, 25-28 (in Russian).

Samorodnitsky, G. and Taqqu, M.S. (1991) Probability laws with 1-stable marginals are 1-stable (unpublished).

Samuels, S.M. (1974) A characterization of the Poisson process. *J. Appl. Prob.*, **11**, 72-85.

Sapatinas, T. (1990) *Damage models and applications in markov chains*. M.Sc. Dissertation, University of Sheffield.

Sapatinas, T. (1993) Characterization and identifiability results in some stochastic models. Ph.D. thesis, University of Sheffield.

Sapatinas, T. and Aly, M.A.H. (1993) Characterizations of some well-known discrete distributions based on variants of the Rao–Rubin condition. *Sankhyā*, Ser. A (to appear).

Sapatinas, T., Rao, C.R. and Shanbhag, D.N. (1993) An identifiability problem in a damage model (unpublished).

Schoenberg, I.J. (1936) Metric spaces and completely monotone functions. *Ann. Math.*, **39**, 811–841.

Schultz, D.M. (1975) Mass-size distributions — a review and a proposed new model. *Statistical Distributions in Scientific Work*, Vol. 2, pp. 275–288, eds Patil, G.P., Kotz, S. and Ord, J.K. Dordrecht, Reidel.

Seneta, E. (1968) The stationary distribution of a branching process allowing immigration. A remark on the critical case. *J. Roy. Statist. Soc.*, Ser. B, **30**, 176–179.

Seneta, E. (1973) *Nonnegative Matrices.* George Allen and Unwin, London.

Seshadri, V. (1983) The inverse Gaussian distribution: some properties and characterizations. *Canad. J. Statist.*, **11**, 131–136.

Seshadri, V. (1993) *The Inverse Gaussian Distribution, A Case Study in Exponential Families*, Clarendon Press, Oxford.

Seth, G.R. (1949) On the variance of estimates. *Ann. Math. Statist.*, **20**, 1–27.

Sethuraman, J. (1965) On a characterization of the three limiting types of the extreme. *Sankhyā*, Ser. A, **27**, 357–364.

Shanbhag, D.N. (1970a) Characterizations for exponential and geometric distributions. *J. Amer. Statist. Assoc.*, **65**, 1256–1259.

Shanbhag, D.N. (1970b) Another characteristic property of the Poisson distribution. *Proc. Camb. Philos. Soc.*, **68**, 167–169.

Shanbhag, D.N. (1971) An extension of Lukacs' result. *Proc. Cam. Phil. Soc.*, **69**, 301–303.

Shanbhag, D.N. (1972a) Some characterizations based on the Bhattacharyya matrix. *J. Appl. Prob.*, **9**, 580–587.

Shanbhag, D.N. (1972b) Characterization under unimodality for exponential distributions. *Research Report 99/DNS3*, Sheffield University.

Shanbhag, D.N. (1972c) Characterizations of some stochastic processes. *Research report 109/DNS 4*, Univ. of Sheffield.

Shanbhag, D.N. (1973a) Characterization for the queueing system $M/G/\infty$. *Proc. Cambridge Philos. Soc.*, **74**, 141–143.

Shanbhag, D.N. (1973b) Characterization of the Yule process. Unpublished manuscript.

shanbhag, D.N. (1973c) Comments on Wang's paper. *Proc. Camb. Philos. Soc.*, **73**, 473–475.

Shanbhag, D.N. (1974) An elementary proof for the Rao–Rubin characterization of the Poisson distribution. *J. Appl. Prob.*, **11**, 211–215.

Shanbhag, D.N. (1977) An extension of the Rao–Rubin characterization of the Poisson distribution. *J. Appl. Prob.*, **14**, No. 3., 640–646.

Shanbhag, D.N. (1978) On some problems in distribution theory. *Sankhyā*, Ser. A, **40**, 208–213.

Shanbhag, D.N. (1979a) Some refinements in distribution theory. *Sankhyā*, Ser. A, **41**, 252–262.

Shanbhag, D.N. (1979b) Diagonality of the Bhattacharyya matrix as a characterization. *Theory Prob. Appl.*, **24**, 430–433.

Shanbhag, D.N. (1983) Review on Wang (1981). *Math. Reviews* **83**m, 62029.

Shanbhag, D.N. (1988) Review on Chen, R.W. and Slud, E. (1984). *Math. Reviews* **88**f, 60024(a,b).

Shanbhag, D.N. (1991) Extended versions of Deny's theorem via de Finettis theorem. *Comput. Statist. Data Analys.*, **12**, 115–126.

Shanbhag, D.N. and Clark, R.M. (1972) Some characterizations of the Poisson distribution starting with a power series distribution. *Proc. Cambridge Philos. Soc.*, **71**, 517–522.

Shanbhag, D.N. and Kapoor, S. (1993) Some questions in characterization theory. *Math. Scientist* (to appear).

Shanbhag, D.N. and Kotz, S. (1987) Some new approaches to multivariate probability distributions. *J. Multivariate Analysis*, **22**(2), 189–211.

Shanbhag, D.N. and Panaretos, J. (1979) Some results related to the Rao–Rubin characterization of the Poisson distribution. *Austral. J. Statist.*, **21**, 78–83.

Shanbhag, D.N. and Rao, M. Bhaskara (1975) A note on characterizations of probability distributions based on conditional expected values. *Sankhyā*, Ser. A, **37**, 297–300.

Shanbhag, D.N. and Taillie, C. (1979) Unpublished.

Shantaram, R. (1978) A characterization of the arc-sine law. *Sankhyā*, Ser. A, **40**, 199–207.

Shantaram, R. (1980) On a conjecture of Norton's. *SIAM J. Appl. Math.*, **42**, 923–925.

Shimizu, R. (1978) Solution to a functional equation and its applications to some characterization problems. *Sankhyā*, Ser. A, **40**, 319–332.

Shimizu, R. (1979) On a lack of memory of the exponential distribution. *Ann. Inst. Statist. Math.*, **31**, 309–313.

Shimizu, R. (1980) Functional equation with an error term and the stability of some characterizations of the exponential distribution. *Ann. Inst. Statist. Math.*, **32**, 1–16.

Shimizu, R. (1986) Inequalities for a distribution with monotone hazard rate. *Ann. Inst. Statist. Math.*, **38**, Part A, 195–204.

Shimizu, R. and Davies, L. (1981) General characterization theorems for the Weibull and the stable distributions. *Sankhyā*, Ser. A, **43**, 282–310.

Shohat, J.A. and Tamarkin, J.D. (1963) The problem of moments. *Math. Surveys*, No. 1, Amer. Math. Soc., New York.

Shorrack, R.W. (1972a) A limit theorem for inter-record times. *J. Appl. Prob.*, **9**, 219–233.

Shorrack, R.W. (1972b) On record values and record times. *J. Appl. Prob.*, **9**, 316–326.

Shorrack, R.W. (1973) Record values and inter-record times. *J. Appl. Prob.*, **10**, 543–555.

Smith, A.F.M. (1981) On random sequences with centered spherical symmetry. *J. Roy. Statist. Soc.*, Ser. B, **43**, 208–209.

Spitzer, F.L. (1960) The Wiener-Hopf equation whose kernel is a probability density. II *Duke. Math. J.*, **27**, 363–372.

Spitzer, F. (1967) Two explicit martin boundary constructions. Symposium on Probability Methods in Analysis. *Lecture Notes in Math.*, **31**, 296–298.

Sreehari, M. (1983) A characterization of the geometric distribution. *J. Appl. Prob.*, **20**, 209–212.

Srivastava, R.C. (1974) Two characterizations of the geometric distribution. *J. Amer. Statist. Assoc.*, **69**, 267–269.

Srivastava, R.C. (1979) Two characterizations of the geometric distribution by record values. *Sankhyā*, Ser. B, **40**, 276–278.

Srivastava, R.C. and Singh, J. (1975) On some characterizations of the binomial and Poisson distributions based on a damage model. *Statistical Distributions in Scientific Work*, Vol. 3, pp. 271–277 eds Patil, G.P., Kotz, S. and Ord, J.K.

Srivastava, R.C. and Srivastava, A.B.L. (1970) On the characterization of the Poisson distribution. *Sankhyā*, Ser. A, **32**, 265–270.

Steutel, F.W. (1970) Preservation of infinite divisibility under mixing and related topics. *Mathematical Centre Tracts*, **33**, Mathematisch Centrum, Amsterdam.

Steutel, F.W. and van Harn, K. (1979) Discrete analogues of self decomposability and stability. *Ann. Prob.*, **7**, 893–899.

Swartz, G.B. (1973) The mean residual life functions. *IEEE Trans. Reliability*, **R22**, 108–109.

Székely, G.J. and Zeng, W. (1990) The Choquet–Deny convolution equation $\mu = \mu * \sigma$ for probability measure on Abelian semigroups. *J. Theoret. Probab.*, **3**, 361–365.

Talwalker, S. (1970) A characterization of the double Poisson distribution. *Sankhyā*, Ser. A, **34**, 191–193.

Talwalker, S. (1977) A note on characterizations by the conditional expectation. *Metrika*, **24**, 129–136.

Talwalker, S. (1980) A note on the generalized Rao–Rubin condition and characterization of certain discrete distributions. *J. Appl. Prob.*, **17**, 563–569.

Tanis, E.A. (1964) Linear forms in the order statistics from an exponential distribution. *Ann. Math. Stat.*, **35**, 270–276.

Targhetta, M.L. (1990) Some characterizations based on Fisher's information. *Metron*, **48**, 421–429.

Tata, M.N. (1969) On outstanding values in a sequence of random variables. *Z. Wahrsch. verw.*, **12**, 9–20.

Titchmarsh, E.C. (1978) *The Theory of Functions*. Clarendon Press, Oxford.

Too, Y.H. and Lin, G.D. (1989) Characterization of uniform and exponential distributions. *Statist. Prob. Letters*, **8**, 490–493.

Tripathi, R.C. and Gupta, R.C. (1985) A generalisation of the log-series distribution. *Commun. Statist. Theory. Meth.*, **A14**, 1779–1799.

Tripathi, R.C. and Gupta, R.C. (1987) A comparison between the ordinary and length-biased modified power series distributions with applications. *Commun. Statist. Theory. Meth.*, **16**, 1195–1206.

Tulcea, I. (1949, 1950) Mesures dans les espaces produits. *Atti. Accad. Naz. Lincei Rend.*, **7**.

Van Harn, K. (1978) Classifying infinitely divisible distributions by functional equations. *Mathematical Centre Tracts 103*, Amsterdam.

Wang, Y.H. (1972) On characterization of certain probability distributions. *Proc. Cambridge Philos. Soc.*, **71**, 347–352.

Wang, Y.H. (1981) Extensions of Lukacs' characterization of the gamma distribution. *Analytical Methods in Probability Theory* (Oberwolfach, 1980). *Lecture Notes in Math.* **861**, 166–177, Springer-Verlag, Berlin.

Wang, Y.H. and Srivastava, R.C. (1980) A characterization of the exponential and related distributions by linear regression. *Ann. Stat.*, **8**(1), 217–220.

Wesolowski, J. (1987) A regression characterization of the normal laws. *Statist. Prob. Letters*, **6**, 11–12.

Wesolowski, J. (1989) A regression characterization of the Poisson distribution stability problems for stochastic models. *Springer Lecture Notes in Maths*, **1412**, 349–351.

Wesolowski, J. (1990) A constant regression characterization of the gamma law. *Adv. Appl. Prob.*, **22**, 488–490.

Westcott, M. (1981) Letter to the editor. *J. Appl. Prob.*, **18**, 568.

Whittaker, J. (1973) The Bhattacharyya matrix for the mixture of two distributions. *Biometrika*, **60**, 201–202.

Widder, D.V. (1946) *The Laplace Transform*. Princeton University Press.

Williams, D. (1979) *Diffusions, Markov Processes, and Martingales*, Vol. 1: *Foundations*. Wiley.

Wishart, D.M.G. (1956) A queueing system with chi-square service time distribution. *Ann. Math. Statist.*, **27**, 768–779.

Witte, H.-J. (1988) Some characterizations of distributions based on the ICFE. *Sankhyā*, Ser. A, **50**, 59–63.

Witte, H.-J. (1990) Characterizations of distributions of exponential or geometric type by the integrated lack of memory property and record values. *Comp. Statist. Data Anal.*, **10**, 283–288.

Witte, H.-J. (1993) Some characterizations of exponential or geometric distributions in a nonstationary record value model. *J. Appl. Prob.*, 30, 373–381.

Wolinska-Welcz, A. and Szynal, D. (1984) On a solution of Dugue's problem for a class of lattice distributions (unpublished).

Yanushkevichius, R. (1988) Convolution equations in problems of the stability of characterizations of probability laws. *Theory Prob. Appl.*, **33**, 668–681.

Yeo, G.F. and Milne, R.K. (1991) One characterization of beta and gamma distributions. *Statist. Prob. Letters*, **11**, 239–242.

Zahedi, H. (1985) Some new classes of multivariate survival functions. *J. Statist. Plann. Inf.*, **11**, 171–188.

Zeng, Wei-Bin (1992) A note on stability of multivariate distributions, *J. Mathematical Research and Exposition*, **2**, 171–175.

Zijlstra, M. (1983) Characterizations of the geometric distribution by distribution properties. *J. Appl. Prob.*, **20**, 843–850.

Zinger, A.A. (1950) On a problem of Kolmogorov's. *Vestnik Leningrad Univ.*, **1**, 53–56.

Zinger, A.A. (1951) On independent sample from normal populations (in Russian). *Uspekhi Matem. Nauk.*, **6**, 172–175.

Zinger, A.A. (1977) On a characterization of the normal law by identically distributed linear statistics. *Sankhyā*, Ser. A, **39**, 232–242.

Zolotarev, V.M. (1964) On the representation of stable laws by integrals. *Trudy Math. Inst. Steklov*, **71**, 46–50 (in Russian) *IMS and AMS Selected Translations in Mathematical Statistics and Probability*, **6**, 84–88.

Zoroa, P., Ruiz, J.M. and Marin, J. (1990) A characterization based on conditional expectations. *Commun. Statist. Theory Meth.*, **91**, 3127–3135.

Index

Index compiled by Geoffrey C. Jones

---

* Now available in a lower priced paperback edition in the Wiley Classics Library.